MATHEMATICS FOR MACHINE TECHNOLOGY

MATHEMATICS FOR MACHINE TECHNOLOGY SECOND EDITION

ROBERT D. SMITH

Cover photo courtesy of DoAll Company, 254 North Laurel Avenue, Des Plaines, IL 60016

DELMAR PUBLISHERS INC.

NOTICE TO THE READER

Publisher does not warrant or guarantee any of the products described herein or perform any independent analysis in connection with any of the product information contained herein. Publisher does not assume, and expressly disclaims, any obligation to obtain and include information other than that provided to it by the manufacturer.

The reader is expressly warned to consider and adopt all safety precautions that might be indicated by the activities described herein and to avoid all potential hazards. By following the instructions contained herein, the reader willingly assumes all risks in connection with such instructions.

The publisher makes no representations or warranties of any kind, including but not limited to, the warranties of fitness for particular purpose or merchantability, nor are any such representations implied with respect to the material set forth herein, and the publisher takes no responsibility with respect to such material. The publisher shall not be liable for any special, consequential or exemplary damages resulting, in whole or in part, from the readers' use of, or reliance upon, this material.

For information, address Delmar Publishers Inc.,
2 Computer Drive West, Box 15-015,
Albany, New York 12212

10   9   8   7   6

LIBRARY OF CONGRESS CATALOG CARD NUMBER: 82-72326
ISBN: 0-8273-2106-6

Printed in the United States of America
Published simultaneously in Canada
by Nelson Canada,
A Division of International Thomson Limited

CONTENTS

SECTION 4   FUNDAMENTALS OF PLANE GEOMETRY

SECTION 5   TRIGONOMETRY

SECTION 6   COMPOUND ANGLES

SECTION 7   NUMERICAL CONTROL

PREFACE

Mathematics for Machine Technology is written to overcome the often mechanical "plug in" approach found in many trade-related mathematics textbooks. An understanding of mathematical concepts is stressed in all topics ranging from general arithmetic processes to oblique trigonometry, compound angles, and numerical control.

Both content and method are those used by the author in teaching applied machine technology mathematics classes for apprentices in the machine, tool-and-die, and tool design trades. Each unit is developed as a learning experience based on preceding units—making prerequisites unnecessary.

Presentation of basic concepts is accompanied by realistic industry-related examples and actual industrial applications. The applications progress from the simple to those with solutions which are relatively complex. Many problems require the student to work with illustrations such as are found in machine trade handbooks and engineering drawings.

An analytical approach to problem solving is emphasized in the geometry, trigonometry, compound angle, and numerical control sections. This approach is necessary in actual practice in translating engineering drawing dimensions to machine working dimensions. Integration of algebraic and geometric principles with trigonometry by careful sequence and treatment of material also helps the student in solving industrial applications. The Instructor's Guide provides answers and solutions for all problems.

Changes from the previous edition have been made to improve the presentation of topics and to update material. The following major changes have been made.

- The metric system is integrated throughout the entire text following the units on common fractions.

- Section II on measurement is greatly expanded with the inclusion of metric as well as English system measuring instruments.

- The number and variety of examples, exercises, and applied problems in Sections III, IV, and V are substantially increased.

- A section on compound angles is now included in the text. An analytical approach is presented which emphasizes student understanding and visualization in the solution of compound angular hole and surface problems.

- An Achievement Review is included at the end of each section to provide a comprehensive review of all material presented within a section.

Robert D. Smith has experience in both the manufacturing industry and in education. He held positions as tool designer, quality control engineer, and chief manufacturing engineer prior to teaching. Mr. Smith taught applied mathematics, physics, and industrial materials and processes on the secondary school level and in Machine Trade Apprentice Programs at the A. I. Prince Vocational-Technical School in Hartford, Connecticut. Mr. Smith is presently Assistant Professor of Vocational-Technical Education at Central Connecticut State University, New Britain, Connecticut. He has membership in several professional organizations in his field of interest: the American Technical Education Association and the Society of Manufacturing Engineers.

SECTION 1 COMMON FRACTIONS AND DECIMAL FRACTIONS

⎺UNIT 1 *INTRODUCTION TO COMMON FRACTIONS AND MIXED NUMBERS* ────────

⎺OBJECTIVES────────────────────────────

After studying this unit you should be able to

- Express fractions in lowest terms.
- Express fractions as equivalent fractions.
- Express mixed numbers as improper fractions.
- Express improper fractions as mixed numbers.

Most measurements and calculations made by a machinist are not limited to whole numbers. Blueprint dimensions are often given as fractions and certain measuring tools are graduated in fractional units. The machinist must be able to make calculations using fractions and to measure fractional values.

⎺FRACTIONAL PARTS────────────────────────

A *fraction* is a value which shows the number of equal parts taken of a whole quantity or unit. The symbols used to indicate a fraction are the bar (─) and the slash (/).

Line segment AB as shown is divided into 4 equal parts.

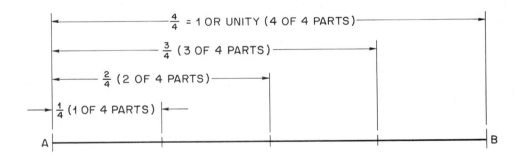

$$1 \text{ part } = \frac{1 \text{ part}}{\text{total parts}} = \frac{1 \text{ part}}{4 \text{ parts}} = \frac{1}{4} \text{ of the length of the line segment.}$$

$$2 \text{ parts} = \frac{2 \text{ parts}}{\text{total parts}} = \frac{2 \text{ parts}}{4 \text{ parts}} = \frac{2}{4} \text{ of the length of the line segment.}$$

$$3 \text{ parts} = \frac{3 \text{ parts}}{\text{total parts}} = \frac{3 \text{ parts}}{4 \text{ parts}} = \frac{3}{4} \text{ of the length of the line segment.}$$

$$4 \text{ parts} = \frac{4 \text{ parts}}{\text{total parts}} = \frac{4 \text{ parts}}{4 \text{ parts}} = \frac{4}{4} = 1, \text{ or unity (4 parts make up the whole).}$$

1

Each of the 4 equal parts of the line segment AB is divided into 8 equal parts. There is a total of 4 x 8 or 32 parts.

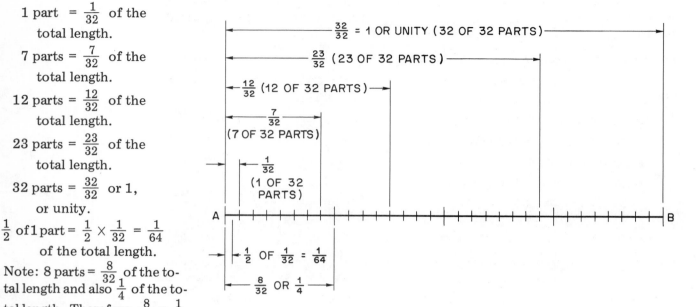

1 part $= \dfrac{1}{32}$ of the total length.

7 parts $= \dfrac{7}{32}$ of the total length.

12 parts $= \dfrac{12}{32}$ of the total length.

23 parts $= \dfrac{23}{32}$ of the total length.

32 parts $= \dfrac{32}{32}$ or 1, or unity.

$\dfrac{1}{2}$ of 1 part $= \dfrac{1}{2} \times \dfrac{1}{32} = \dfrac{1}{64}$ of the total length.

Note: 8 parts $= \dfrac{8}{32}$ of the total length and also $\dfrac{1}{4}$ of the total length. Therefore, $\dfrac{8}{32} = \dfrac{1}{4}$.

DEFINITIONS OF FRACTIONS

A *fraction* is a value which shows the number of equal parts taken of a whole quantity or unit.

The *denominator* of a fraction is the number that shows how many equal parts are in the whole quantity. The denominator is written below the bar.

The *numerator* of a fraction is the number that shows how many equal parts of the whole are taken. The numerator is written above the bar.

The numerator and denominator are called the *terms* of the fraction.

$\dfrac{3}{4}$ $\leftarrow$ numerator
$\leftarrow$ denominator

An *improper* fraction is a fraction in which the numerator is larger than or equal to the denominator, as $\dfrac{3}{2}$, $\dfrac{5}{4}$, $\dfrac{15}{8}$, $\dfrac{6}{6}$, $\dfrac{17}{17}$.

A *mixed number* is a number composed of a whole number and a fraction, as $3\dfrac{7}{8}$, $7\dfrac{1}{2}$.

Note: $3\dfrac{7}{8}$ means $3 + \dfrac{7}{8}$. It is read as three and seven-eighths. $7\dfrac{1}{2}$ means $7 + \dfrac{1}{2}$. It is read as seven and one-half.

A *complex fraction* is a fraction in which one or both of the terms are fractions or mixed numbers, as $\dfrac{\frac{3}{4}}{6}$, $\dfrac{32}{\frac{15}{4}}$, $\dfrac{8\frac{3}{4}}{3}$, $\dfrac{\frac{7}{16}}{2\frac{2}{5}}$, $\dfrac{4\frac{1}{4}}{7\frac{5}{8}}$.

EXPRESSING FRACTIONS AS EQUIVALENT FRACTIONS

The numerator and denominator of a fraction can be multiplied or divided by the same number without changing the value. For example, $\dfrac{1}{2} = \dfrac{1 \times 4}{2 \times 4} = \dfrac{4}{8}$. Both the numerator and denominator are multiplied by 4. Because $\dfrac{1}{2}$ and $\dfrac{4}{8}$ have the same value, they are *equivalent*. Also, $\dfrac{8}{12} = \dfrac{8 \div 4}{12 \div 4} = \dfrac{2}{3}$. Both numerator and denominator are divided by 4. Since $\dfrac{8}{12}$ and $\dfrac{2}{3}$ have the same value, they are *equivalent*.

A fraction is in its *lowest terms* when the numerator and denominator do not contain a common factor, as $\frac{5}{9}$, $\frac{7}{8}$, $\frac{3}{4}$, $\frac{11}{12}$, $\frac{15}{32}$, $\frac{9}{11}$. *Factors* are the numbers used in multiplying. For example, 2 and 5 are each factors of 10; 2 × 5 = 10. Expressing a fraction in lowest terms is often called *reducing* a fraction to lowest terms.

Procedure: To reduce a fraction to lowest terms

- Divide both numerator and denominator by the greatest common factor.

Example: Reduce $\frac{12}{42}$ to lowest terms.

Both terms can be divided by 2. $\frac{12 \div 2}{42 \div 2} = \frac{6}{21}$

Note: The fraction is reduced, but not to lowest terms.

Further reduce $\frac{6}{21}$.

Both terms can be divided by 3. $\frac{6 \div 3}{21 \div 3} = \frac{2}{7}$ Ans

Note: The value $\frac{2}{7}$ may be obtained in one step if
each term of $\frac{12}{42}$ is divided by 2 × 3 or 6. $\frac{12 \div 6}{42 \div 6} = \frac{2}{7}$ Ans

Procedure: To express a fraction as an equivalent fraction with an indicated denominator which is larger than the denominator of the fraction

- Divide the indicated denominator by the denominator of the fraction.
- Multiply both the numerator and denominator of the fraction by the value obtained.

Example: Express $\frac{3}{4}$ as an equivalent fraction with 12 as the denominator.
Divide 12 by 4. $\frac{3 \times 3}{4 \times 3} = \frac{9}{12}$ Ans
12 ÷ 4 = 3
Multiply both 3 and 4 by 3.

⌐EXPRESSING MIXED NUMBERS AS IMPROPER FRACTIONS_____

Procedure: To express a mixed number as an improper fraction

- Multiply the whole number by the denominator.
- Add the numerator to obtain the numerator of the improper fraction.
- The denominator is the same as that of the original fraction.

Example 1: Express $4\frac{1}{2}$ as an improper fraction.
Multiply the whole number by the denominator. $\frac{4 \times 2 + 1}{2} = \frac{9}{2}$ Ans
Add the numerator to obtain numerator for the
improper fraction.

The denominator is the same as that of the original fraction.

Example 2: Express $12\frac{3}{16}$ as an improper fraction.

$\frac{12 \times 16 + 3}{16} = \frac{195}{16}$ Ans

⌐EXPRESSING IMPROPER FRACTIONS AS MIXED NUMBERS_____

Procedure: To express an improper fraction as a mixed number

- Divide the numerator by the denominator.

Examples: Express the following improper fractions as mixed numbers.

$\frac{11}{4} = 11 \div 4 = 2\frac{3}{4}$ Ans

$\frac{43}{3} = 43 \div 3 = 14\frac{1}{3}$ Ans

$\frac{931}{8} = 931 \div 8 = 116\frac{3}{8}$ Ans

4   Section 1 Common Fractions and Decimal Fractions

APPLICATION

Fractional Parts

1. Write the fractional part which each length, A through F, represents of the total shown on the scale.

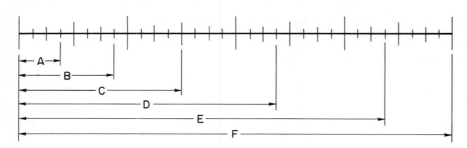

A = _____

B = _____

C = _____

D = _____

E = _____

F = _____

2. The circle is divided into equal parts. Write the fractional part each of the following represents.

a. 1 part _____

b. 3 parts _____

c. 9 parts _____

d. 5 parts _____

e. 16 parts _____

f. $\frac{1}{2}$ of 1 part _____

g. $\frac{1}{3}$ of 1 part _____

h. $\frac{1}{4}$ of 1 part _____

i. $\frac{1}{10}$ of 1 part _____

j. $\frac{1}{16}$ of 1 part _____

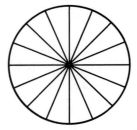

Expressing Fractions as Equivalent Fractions

3. Reduce to halves.

a. $\frac{4}{8}$ _____

b. $\frac{9}{18}$ _____

c. $\frac{100}{200}$ _____

d. $\frac{121}{242}$ _____

e. $\frac{15}{10}$ _____

f. $\frac{18}{12}$ _____

g. $\frac{54}{36}$ _____

h. $\frac{125}{50}$ _____

4. Reduce to lowest terms.

a. $\frac{6}{8}$ _____

b. $\frac{12}{4}$ _____

c. $\frac{6}{10}$ _____

d. $\frac{35}{5}$ _____

e. $\frac{11}{44}$ _____

f. $\frac{14}{6}$ _____

g. $\frac{27}{6}$ _____

h. $\frac{65}{15}$ _____

i. $\frac{25}{150}$ _____

j. $\frac{8}{128}$ _____

5. Express as thirty-seconds.

a. $\frac{1}{4}$ _____

b. $\frac{3}{4}$ _____

c. $\frac{9}{8}$ _____

d. $\frac{7}{16}$ _____

e. $\frac{21}{16}$ _____

f. $\frac{17}{2}$ _____

g. $\frac{197}{16}$ _____

h. $\frac{33}{8}$ _____

6. Express as equivalent fractions as indicated.

a. $\frac{3}{4} = \frac{?}{8}$ _____

b. $\frac{5}{12} = \frac{?}{36}$ _____

c. $\frac{6}{15} = \frac{?}{60}$ _____

d. $\frac{17}{14} = \frac{?}{42}$ _____

e. $\frac{22}{9} = \frac{?}{45}$ _____

f. $\frac{14}{3} = \frac{?}{18}$ _____

g. $\frac{7}{16} = \frac{?}{128}$ _____

h. $\frac{13}{8} = \frac{?}{48}$ _____

i. $\frac{19}{16} = \frac{?}{160}$ _____

Mixed Numbers and Improper Fractions

7. Express the following mixed numbers as improper fractions.

a. $2\frac{2}{3}$ _____

b. $1\frac{7}{8}$ _____

c. $6\frac{4}{5}$ _____

d. $3\frac{3}{8}$ _____

e. $5\frac{9}{32}$ _____

f. $7\frac{4}{7}$ _____

g. $10\frac{1}{3}$ _____

h. $8\frac{2}{5}$ _____

i. $100\frac{1}{2}$ _____

j. $4\frac{63}{64}$ _____

k. $51\frac{3}{4}$ _____

l. $408\frac{13}{16}$ _____

8. Express the following improper fractions as mixed numbers.

   a. $\dfrac{10}{3}$ _____   d. $\dfrac{87}{4}$ _____   g. $\dfrac{127}{32}$ _____   j. $\dfrac{235}{16}$ _____

   b. $\dfrac{21}{2}$ _____   e. $\dfrac{63}{18}$ _____   h. $\dfrac{43}{13}$ _____   k. $\dfrac{514}{4}$ _____

   c. $\dfrac{9}{8}$ _____   f. $\dfrac{127}{124}$ _____   i. $\dfrac{150}{9}$ _____   l. $\dfrac{337}{64}$ _____

9. Express the following mixed numbers as improper fractions. Then express the improper fractions as the equivalent fractions indicated.

   a. $2\dfrac{1}{2} = \dfrac{?}{8}$ _____   c. $6\dfrac{4}{5} = \dfrac{?}{15}$ _____   e. $9\dfrac{7}{8} = \dfrac{?}{64}$ _____

   b. $1\dfrac{3}{4} = \dfrac{?}{16}$ _____   d. $12\dfrac{2}{3} = \dfrac{?}{18}$ _____   f. $15\dfrac{1}{2} = \dfrac{?}{128}$ _____

10. Sketch and redimension this plate. Reduce all proper fractions to lowest terms. Reduce all improper fractions to lowest terms and express as mixed numbers. All dimensions are in inches.

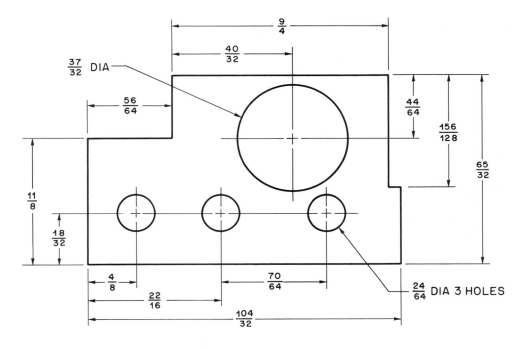

UNIT 2 ADDITION OF COMMON FRACTIONS AND MIXED NUMBERS _____

OBJECTIVES _____

After studying this unit you should be able to
- Determine least common denominators.
- Express fractions as equivalent fractions having least common denominators.
- Add fractions and mixed numbers.

A machinist must be able to add fractions and mixed numbers in order to determine the length of stock required for a job, the distances between various parts of a machined piece, and the depth of holes and cutouts in a workpiece.

⎯LEAST COMMON DENOMINATORS⎯⎯⎯⎯⎯⎯⎯⎯⎯⎯⎯⎯⎯⎯⎯⎯⎯

Fractions cannot be added unless they have a common denominator. *Common denominator* means that the denominators of each of the fractions are the same, as $\frac{5}{8}$, $\frac{7}{8}$, $\frac{15}{8}$.

In order to add fractions which do not have common denominators, such as $\frac{3}{8} + \frac{1}{4} + \frac{7}{16}$, it is necessary to determine the least common denominator.

The *least common denominator* is the smallest denominator which is evenly divisible by each of the denominators of the fractions being added. Or, stated in another way, the *least common denominator* is the smallest denominator into which each denominator can be divided without leaving a remainder.

Procedure: To find the least common denominator
- Determine the smallest number into which all denominators can be divided without leaving a remainder.
- Use this number as a common denominator.

Example 1: Find the least common denominator of $\frac{3}{8}$, $\frac{1}{4}$, and $\frac{7}{16}$.

The smallest number into which 8, 4, and 16 can be divided without leaving a remainder is 16.

Write 16 as the least common denominator.

Example 2: Find the least common denominator of $\frac{3}{4}$, $\frac{1}{3}$, $\frac{7}{8}$, and $\frac{5}{12}$.

The smallest number into which 4, 3, 8, and 12 can be divided is 24.

Write 24 as the least common denominator.

Note: In this example, denominators such as 48, 72, and 96 are common denominators because 4, 3, 8, and 12 divide evenly into these numbers, but they are not the least common denominators.

Although any common denominator can be used when adding fractions, it is generally easier and faster to use the least common denominator.

⎯EXPRESSING FRACTIONS AS EQUIVALENT FRACTIONS WITH THE LEAST COMMON DENOMINATOR ⎯⎯⎯⎯⎯⎯⎯⎯⎯⎯⎯⎯⎯⎯

Procedure: To change fractions into equivalent fractions having the least common denominator
- Divide the least common denominator by each denominator.
- Multiply both the numerator and denominator of each fraction by the value obtained.

Example 1: Express $\frac{2}{3}$, $\frac{7}{15}$, and $\frac{1}{2}$ as equivalent fractions having a least common denominator.

The least common denominator is 30. $30 \div 3 = 10;$ $\frac{2 \times 10}{3 \times 10} = \frac{20}{30}$ Ans

Divide 30 by each denominator.

Multiply each term of the fraction by $30 \div 15 = 2;$ $\frac{7 \times 2}{15 \times 2} = \frac{14}{30}$ Ans
the value obtained.

$30 \div 2 = 15;$ $\frac{1 \times 15}{2 \times 15} = \frac{15}{30}$ Ans

Example 2: Change $\frac{5}{8}$, $\frac{15}{32}$, $\frac{3}{4}$, and $\frac{9}{16}$ to equivalent fractions having a least common denominator.

The least common denominator is 32.

$32 \div 8 = 4$; $\frac{5 \times 4}{8 \times 4} = \frac{20}{32}$ Ans $\qquad$ $32 \div 4 = 8$; $\frac{3 \times 8}{4 \times 8} = \frac{24}{32}$ Ans

$32 \div 32 = 1$; $\frac{15 \times 1}{32 \times 1} = \frac{15}{32}$ Ans $\qquad$ $32 \div 16 = 2$; $\frac{9 \times 2}{16 \times 2} = \frac{18}{32}$ Ans

ADDING FRACTIONS

Procedure: To add fractions

- Express the fractions as equivalent fractions having the least common denominator.
- Add the numerators and write their sum over the least common denominator.
- Express an improper fraction as a mixed number when necessary and reduce the fractional part to lowest terms.

Example 1: Add $\frac{1}{2} + \frac{3}{5} + \frac{7}{10} + \frac{5}{6}$.

Express the fractions as equivalent fractions with 30 as the denominator.

Add the numerators and write their sum over the least common denominator, 30.

Express the fraction as a mixed number.

$$\frac{1}{2} = \frac{15}{30}$$
$$\frac{3}{5} = \frac{18}{30}$$
$$\frac{7}{10} = \frac{21}{30}$$
$$+ \frac{5}{6} = \frac{25}{30}$$
$$\frac{79}{30} = 2\frac{19}{30} \text{ Ans}$$

Example 2: Determine the total length of the shaft shown. All dimensions are in inches.

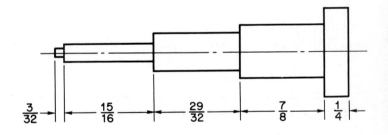

Express the fractions as equivalent fractions with 32 as the denominator.

Add the numerators and write their sum over the least common denominator, 32.

Express $\frac{98}{32}$ as a mixed number and reduce to lowest terms.

Total Length = $3\frac{1}{16}''$ Ans

$$\frac{3}{32} = \frac{3}{32}$$
$$\frac{15}{16} = \frac{30}{32}$$
$$\frac{29}{32} = \frac{29}{32}$$
$$\frac{7}{8} = \frac{28}{32}$$
$$+ \frac{1}{4} = \frac{8}{32}$$
$$\frac{98}{32} = 3\frac{2}{32} = 3\frac{1}{16}$$

ADDING FRACTIONS, MIXED NUMBERS, AND WHOLE NUMBERS

Procedure: To add fractions, mixed numbers, and whole numbers

- Add the whole numbers.
- Add the fractions.
- Combine whole number and fraction.

Example 1:   Add $\frac{1}{3} + 7 + 3\frac{1}{2} + \frac{5}{12} + 2\frac{19}{24}$.

Express the fractional parts as equivalent fractions with 24 as the denominator.

Add the whole numbers.

Add the fractions.

Combine the whole number and the fraction. Express the answer in lowest terms.

$$\frac{1}{3} = \frac{8}{24}$$

$$7 = 7$$

$$3\frac{1}{2} = 3\frac{12}{24}$$

$$\frac{5}{12} = \frac{10}{24}$$

$$+\ 2\frac{19}{24} = 2\frac{19}{24}$$

$$12\frac{49}{24} = 14\frac{1}{24}\ \text{Ans}$$

Example 2:   Find the distance between the two $\frac{1}{2}$-inch diameter holes in the plate shown. All dimensions are in inches.

$$1 = 1$$

$$\frac{13}{32} = \frac{26}{64}$$

$$1\frac{47}{64} = 1\frac{47}{64}$$

$$+\ \frac{3}{16} = \frac{12}{64}$$

$$2\frac{85}{64} = 3\frac{21}{64}$$

Distance $= 3\frac{21}{64}''$ Ans

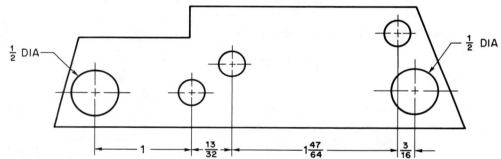

APPLICATION

Least Common Denominators

Determine the least common denominators of the following sets of fractions.

1. $\frac{2}{3}$, $\frac{1}{6}$, $\frac{5}{12}$ _____

2. $\frac{3}{5}$, $\frac{9}{10}$, $\frac{1}{3}$ _____

3. $\frac{5}{6}$, $\frac{7}{12}$, $\frac{3}{16}$, $\frac{19}{24}$ _____

4. $\frac{4}{5}$, $\frac{3}{4}$, $\frac{7}{10}$, $\frac{1}{2}$ _____

Equivalent Fractions with Least Common Denominators

Express these fractions as equivalent fractions having the least common denominator.

5. $\frac{1}{2}$, $\frac{2}{3}$, $\frac{5}{12}$ _____

6. $\frac{7}{16}$, $\frac{3}{8}$, $\frac{1}{2}$ _____

7. $\frac{9}{10}$, $\frac{1}{4}$, $\frac{3}{5}$, $\frac{1}{2}$ _____

8. $\frac{3}{16}$, $\frac{9}{32}$, $\frac{17}{64}$, $\frac{3}{4}$ _____

Adding Fractions

9. Determine the dimensions A, B, C, D, E, and F of this profile gage. All dimensions are in inches.

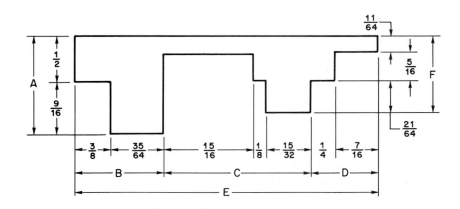

A = _____
B = _____
C = _____
D = _____
E = _____
F = _____

10. Determine the length, width, and height of this casting. All dimensions are in inches.

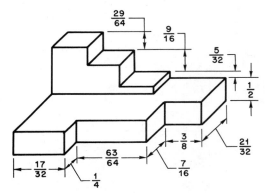

length = _____
width = _____
height = _____

Adding Fractions, Mixed Numbers, and Whole Numbers

11. Determine dimensions A, B, C, D, E, F, and G of this plate. Reduce to lowest terms where necessary. All dimensions are in inches.

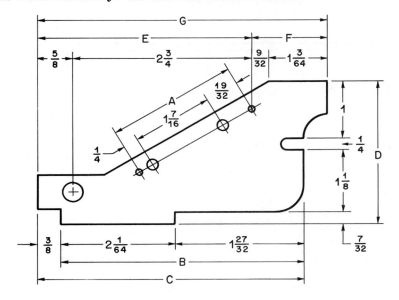

A = _____
B = _____
C = _____
D = _____
E = _____
F = _____
G = _____

12. Determine dimensions A, B, C, and D of this pin. All dimensions are in inches.

A = _____

B = _____

C = _____

D = _____

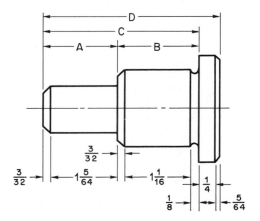

13. The operation sheet for machining an aluminum housing specifies 1 hour for facing, 2 3/4 hours for milling, 5/6 hour for drilling, 3/10 hour for tapping, and 2/5 hour for setting up. What is the total time allotted for this job?

_____

‾UNIT 3 *SUBTRACTION OF COMMON FRACTIONS AND MIXED NUMBERS* _____

‾OBJECTIVES _____

After studying this unit you should be able to

• Subtract fractions.

• Subtract mixed numbers.

While making a part from a blueprint, a machinist often finds it necessary to express blueprint dimensions as working dimensions. Subtraction of fractions and mixed numbers is required in order to properly position a part on a machine, to establish hole locations, and to determine depths of cut.

‾SUBTRACTING FRACTIONS _____

Procedure: To subtract fractions

• Express the fractions as equivalent fractions having the least common denominator.

• Subtract the numerators.

• Write their difference over the least common denominator.

• Reduce the fraction to lowest terms.

Example 1: Subtract $\frac{3}{8}$ from $\frac{9}{16}$.

The least common denominator is 16. Express $\frac{3}{8}$ as 16ths.

Subtract the numerators.

Write their difference over the least common denominator.

$$\frac{9}{16} = \frac{9}{16}$$
$$-\ \frac{3}{8} = \frac{6}{16}$$
$$\overline{\qquad\quad\frac{3}{16}}\ \ \text{Ans}$$

Example 2: Subtract $\frac{2}{5}$ from $\frac{3}{4}$.

$$\frac{3}{4} = \frac{15}{20}$$
$$-\frac{2}{5} = \frac{8}{20}$$
$$\frac{7}{20} \quad \text{Ans}$$

Example 3: Find the distances x and y between the centers of the pairs of holes in the strap shown. All dimensions are in inches.

To find distance x:

$$\frac{7}{8} = \frac{28}{32}$$
$$-\frac{11}{32} = \frac{11}{32}$$
$$\frac{17}{32}$$

$$x = \frac{17}{32}'' \quad \text{Ans}$$

To find distance y:

$$\frac{63}{64} = \frac{63}{64}$$
$$-\frac{1}{4} = \frac{16}{64}$$
$$\frac{47}{64}$$

$$y = \frac{47}{64}'' \quad \text{Ans}$$

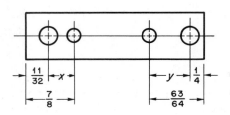

SUBTRACTING MIXED NUMBERS

Procedure: To subtract mixed numbers
- Subtract the whole numbers.
- Subtract the fractions.
- Combine whole number and fraction.

Example 1: Subtract $2\frac{1}{4}$ from $9\frac{3}{8}$.

Subtract the whole numbers.
Subtract the fractions.
Combine.

$$9\frac{3}{8} = 9\frac{3}{8}$$
$$-2\frac{1}{4} = 2\frac{2}{8}$$
$$7\frac{1}{8} \quad \text{Ans}$$

Example 2: Find the length of thread x of the bolt shown. All dimensions are in inches.

$$2\frac{7}{8} = 2\frac{28}{32}$$
$$-1\frac{3}{32} = 1\frac{3}{32}$$
$$1\frac{25}{32}$$

$$x = 1\frac{25}{32}'' \quad \text{Ans}$$

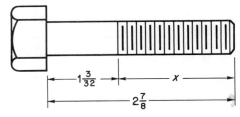

Example 3: Subtract $7\frac{15}{16}$ from $12\frac{5}{8}$.

$$12\frac{5}{8} = 12\frac{10}{16} = 11\frac{26}{16}$$
$$-7\frac{15}{16} = 7\frac{15}{16} = 7\frac{15}{16}$$
$$4\frac{11}{16} \quad \text{Ans}$$

Note: Since 15/16 cannot be subtracted from 10/16, one unit of the whole number 12 is expressed as a fraction with the common denominator 16.

Example 4: Subtract $52\frac{31}{64}$ from 75.

$$75 \quad = 74\frac{64}{64}$$

$$- 52\frac{31}{64} = 52\frac{31}{64}$$

$$\overline{\phantom{-52\frac{31}{64}} 22\frac{33}{64}} \quad \text{Ans}$$

Example 5: Find dimension *y* of the counterbored block shown. All dimensions are in inches.

$$2\frac{3}{8} = 2\frac{12}{32} = 1\frac{44}{32}$$

$$- \frac{29}{32} = \frac{29}{32} = \frac{29}{32}$$

$$\overline{\phantom{-\frac{29}{32}} 1\frac{15}{32}}$$

$$y = 1\frac{15}{32}'' \quad \text{Ans}$$

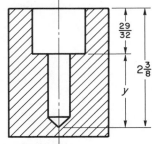

APPLICATION

Subtracting Fractions

1. Subtract each of the following fractions. Reduce to lowest terms where necessary.

   a. $\frac{5}{8} - \frac{11}{32}$ _____

   b. $\frac{7}{8} - \frac{5}{8}$ _____

   c. $\frac{9}{10} - \frac{21}{50}$ _____

   d. $\frac{5}{8} - \frac{9}{64}$ _____

   e. $\frac{9}{16} - \frac{13}{64}$ _____

   f. $\frac{19}{24} - \frac{1}{16}$ _____

2. Determine dimensions A, B, C, and D of this casting. All dimensions are in inches.

   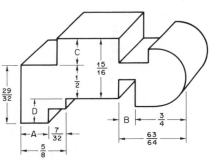

   A = _____
   B = _____
   C = _____
   D = _____

3. Determine dimensions A, B, C, D, E, and F of this drill jig. All dimensions are in inches.

   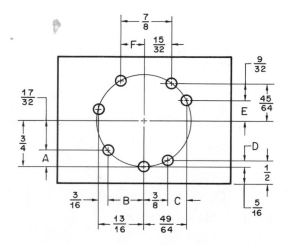

   A = _____
   B = _____
   C = _____
   D = _____
   E = _____
   F = _____

Subtracting Mixed Numbers

4. Determine dimensions A, B, C, D, E, F, and G of this tapered pin. All dimensions are in inches.

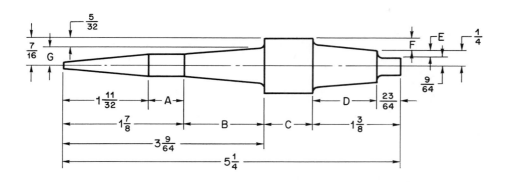

A = _____

B = _____

C = _____

D = _____

E = _____

F = _____

G = _____

5. Determine dimensions A, B, C, D, E, F, G, H, and I of this plate. All dimensions are in inches.

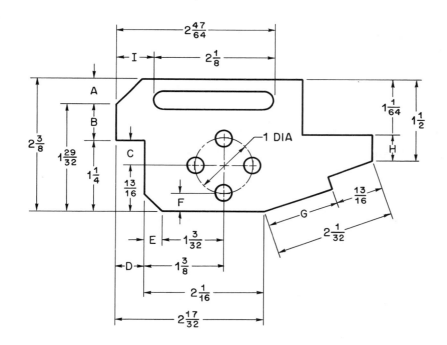

A = _____

B = _____

C = _____

D = _____

E = _____

F = _____

G = _____

H = _____

I = _____

6. Three holes are bored in a checking gage. The lower left edge of the gage is the reference point for the hole locations. Sketch the hole locations and determine the missing distances. From the reference point:

        Hole #1 is 1 3/32″ to the right, and 1 5/8″ up.
        Hole #2 is 2 1/64″ to the right, and 2 3/16″ up.
        Hole #3 is 3 1/4″ to the right, and 3 1/2″ up.

Determine:

a. The horizontal distance between hole #1 and hole #2. _____

b. The horizontal distance between hole #2 and hole #3. _____

c. The horizontal distance between hole #1 and hole #3. _____

d. The vertical distance between hole #1 and hole #2. _____

e. The vertical distance between hole #2 and hole #3. _____

f. The vertical distance between hole #1 and hole #3. _____

UNIT 4 MULTIPLICATION OF COMMON FRACTIONS AND MIXED NUMBERS

OBJECTIVES

After studying this unit you should be able to

- Multiply fractions.
- Multiply mixed numbers.
- Divide by common factors (cancellation).

MULTIPLYING FRACTIONS

Procedure: To multiply two or more fractions

- Multiply the numerators and the denominators separately.
- Write the product of the numerators over the product of the denominators.
- Reduce the resulting fraction to lowest terms.

Example 1: Multiply $\frac{3}{4}$ by $\frac{8}{9}$.

Multiply the numerators.

Multiply the denominators.

$$\frac{3 \times 8}{4 \times 9} = \frac{24}{36} = \frac{24 \div 12}{36 \div 12} = \frac{2}{3} \text{ Ans}$$

Write the product of the numerators over the product of the denominators.

Reduce the resulting fraction to lowest terms.

Example 2: Multiply $\frac{2}{3} \times \frac{5}{6} \times \frac{3}{10}$.

$$\frac{2 \times 5 \times 3}{3 \times 6 \times 10} = \frac{30}{180} = \frac{30 \div 30}{180 \div 30} = \frac{1}{6} \text{ Ans}$$

Example 3: Find the distance between centers of the first and last holes shown in this figure. All dimensions are in inches.

Multiply $6 \times \frac{7}{16} = \frac{6}{1} \times \frac{7}{16} = \frac{6 \times 7}{1 \times 16} = \frac{42}{16}$

Reduce $\frac{42}{16} = 2\frac{10}{16} = 2\frac{5}{8}$

Distance $= 2\frac{5}{8}''$ Ans

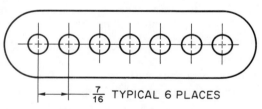

$\frac{7}{16}$ TYPICAL 6 PLACES

Note: The value of a number remains unchanged when the number is placed over a denominator of 1.

DIVIDING BY COMMON FACTORS (CANCELLATION)

Problems involving multiplication of fractions are generally solved more quickly and easily if a numerator and denominator are divided by any common factors before the fractions are multiplied. This process of first dividing by common factors is commonly called *cancellation*.

14

Example 1: Multiply by cancellation method. $\frac{3}{4} \times \frac{8}{9}$

Divide by 3 which is the factor common to both the numerator 3 and the denominator 9.
$3 \div 3 = 1$
$9 \div 3 = 3$

$$\frac{3}{4} \times \frac{8}{9} = \frac{\overset{1}{\cancel{3}}}{\underset{1}{\cancel{4}}} \times \frac{\overset{2}{\cancel{8}}}{\underset{3}{\cancel{9}}} = \frac{1 \times 2}{1 \times 3} = \frac{2}{3} \quad \text{Ans}$$

Divide by 4 which is the factor common to both the denominator 4 and the numerator 8.
$4 \div 4 = 1$
$8 \div 4 = 2$
Multiply reduced fractions.

Example 2: Multiply. $\frac{4}{7} \times \frac{5}{18} \times \frac{14}{15}$

Divide 4 and 18 by 2.
Divide 7 and 14 by 7.
Divide 5 and 15 by 5.
Multiply.

$$\frac{\overset{2}{\cancel{4}}}{\underset{1}{\cancel{7}}} \times \frac{\overset{1}{\cancel{5}}}{\underset{9}{\cancel{18}}} \times \frac{\overset{2}{\cancel{14}}}{\underset{3}{\cancel{15}}} = \frac{2 \times 1 \times 2}{1 \times 9 \times 3} = \frac{4}{27} \quad \text{Ans}$$

Example 3: Multiply. $\frac{5}{14} \times \frac{8}{9} \times \frac{7}{10}$

Divide 5 and 10 by 5.
Divide 14 and 8 by 2.
The process is continued by dividing 7 and 7 by 7 and dividing 2 and 4 by 2.
Multiply.

$$\frac{\overset{1}{\cancel{5}}}{\underset{7}{\underset{1}{\cancel{14}}}} \times \frac{\overset{\overset{2}{\cancel{4}}}{\cancel{8}}}{9} \times \frac{\overset{1}{\cancel{7}}}{\underset{2}{\underset{1}{\cancel{10}}}} = \frac{1 \times 2 \times 1}{1 \times 9 \times 1} = \frac{2}{9} \quad \text{Ans}$$

⌐MULTIPLYING MIXED NUMBERS

Procedure: To multiply mixed numbers
  • Express the mixed numbers as improper fractions.
  • Follow the procedure for multiplying proper fractions.

Example 1: Multiply. $2\frac{2}{5} \times 6\frac{7}{8}$

Express $2\frac{2}{5}$ and $6\frac{7}{8}$ as improper fractions.
Divide 5 and 55 by 5.
Divide 12 and 8 by 4.
Multiply and express the product as a mixed number.

$$\frac{\overset{3}{\cancel{12}}}{\underset{1}{\cancel{5}}} \times \frac{\overset{11}{\cancel{55}}}{\underset{2}{\cancel{8}}} = \frac{3 \times 11}{1 \times 2} = \frac{33}{2} = 16\frac{1}{2} \quad \text{Ans}$$

Example 2: The block of steel shown is to be machined. The block measures $8\frac{3}{4}$ inches long, $4\frac{9}{16}$ inches wide, and $\frac{7}{8}$ inch thick. Find the volume of the block. All dimensions are in inches. (Volume = length × width × thickness.)

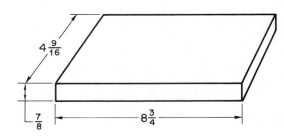

$$8\frac{3}{4} \times 4\frac{9}{16} \times \frac{7}{8} = \frac{35}{4} \times \frac{73}{16} \times \frac{7}{8} = \frac{35 \times 73 \times 7}{4 \times 16 \times 8} = \frac{17885}{512} = 34\frac{477}{512}$$

Volume = $34\frac{477}{512}$ cubic inches Ans

APPLICATION

Multiplying Fractions

1. Multiply these fractions. Reduce to lowest terms where necessary.

a. $\frac{2}{3} \times \frac{1}{6}$ _____   c. $\frac{5}{8} \times \frac{13}{64}$ _____   e. $7 \times \frac{9}{14} \times 3$ _____

b. $\frac{1}{2} \times \frac{1}{8}$ _____   d. $\frac{3}{4} \times \frac{3}{5} \times \frac{2}{3}$ _____   f. $\frac{8}{25} \times \frac{5}{8} \times \frac{3}{7}$ _____

2. Determine dimensions A, B, C, D, and E of the template shown. All dimensions are in inches.

A = _____
B = _____
C = _____
D = _____
E = _____

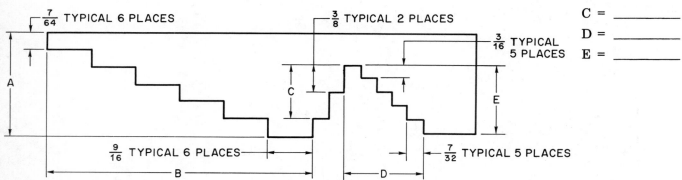

3. A special washer-faced nut is shown. All dimensions are in inches.

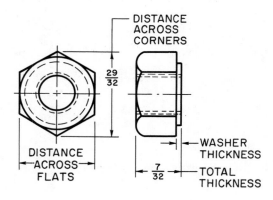

a. Determine the distance across flats. _____

Distance across flats $= \frac{55}{64} \times$ Distance across corners

b. Determine the washer thickness. _____

Washer thickness $= \frac{1}{8} \times$ Total thickness

4. The Unified Thread may have either a flat or rounded crest or root. If the sides of the Unified Thread are extended a sharp V-thread is formed. H is the height of a sharp V-thread. The pitch, P, is the distance between two adjacent threads.

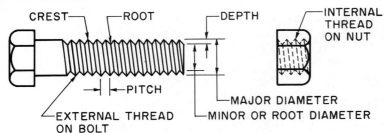

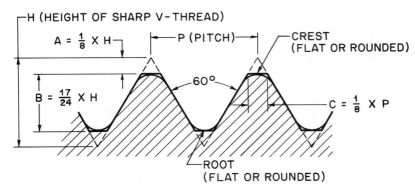

Find dimensions A, B, and C as indicated.

a. H $= \frac{5''}{16}$,   A = _____ ,   B = _____   f. P $= \frac{1''}{4}$,   C = _____

b. H $= \frac{3''}{8}$,   A = _____ ,   B = _____   g. P $= \frac{1''}{6}$,   C = _____

c. H $= \frac{15''}{16}$,   A = _____ ,   B = _____   h. P $= \frac{1''}{20}$,   C = _____

d. H $= \frac{1''}{2}$,   A = _____ ,   B = _____   i. P $= \frac{1''}{28}$,   C = _____

e. H $= \frac{3''}{4}$,   A = _____ ,   B = _____   j. P $= \frac{1''}{32}$,   C = _____

Multiplying Mixed Numbers

5. Multiply these mixed numbers. Reduce to lowest terms where necessary.

a. $1\frac{2}{3} \times 6\frac{3}{10}$ _____   d. $1\frac{2}{3} \times 10\frac{1}{4} \times \frac{3}{8}$ _____

b. $3\frac{5}{16} \times 7\frac{3}{4}$ _____   e. $1\frac{7}{32} \times 2 \times \frac{1}{10}$ _____

c. $3\frac{7}{8} \times 2\frac{1}{2}$ _____   f. $2\frac{2}{3} \times 2\frac{2}{3} \times 5\frac{1}{4}$ _____

6. How many inches of drill rod are required in order to make 20 drills each 3 3/16″ long? Allow 3/32″ waste for each drill.   _____

UNIT 5 DIVISION OF COMMON FRACTIONS AND MIXED NUMBERS

OBJECTIVES

After studying this unit you should be able to

- Divide fractions.
- Divide mixed numbers.

In machine technology, division of fractions and mixed numbers is used in determining production times and costs per machined unit, in calculating the pitch of screw threads, and in computing the number of parts that can be manufactured from a given amount of raw material.

DIVIDING FRACTIONS AS THE INVERSE OF MULTIPLYING FRACTIONS

Division is the inverse of multiplication. Dividing by 2 is the same as multiplying by $\frac{1}{2}$.

$$5 \div 2 = 2\frac{1}{2}$$
$$5 \times \frac{1}{2} = 2\frac{1}{2}$$
$$5 \div 2 = 5 \times \frac{1}{2}$$

Two is the *inverse* of $\frac{1}{2}$, and $\frac{1}{2}$ is the *inverse* of 2. *Inverting* a fraction means turning the fraction upside down, such as, $\frac{1}{3}$ inverted is $\frac{3}{1}$, $\frac{8}{7}$ inverted is $\frac{7}{8}$, $\frac{63}{64}$ inverted is $\frac{64}{63}$, and $\frac{9}{16}$ inverted is $\frac{16}{9}$.

Procedure: To divide fractions

- Invert the divisor.
- Change the division operation to a multiplication operation.
- Follow the procedure for multiplying fractions.

Example 1: Divide $\frac{5}{8}$ by $\frac{3}{4}$.

Invert the divisor.

Change the division operation to a multiplication operation.

$$\frac{5}{8} \div \frac{3}{4} = \frac{5}{\underset{2}{8}} \times \frac{\overset{1}{4}}{3} = \frac{5}{6} \quad \text{Ans}$$

Follow the procedure for multiplication.

Example 2: The machine bolt shown has a pitch of $\frac{1''}{16}$. The pitch is the distance between 2 adjacent threads or the thickness of one thread. Find the number of threads in $\frac{7''}{8}$. All dimensions are in inches.

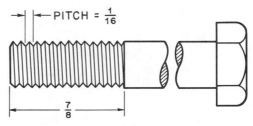

Divide $\frac{7}{8}$ by $\frac{1}{16}$.

$$\frac{7}{8} \div \frac{1}{16} = \frac{7}{\underset{1}{8}} \times \frac{\overset{2}{16}}{1} = 14 \quad \text{Ans}$$

DIVIDING MIXED NUMBERS

Procedure: To divide mixed numbers

- Express the mixed numbers as improper fractions.
- Follow the procedure for dividing fractions.

18

Example 1: Divide $7\frac{1}{2}$ by $2\frac{3}{8}$.

Express $7\frac{1}{2}$ and $2\frac{3}{8}$ as improper fractions. $7\frac{1}{2} \div 2\frac{3}{8} = \frac{15}{2} \div \frac{19}{8} =$

Invert the divisor.

Change the division operation to a multiplication operation. $\frac{15}{\underset{1}{2}} \times \frac{\overset{4}{8}}{19} = \frac{60}{19} = 3\frac{3}{19}$ Ans

Multiply.

Example 2: A section of strip stock is shown with 5 equally spaced holes. Determine the distance between two consecutive holes. All dimensions are in inches.

Note: The number of spaces between the holes is one less than the number of holes.

Express as improper fractions. $4\frac{3}{8} \div 4 = \frac{35}{8} \div \frac{4}{1} =$

Invert the divisor and multiply.

$x = 1\frac{3}{32}''$ Ans $\frac{35}{8} \times \frac{1}{4} = \frac{35}{32} = 1\frac{3}{32}$

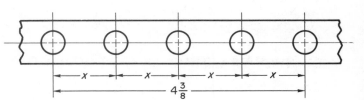

APPLICATION

Inverting Fractions.

Invert each of the following.

1. $\frac{7}{8}$ _____   2. $\frac{1}{4}$ _____   3. $\frac{97}{8}$ _____   4. 2 _____

Dividing Fractions

5. This casting shows seven tapped holes, A–G. The number of threads is determined by dividing the depth of thread by the thread pitch. Find the number of threads in each of the tapped holes. All dimensions are in inches.

A = _____
B = _____
C = _____
D = _____
E = _____
F = _____
G = _____

HOLE A $\frac{1}{12}$ PITCH

HOLE B $\frac{1}{9}$ PITCH

HOLE C $\frac{1}{14}$ PITCH

HOLE G $\frac{1}{5}$ PITCH

HOLE D $\frac{1}{8}$ PITCH

HOLE F $\frac{1}{16}$ PITCH

HOLE E $\frac{1}{11}$ PITCH

$\frac{1}{2}$ $\frac{1}{4}$ $\frac{15}{16}$ $\frac{13}{32}$ $\frac{5}{8}$ $\frac{9}{16}$ $\frac{7}{8}$

6. Bar stock is being cut on a lathe. The tool feeds (advances) 1/32 inch each time the stock turns once (1 revolution). How many revolutions will the stock make when the tool advances 7/8 inch?   _____

7. A groove 15/16 inch deep is to be milled in a steel plate. How many cuts are required if each cut is 3/16 inch deep?   _____

Dividing Mixed Numbers

8. This sheet metal section has 5 sets of drilled holes: A, B, C, D, and E. The holes within a set are equally spaced in the horizontal direction. Compute the horizontal distance between 2 consecutive holes for each set. All dimensions are in inches.

Set A: _____

Set B: _____

Set C: _____

Set D: _____

Set E: _____

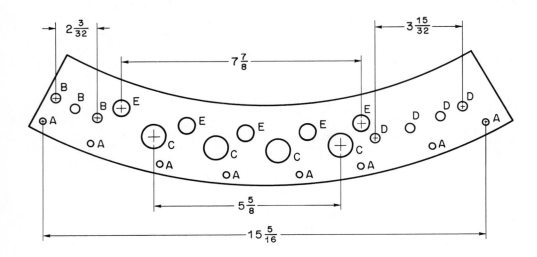

9. The feed on a boring mill is set for 1/64 inch. How many revolutions does the work make when the tool advances 4 1/2 inches?   _____

10. How many complete pieces can be blanked from a strip of steel 27 1/4 feet long if each stamping requires 2 3/16 inches of material plus an allowance of 5/16 inch at one end of the strip? (12 inches = 1 foot)   _____

11. A groove is milled the full length of a steel plate which is 3 1/4 feet long. This operation takes a total of 4 1/16 minutes. How many feet of steel are cut in one minute?   _____

12. How many binding posts can be cut from a brass rod 37 1/2 inches long if each post is 1 7/8 inches long? Allow 3/32 inch waste for each cut.   _____

13. A bar of steel 22 3/4 feet long weighs 107 11/16 pounds. How much does a one-foot length of bar weigh?   _____

14. A single-threaded square thread screw is shown. The lead of a screw is the distance that the screw advances in one turn (revolution). The lead is equal to the pitch in a single-threaded screw. Given the number of turns and the amount of screw advance, determine the leads.

	Screw Advance	Number of Turns	Lead
a.	2 1/4″	10	
b.	7 37/64″	24 1/4	
c.	2 7/16″	6 1/2	
d.	1 1/2″	15	
e.	6 3/10″	12 3/5	

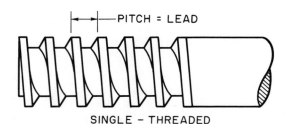

SINGLE – THREADED
SQUARE THREAD SCREW

UNIT 6 COMBINED OPERATIONS OF COMMON FRACTIONS AND MIXED NUMBERS

OBJECTIVES

After studying this unit you should be able to

- Solve problems which involve combined operations of fractions and mixed numbers.
- Solve complex fractions.

Before a part is machined, the sequence of machining operations, the machine setup, and the working dimensions needed to produce the part must be determined. In actual practice, calculations of machine setup and working dimensions require not only the individual operations of addition, subtraction, multiplication, and division, but a combination of two or more of these operations.

ORDER OF OPERATIONS FOR COMBINED OPERATIONS

Procedure:

- **Do all the work in the parentheses first.** Parentheses are used to group numbers. In a problem expressed in fractional form, the numerator and the denominator are each considered as being enclosed in parentheses.

$$\frac{4\frac{3}{4} - \frac{1}{2}}{10 + 6\frac{5}{8}} = \left(4\frac{3}{4} - \frac{1}{2}\right) \div \left(10 + 6\frac{5}{8}\right)$$

If an expression contains parentheses with brackets, do the work within the innermost parentheses first.

- **Do multiplication and division next.** Perform multiplication and division in order from left to right.
- **Do addition and subtraction last.** Perform addition and subtraction in order from left to right.

Combining Addition and Subtraction

Example 1: Find the value of $3\frac{1}{2} - \frac{3}{8} + \frac{5}{16}$.

Subtract $\frac{3}{8}$ from $3\frac{1}{2}$. $3\frac{1}{2} - \frac{3}{8} = 3\frac{1}{8}$

Add $3\frac{1}{8}$ to $\frac{5}{16}$. $3\frac{1}{8} + \frac{5}{16} = 3\frac{7}{16}$ Ans

Example 2: Find x, the distance from the base of the plate to the center of hole #2. All dimensions are in inches.

$x = \frac{9''}{16} + 2\frac{1''}{8} - \frac{13''}{32}$

Add. $\frac{9''}{16} + 2\frac{1''}{8} = 2\frac{11''}{16}$

Subtract. $2\frac{11''}{16} - \frac{13''}{32} = 2\frac{9''}{32}$ Ans

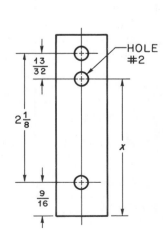

Combining Multiplication and Division

Example 1: Find the value of $\frac{2}{3} \times 8 \div 2\frac{1}{2}$.

Multiply. $\frac{2}{3} \times 8 = \frac{2 \times 8}{3 \times 1} = \frac{16}{3}$

Divide. $\frac{16}{3} \div 2\frac{1}{2} = \frac{16}{3} \times \frac{2}{5} = \frac{32}{15} = 2\frac{2}{15}$ Ans

Example 2: The stainless steel plate shown has grooves which are of uniform length and equally spaced within a distance of $33\frac{1}{2}$ inches. The time required to rough and finish mill a one-inch length of groove is $\frac{7}{10}$ minute. How many minutes are required to cut all the grooves? Disregard the time required to reposition the part. All dimensions are in inches.

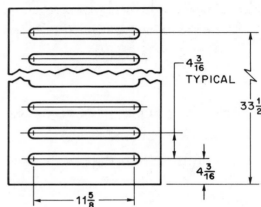

The number of grooves in $33\frac{1}{2}'' = 33\frac{1}{2} \div 4\frac{3}{16}$.

The time required to cut 1 groove $= \frac{7}{10} \times 11\frac{5}{8}$.

Total time equals the number of grooves multiplied by the time for each groove.

$$33\frac{1}{2} \div 4\frac{3}{16} \times \frac{7}{10} \times 11\frac{5}{8}$$

Divide. $33\frac{1}{2} \div 4\frac{3}{16} = \frac{67}{2} \times \frac{16}{67} = 8$

Multiply. $8 \times \frac{7}{10} \times 11\frac{5}{8} = \frac{8}{1} \times \frac{7}{10} \times \frac{93}{8} = 65\frac{1}{10}$

Total Time $= 65\frac{1}{10}$ minutes Ans

Combining Addition, Subtraction, Multiplication, and Division

Example 1: Find the value of $7\frac{5}{6} + 5\frac{1}{2} \div \frac{3}{4} - 10 \times \frac{7}{16}$.

First divide and multiply. $5\frac{1}{2} \div \frac{3}{4} = \frac{11}{2} \times \frac{4}{3} = 7\frac{1}{3}$

$10 \times \frac{7}{16} = \frac{10}{1} \times \frac{7}{16} = 4\frac{3}{8}$

Next add and subtract. $7\frac{5}{6} + 7\frac{1}{3} = 15\frac{1}{6}$

$15\frac{1}{6} - 4\frac{3}{8} = 10\frac{19}{24}$ Ans

Example 2: Find the value of $\left(7\frac{5}{6} + 5\frac{1}{2}\right) \div \frac{3}{4} - 10 \times \frac{7}{16}$.

First do the work in parentheses. $\left(7\frac{5}{6} + 5\frac{1}{2}\right) = 7\frac{5}{6} + 5\frac{3}{6} = 13\frac{1}{3}$

Next divide and multiply. $13\frac{1}{3} \div \frac{3}{4} = \frac{40}{3} \times \frac{4}{3} = \frac{160}{9} = 17\frac{7}{9}$

$10 \times \frac{7}{16} = 4\frac{3}{8}$

Then add and subtract. $17\frac{7}{9} - 4\frac{3}{8} = 13\frac{29}{72}$ Ans

Note: This example is the same as the preceding example except for the parentheses.

COMPLEX FRACTIONS

A *complex fraction* is an expression in which either the numerator or denominator or both are fractions or mixed numbers. A fraction indicates a division operation. Therefore, complex fractions can be solved by dividing the numerator by the denominator.

$$\frac{\frac{5}{9}}{\frac{1}{3}} = \frac{5}{9} \div \frac{1}{3}$$

Example: Find the value of $\dfrac{5\frac{7}{8} + 2\frac{3}{4}}{3\frac{15}{16} - 1\frac{1}{8}}$.

Note: The complete numerator is divided by the complete denominator. Therefore, parentheses are used to indicate that addition in the numerator and subtraction in the denominator must be performed before division.

$$\dfrac{5\frac{7}{8} + 2\frac{3}{4}}{3\frac{15}{16} - 1\frac{1}{8}} \qquad \left(5\frac{7}{8} + 2\frac{3}{4}\right) \div \left(3\frac{15}{16} - 1\frac{1}{8}\right) = 8\frac{5}{8} \div 2\frac{13}{16} = 3\frac{1}{15} \text{ Ans}$$

APPLICATION

Order of Operations for Combined Operations

1. Solve the following examples of combined operations.

    a. $\frac{1}{2} + \frac{3}{16} - \frac{1}{8}$ _____

    b. $3\frac{7}{8} - 2\frac{3}{16} + \frac{3}{8}$ _____

    c. $\frac{3}{10} + 8\frac{2}{5} - 3\frac{1}{25}$ _____

    d. $18 - 2\frac{2}{3} + 4\frac{1}{6}$ _____

    e. $32\frac{1}{8} + 2\frac{3}{16} \times \frac{3}{4}$ _____

    f. $\frac{7}{9} \times \left(\frac{2}{3} + 3\frac{5}{6}\right)$ _____

    g. $16 - 4\frac{1}{2} \div \frac{1}{2} + 2\frac{1}{8}$ _____

    h. $\left(16 - 4\frac{1}{2}\right) \div \frac{1}{2} + 2\frac{1}{8}$ _____

    i. $\left(16 - 4\frac{1}{2}\right) \div \left(\frac{1}{2} + 2\frac{1}{8}\right)$ _____

    j. $15\frac{1}{4} \times 1\frac{1}{3} + 2\frac{2}{3} \div 4\frac{5}{6}$ _____

Complex Fractions

2. Find the value of the following complex fractions.

    a. $\dfrac{\frac{1}{3}}{\frac{1}{2}}$ _____

    b. $\dfrac{3\frac{7}{8}}{5}$ _____

    c. $\dfrac{\frac{15}{16}}{2\frac{1}{8}}$ _____

    d. $\dfrac{\frac{1}{3} + \frac{5}{6}}{7\frac{1}{2}}$ _____

    e. $\dfrac{6\frac{3}{4} - 2\frac{7}{8}}{3\frac{1}{2} + 1\frac{1}{16}}$ _____

    f. $\dfrac{10\frac{1}{2} \times \frac{1}{2}}{4 \div 2\frac{1}{4}}$ _____

Related Problems

3. Refer to the shaft shown. Determine the missing dimensions in the table using the dimensions given. All dimensions are in inches.

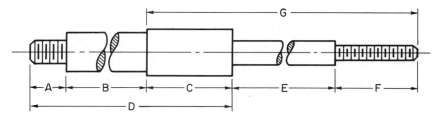

	A	B	C	D	E	F	G
a.	1/2		1 3/8	6 3/4		15/16	8 1/8
b.		3 13/16	1 5/8	5 37/64	4 3/8	3/4	
c.	9/16	4 3/32			5 1/8	27/32	7 1/32
d.	5/8		1 7/16	5 31/32		7/8	7 15/16
e.		3 3/4	1 11/16	6 1/32	4 61/64	25/32	
f.	11/16	4 1/8			5 3/16	7/8	7 3/64

4. The outside diameter of an aluminum tube is 3 1/16 inches. The wall thickness is 5/32 inch. What is the inside diameter? _____

5. Four studs of the following lengths in inches are to be machined from bar stock: 2 3/4", 1 7/8", 2 5/16", and 1 13/32". Allow 1/8 inch waste for each cut and 1/32 inch on each end of each stud for facing. What is the total length of bar stock required? _____

6. Find dimensions A, B, C, and D of the idler bracket in the figure. All dimensions are in inches.

    A = _____

    B = _____

    C = _____

    D = _____

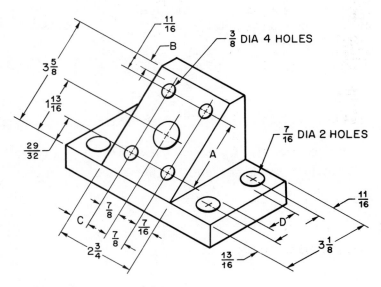

7. How long does it take to cut a distance of 2 1/2 feet along a shaft that turns 150 revolutions per minute with a tool feed of 1/32 inch per revolution? _____

8. An angle iron 47 1/2 inches long has two drilled holes which are equally spaced from the center of the piece. The center distance between the two holes is 19 7/8 inches. What is the distance from each end of the piece to the closest hole? _____

9. A tube has an inside diameter of 3/4 inch and a wall thickness of 1/16 inch. The tube is to be fitted in a drilled hole in a block. What diameter hole should be drilled in the block to give 1/64 inch total clearance? _____

10. Two views of a mounting block are shown. Determine dimensions A–G. All dimensions are in inches.

    A = _____

    B = _____

    C = _____

    D = _____

    E = _____

    F = _____

    G = _____

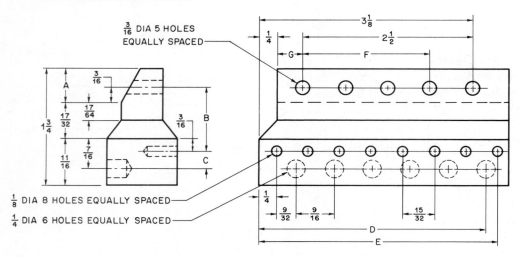

11. The composition of an aluminum alloy by weight is 24/25 aluminum and 1/40 copper. The only other element in the alloy is magnesium. How many pounds of magnesium are required for casting 125 pounds of alloy? _____

12. Pieces of the following lengths are cut from an 18-inch steel bar: 2 1/2", 1 3/4", 1 7/8", and 5/16". Allowing 1/8-inch waste for each cut, what is the length of bar left after the pieces are cut? _____

UNIT 7 INTRODUCTION TO DECIMAL FRACTIONS

OBJECTIVES

After studying this unit you should be able to
- Locate decimal fractions on a number line.
- Express common fractions having denominators of powers of ten as equivalent decimal fractions.
- Write decimal numbers in word form.
- Write numbers expressed in word form as decimal fractions.

Most blueprints are dimensioned with decimal fractions rather than common fractions. The dials which are used in establishing machine settings and movement, in determining tool speeds and travel, and in measuring dimensions of parts are usually graduated in decimal units.

EXPLANATION OF DECIMAL FRACTIONS

A decimal fraction is not written as a common fraction with a numerator and denominator. The denominator is omitted and replaced by a decimal point placed to the left of the numerator. *Decimal fractions* are equivalent to common fractions having denominators which are powers of 10, such as 10; 100; 1000; 10,000; 100,000; and 1,000,000. *Powers of 10* are numbers which are obtained by multiplying 10 by itself a certain number of times.

MEANING OF FRACTIONAL PARTS

The line segment shown is 1 unit long. It is divided into 10 equal smaller parts. The locations of common fractions and their decimal fraction equivalents are shown on the line.

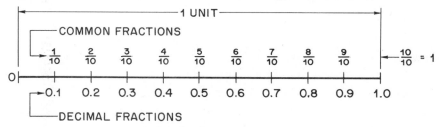

1 UNIT LINE

One of the ten equal small parts, 1/10 (0.1) of the 1 unit line, is shown enlarged. The 1/10 or 0.1 unit is divided into 10 equal smaller units. The locations of common fractions and their decimal fraction equivalents are shown on this line.

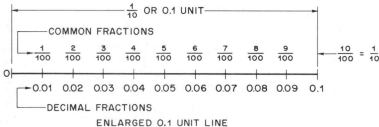

ENLARGED 0.1 UNIT LINE

If the 1/100 (0.01) division is divided into 10 equal smaller parts, the resulting parts are 1/1000 (0.001); 2/1000 (0.002); 3/1000 (0.003); . . . 9/1000 (0.009); 10/1000 = 1/100 (0.01).
- Each time the decimal point is moved one place to the left, a value 1/10 (0.1) times the previous value is obtained.
- Each time a decimal point is moved one place to the right, a value 10 times greater than the previous value is obtained.

Each time a decimal fraction is multiplied by 10 the decimal point is moved one place to the right. Each step in the following table shows both the decimal fraction and its equivalent common fraction.

Decimal Fraction	Common Fraction
0.000003 × 10 = 0.00003	3/1,000,000 × 10 = 3/100,000
0.00003 × 10 = 0.0003	3/100,000 × 10 = 3/10,000
0.0003 × 10 = 0.003	3/10,000 × 10 = 3/1,000
0.003 × 10 = 0.03	3/1000 × 10 = 3/100
0.03 × 10 = 0.3	3/100 × 10 = 3/10
0.3 × 10 = 3.	3/10 × 10 = 3

READING AND WRITING DECIMAL FRACTIONS

The following chart gives the names of the parts of a number with respect to the positions from the decimal point.

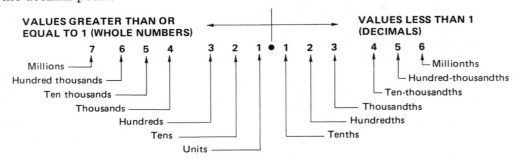

To read a decimal, read the number as a whole number. Then say the name of the decimal place of the last digit to the right.

Examples: 1. 0.5 is read as five tenths.
            2. 0.07 is read as seven hundredths.
            3. 0.011 is read as eleven thousandths.

To write a decimal fraction from a word statement, write the number using a decimal point and zeros before the number as necessary for the given place value.

Examples: 1. Two hundred nineteen ten-thousandths is written as 0.0219.
            2. Forty-three hundred-thousandths is written as 0.00043.
            3. Eight hundred seventeen millionths is written as 0.000817.

A number that consists of a whole number and a decimal fraction is called a *mixed decimal*. To read a mixed decimal, read the whole number, read the word *and* at the decimal point, and read the decimal.

Examples: 1. 3.4 is read as three and four tenths.
            2. 1.002 is read as one and two thousandths.
            3. 16.0793 is read as sixteen and seven hundred ninety-three ten-thousandths.
            4. 8.00032 is read as eight and thirty-two hundred-thousandths.

SIMPLIFIED METHOD OF READING DECIMAL FRACTIONS

Usually a simplified method of reading decimal fractions is used in the machine trades. This method is generally quicker, easier, and less likely to be misinterpreted. A tool-and-die maker reads 0.0265 inches as point zero, two, six, five inches. A machinist reads 4.172 millimeters as four, point one, seven, two millimeters.

WRITING DECIMAL FRACTIONS FROM COMMON FRACTIONS HAVING DENOMINATORS WHICH ARE POWERS OF TEN

A common fraction with a denominator which is a power of ten can be written as a decimal fraction. For a common fraction with a numerator smaller than the denominator, replace the denominator with a decimal point. The decimal point is placed to the left of the first digit of the numerator. There are as many decimal places as there are zeros in the denominator. When writing a decimal fraction it is advisable to place a zero to the left of the decimal point.

Examples:

1. $\frac{9}{10}$ = 0.9 Ans   There is 1 zero in 10 and 1 decimal place in 0.9.

2. $\frac{381}{1000}$ = 0.381 Ans   There are 3 zeros in 1000 and 3 decimal places in 0.381.

3. $\frac{7}{10,000}$ = 0.0007 Ans   There are 4 zeros in 10,000 and 4 decimal places in 0.0007. In order to maintain proper place value, 3 zeros are written between the decimal point and the 7.

APPLICATION

Meaning of Fractional Parts

1. Find the decimal value of each of the distances A, B, C, D, and E. Note the total unit value of the line.

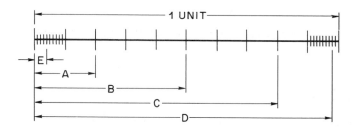

A = _____

B = _____

C = _____

D = _____

E = _____

2. Find the decimal value of each of the distances A, B, C, D, and E. Note the total unit value of the line.

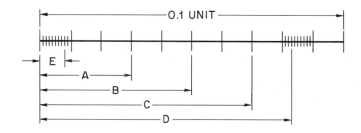

A = _____

B = _____

C = _____

D = _____

E = _____

3. Find the decimal value of each of the distances A, B, C, D, and E. Note the total unit value of the line.

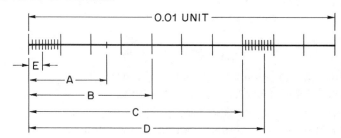

A = _____

B = _____

C = _____

D = _____

E = _____

In each of the following problems, the value on the left must be multiplied by one of the following numbers: 0.0001; 0.001; 0.01; 0.1; 10; 100; 1000; or 10,000 in order to obtain the value on the right of the equal sign. Determine the proper number.

4. 0.9 × _____ = 0.0009   8. 0.135 × _____ = 0.00135   11. 0.0643 × _____ = 6.43

5. 0.7 × _____ = 0.007   9. 4 × _____ = 0.4   12. 0.00643 × _____ = 64.3

6. 0.03 × _____ = 0.3   10. 0.0643 × _____ = 0.000643   13. 643 × _____ = 0.643

7. 0.0003 × _____ = 0.003

Reading and Writing Decimal Fractions

Write these numbers as words.

14. 0.032 _____   19. 1.5 _____

15. 0.007 _____   20. 10.37 _____

16. 0.132 _____   21. 17.0009 _____

17. 0.0075 _____   22. 4.0012 _____

18. 0.108 _____   23. 13.103 _____

Write these words as numbers.

24. eighty-four ten-thousandths _____   28. thirty-five ten-thousandths _____

25. seven tenths _____   29. ten and two tenths _____

26. forty-three and eight hundredths _____   30. five and one ten-thousandths _____

27. two and seven hundred-thousandths _____   31. twenty and seventy-one hundredths _____

Each of the following common fractions has a denominator which is a power of 10. Write the equivalent decimal fraction for each.

32. $\frac{9}{10}$ _____   35. $\frac{43}{100}$ _____   38. $\frac{73}{1000}$ _____

33. $\frac{3}{10,000}$ _____   36. $\frac{99}{1000}$ _____   39. $\frac{1973}{100,000}$ _____

34. $\frac{17}{100}$ _____   37. $\frac{999}{10,000}$ _____   40. $\frac{63,917}{100,000}$ _____

UNIT 8 ROUNDING DECIMAL FRACTIONS AND EQUIVALENT DECIMAL AND COMMON FRACTIONS

OBJECTIVES

After studying this unit you should be able to
- Round decimal fractions to any required number of places.
- Express common fractions as decimal fractions.
- Express decimal fractions as common fractions.

When blueprint dimensions of a part are given in fractional units, a machinist is usually required to express these fractional values as decimal working dimensions. In computing material requirements and in determining stock waste and scrap allowances, it is sometimes more convenient to express decimal values as approximate fractional equivalents.

ROUNDING DECIMAL FRACTIONS

When working with decimals, the computations and answers may contain more decimal places than are required. The number of decimal places needed depends on the degree of precision desired. The degree of precision depends on how the decimal value is going to be used. The tools, machines, equipment, and materials determine the degree of precision obtainable. For example, a length of 0.875376 inch cannot be cut on a milling machine. In cutting to the nearer thousandths of an inch, the machinist would consider 0.875376 inch as 0.875 inch. *Rounding a decimal* means expressing the decimal with a fewer number of decimal places.

Procedure: To round a decimal fraction
- Determine the number of decimal places required in an answer.
- If the digit directly following the last decimal place required is less than 5, drop all digits which follow the required number of decimal places.
- If the digit directly following the last decimal place required is 5 or larger, add one to the last required digit and drop all digits which follow the required number of decimal places.

Example 1: Round 0.873429 to three decimal places.

The digit following the third decimal place is 4.   0.873 ④ 29

Because 4 is less than 5, drop all digits after the third decimal place.   0.873 Ans

Example 2: Round 0.36845 to two decimal places.

The digit following the second decimal place is 8.   0.36 ⑧ 45

Because 8 is greater than 5, add 1 to the 6.   0.37 Ans

Example 3: Round 18.738257 to four decimal places.

The digit following the fourth decimal place is 5.   18.7382 ⑤ 7

Add 1 to the 2.   18.7383 Ans

EXPRESSING COMMON FRACTIONS AS DECIMAL FRACTIONS

A common fraction is an indicated division. For example, 3/4 is the same as $3 \div 4$; 5/16 is the same as $5 \div 16$; 99/171 is the same as $99 \div 171$.

Because both the numerator and the denominator of a common fraction are whole numbers, expressing a common fraction as a decimal fraction requires division with whole numbers.

Procedure: To express a common fraction as a decimal fraction
- Divide the numerator by the denominator.

A common fraction which divides evenly is expressed as an even or *terminating decimal.* A common fraction which will not divide evenly is expressed as a repeating or *nonterminating decimal.*

The division should be carried out to one more place than the number of places required in the answer, then rounded one place.

Example 1: Express $\frac{2}{3}$ as a 4-place decimal.

Divide the numerator by the denominator.

After the 2, add one more zero than the required number of decimal places. (Add 5 zeros.)

$$\begin{array}{r} 0.66666 \\ 3\overline{)2.00000} \end{array}$$

Round 0.66666 to 4 places.   0.6667   Ans

Example 2: Express $\frac{5}{7}$ as a 2-place decimal.

Add 3 zeros after the 5.

$$\begin{array}{r} 0.714 \\ 7\overline{)5.000} \end{array}$$

Round to 2 places.   0.71 Ans

EXPRESSING DECIMAL FRACTIONS AS COMMON FRACTIONS

Procedure: To express a decimal fraction as a common fraction

- Make the numerator of the common fraction the decimal with the decimal point omitted.
- The denominator is 1 followed by the same number of zeros as there are decimal places in the decimal fraction.
- Reduce to lowest terms.

Example 1:   Express 0.375 as a common fraction.

The decimal 0.375 without the decimal point is the numerator. The numerator is 375.

$$0.375 = \frac{375}{1000} = \frac{3}{8} \text{ Ans}$$

The denominator is 1 with the same number of zeros as there are decimal places in the decimal fraction. There are three zeros. The denominator is 1000.

Reduce $\frac{375}{1000}$ to lowest terms.

Example 2:   Express 0.27 as a common fraction.

The numerator is 27.

The denominator is 100.

$$0.27 = \frac{27}{100} \text{ Ans}$$

Example 3:   Express 0.03125 as a common fraction.

The numerator is 3125.

The denominator is 100,000.

$$0.03125 = \frac{3125}{100,000} = \frac{1}{32} \text{ Ans}$$

Reduce $\frac{3125}{100,000}$ to lowest terms.

APPLICATION

Rounding Decimal Fractions

Round the following decimals to the indicated number of decimal places.

1. 0.78273 (3 places) _____
2. 0.1247 (2 places) _____
3. 0.23975 (3 places) _____
4. 0.01723 (3 places) _____
5. 0.01723 (2 places) _____

6. 0.90039 (2 places) _____
7. 0.72008 (4 places) _____
8. 0.0001 (3 places) _____
9. 0.0005 (3 places) _____
10. 0.099 (2 places) _____

Expressing Common Fractions as Decimal Fractions

Express the common fractions as decimal fractions. Express the answer to 4 decimal places.

11. $\frac{13}{32}$ _____
12. $\frac{7}{8}$ _____
13. $\frac{5}{8}$ _____
14. $\frac{1}{4}$ _____

15. $\frac{2}{3}$ _____
16. $\frac{10}{11}$ _____
17. $\frac{1}{25}$ _____
18. $\frac{47}{64}$ _____

19. $\frac{7}{32}$ _____
20. $\frac{1}{2}$ _____
21. $\frac{4}{7}$ _____
22. $\frac{3}{8}$ _____

Solve the following.

23. What decimal fraction of distance B is distance A? Express the answer to 4 decimal places. All dimensions are in inches.

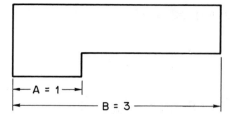

24. What decimal fraction of distance D is distance C? All dimensions are in inches.

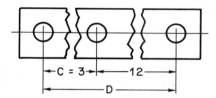

25. Dimensions in this figure are in feet and inches.

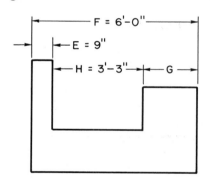

   a. What decimal fraction of distance F is distance E? Note: Both the numerator and denominator of a common fraction must be in the same units before the value is expressed as a decimal fraction. Use 1 foot = 12 inches.

   b. What decimal fraction of distance H is distance G? Express the answer to 4 decimal places.

Expressing Decimal Fractions as Common Fractions

Express the following decimal fractions as common fractions. Reduce to lowest terms.

26. 0.625	_____	33. 0.003	_____	40. 0.0005	_____
27. 0.125	_____	34. 0.004	_____	41. 0.06	_____
28. 0.4	_____	35. 0.502	_____	42. 0.09375	_____
29. 0.75	_____	36. 0.99	_____	43. 0.753	_____
30. 0.8	_____	37. 0.4375	_____	44. 0.45	_____
31. 0.6875	_____	38. 0.1111	_____	45. 0.045	_____
32. 0.67	_____	39. 0.8717	_____	46. 0.0045	_____

Solve the following.

47. What common fractional part of distance B is distance A? All dimensions
are in inches.   _____

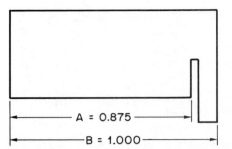

48. What common fractional part of diameter C is diameter D? All dimensions
are in feet.   _____

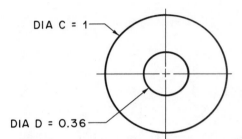

49. What common fractional part of distance A is each distance listed. All dimensions are in inches.

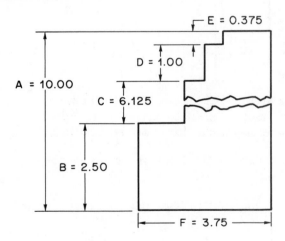

a. Distance B   _____

b. Distance C   _____

c. Distance D   _____

d. Distance E   _____

e. Distance F   _____

UNIT 9 ADDITION AND SUBTRACTION OF DECIMAL FRACTIONS

OBJECTIVES

After studying this unit you should be able to
- Add decimal fractions.
- Add combinations of decimals, mixed decimals, and whole numbers.
- Subtract decimal fractions.
- Subtract combinations of decimals, mixed decimals, and whole numbers.

Adding and subtracting decimal fractions are required at various stages in the production of most products and parts. It is necessary to add and subtract decimals in order to estimate machining costs and production times, to compute stock allowances and tolerances, to determine locations and lengths of cuts, and to inspect finished parts.

ADDING DECIMAL FRACTIONS

Procedure: To add decimal fractions
- Arrange the numbers so that the decimal points are directly under each other.
- Proceed with addition as with whole numbers.
- Place the decimal point in the sum directly under the other decimal points.

Example 1: Add. 7.35 + 114.075 + 0.3422 + 0.003 + 218.7

Note: To reduce the possibility of error, add zeros to decimals so that all the values have the same number of places to the right of the decimal point. Zeros added in this manner do not affect the value of the number.

Arrange the numbers so that the decimal points are directly under each other.

Proceed with addition as with whole numbers.

The decimal point of the sum is placed in the same position as the other decimal points.

$$
\begin{array}{r}
7.3500 \\
114.0750 \\
0.3422 \\
0.0030 \\
+\ 218.7000 \\
\hline
340.4702 \quad \text{Ans}
\end{array}
$$

The decimal point location of a whole number is directly to the right of the last digit.

Example 2: Find the length x of the swivel bracket shown. All dimensions are in millimeters.

Add.
$$
\begin{array}{r}
8.78 \\
25.40 \\
12.80 \\
30.00 \\
3.90 \\
+\ 9.25 \\
\hline
90.13
\end{array}
$$

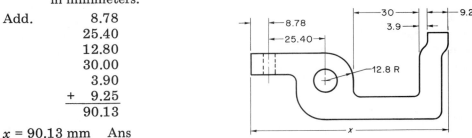

$x = 90.13$ mm   Ans

SUBTRACTING DECIMAL FRACTIONS

Procedure: To subtract decimal fractions
- Arrange the numbers so that the decimal points are directly under each other.
- Proceed with subtraction as with whole numbers.
- Place the decimal point in the difference under the other decimal points.

Example 1: Subtract 13.261 from 25.6.

Arrange the numbers so that the decimal points are directly under each other.

Add two zeros to 25.6 so that it has the same number of decimal places as 13.261. Place the decimal point of the answer in the same position as the other decimal points.

Subtract.

$$\begin{array}{r} 25.600 \\ -\ 13.261 \\ \hline 12.339 \quad \text{Ans} \end{array}$$

Example 2: Determine dimensions A, B, C, and D of the support bracket shown. All dimensions are given in inches.

Solve for A:
A = 0.505 – 0.18
A = 0.325″   Ans

$$\begin{array}{r} 0.505 \\ -\ 0.180 \\ \hline 0.325 \end{array}$$

Solve for B:
B = 1.4 – 0.301
B = 1.099″   Ans

$$\begin{array}{r} 1.400 \\ -\ 0.301 \\ \hline 1.099 \end{array}$$

Solve for C:
C = 1.74 – 0.365
C = 1.375″   Ans

$$\begin{array}{r} 1.740 \\ -\ 0.365 \\ \hline 1.375 \end{array}$$

Solve for D:
D = 0.746 – 0.46
D = 0.286″   Ans

$$\begin{array}{r} 0.746 \\ -\ 0.460 \\ \hline 0.286 \end{array}$$

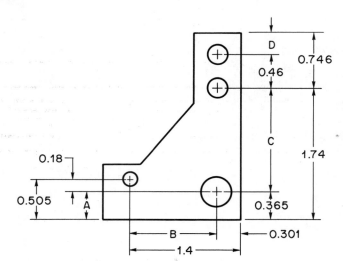

APPLICATION

Adding Decimal Fractions

1. Add the following numbers

    a. 0.132 + 12.9 + 5   _____

    b. 0.003 + 0.13795   _____

    c. 0.375 + 0.8 + 0.12   _____

    d. 4.187 + 0.932 + 0.01   _____

    e. 873.14 + 19.3 + 0.137   _____

    f. 4 + 0.4 + 0.04 + 0.004   _____

    g. 87 + 0.0239 + 7.23   _____

    h. 0.0001 + 0.1 + 0.01   _____

    i. 4.705 + 0.0937 + 0.98   _____

    j. 0.057 + 5.7 + 570   _____

2. Determine dimensions A, B, C, D, E, and F of the profile gage shown. All dimensions are in inches.

    A = _____

    B = _____

    C = _____

    D = _____

    E = _____

    F = _____

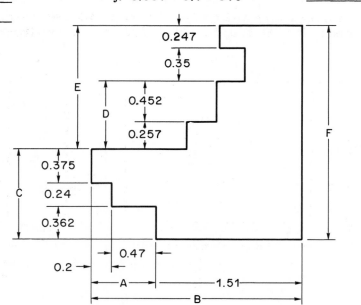

3. A sine plate is to be set to a desired angle by using size blocks of the following thicknesses: 2.000 inches, 0.500 inch, 0.250 inch, 0.125 inch, 0.100 inch, 0.1007 inch, and 0.1001 inch. Determine the total height that the sine plate is raised. _____

4. Three cuts are required to turn a steel shaft. The depths of the cuts, in millimeters, are 6.25, 3.18, and 0.137. How much stock has been removed per side? Round answer to 2 decimal places. _____

5. Determine dimensions A, B, C, D, and E of the gear arm shown. All dimensions are given in millimeters.

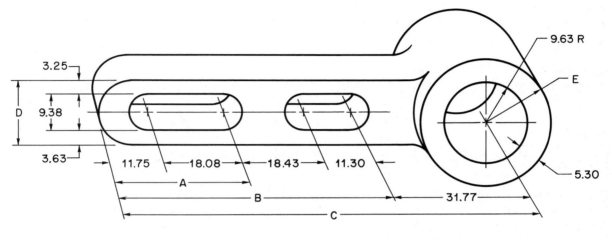

A = _____   B = _____   C = _____   D = _____   E = _____

Subtracting Decimal Fractions

6. Subtract the following numbers. Where necessary, round answers to 3 decimal places.

   a. 0.617 – 0.4136 _____   d. 0.3 – 0.299 _____   g. 0.313 – 0.2323 _____
   b. 0.319 – 0.0127 _____   e. 0.4327 – 0.232 _____   h. 3.872 – 0.0002 _____
   c. 2.308 – 0.7859 _____   f. 23.062 – 0.973 _____   i. 5.923 – 3.923 _____

7. The front and right side views of a sliding shoe are shown. Determine dimensions A, B, C, D, E, and F. All dimensions are in millimeters.

   A = _____
   B = _____
   C = _____
   D = _____
   E = _____
   F = _____

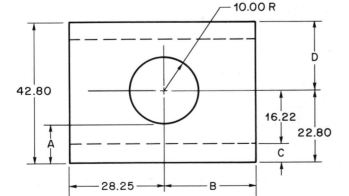

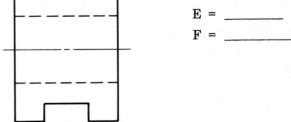

8. Refer to the plate shown and determine the following distances. All dimensions are in inches.

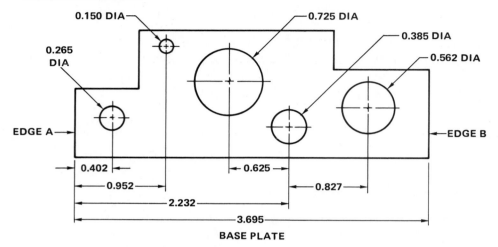

BASE PLATE

a. The center distance between the 0.265″ diameter hole and the 0.150″ diameter hole. _____

b. The center distance between the 0.385″ diameter hole and the 0.150″ diameter hole. _____

c. The distance between edge A and the center of the 0.725″ diameter hole. _____

d. The distance between edge B and the center of the 0.385″ diameter hole. _____

e. The distance between edge B and the center of the 0.562″ diameter hole. _____

UNIT 10 *MULTIPLICATION OF DECIMAL FRACTIONS*

OBJECTIVES

After studying this unit you should be able to
* Multiply decimal fractions.
* Multiply combinations of decimals, mixed decimals, and whole numbers.

A machinist must readily be able to multiply decimal fractions for computing machine feeds and speeds, for determining tapers, and for determining lengths and stock sizes. Multiplication of decimal fractions is also required in order to solve problems which involve geometry and trigonometry.

MULTIPLYING DECIMAL FRACTIONS

Procedure: To multiply decimal fractions
* Multiply using the same procedure as with whole numbers.
* Beginning at the right of the product, point off the same number of decimal places as there are in the multiplicand and the multiplier combined.

Example 1: Multiply 50.123 by 0.87.

Multiply the same as with whole numbers.

Beginning at the right of the product, point off as many decimal places as there are in both the multiplicand and the multiplier.

Multiplicand → 50.123 (3 places)
Multiplier → ✕ 0.87 (2 places)
3 50861
40 0984
Product → 43.60701 (5 places) Ans

Example 2: Compute the lengths of thread on each end of this shaft. All dimensions are in inches.

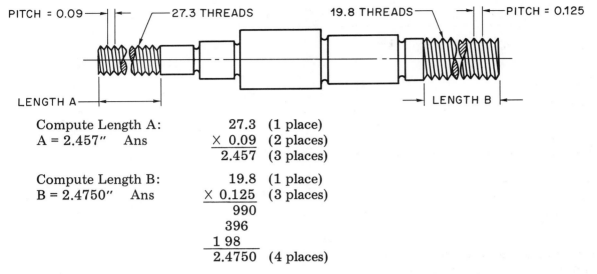

Compute Length A: 27.3   (1 place)
A = 2.457″   Ans   X 0.09   (2 places)
                          2.457   (3 places)

Compute Length B: 19.8   (1 place)
B = 2.4750″   Ans   X 0.125   (3 places)
                           990
                           396
                      1 98
                       2.4750   (4 places)

When multiplying certain decimal fractions, the product has a smaller number of digits than the number of decimal places required. For these products add as many zeros to the left of the product as are necessary to give the required number of decimal places.

Example: Multiply 0.0237 by 0.04. Round the answer to 5 decimal places.

Multiply.   0.0237   (4 places)

The multiplicand, 0.0237, has four decimal   X   0.04   (2 places)
places, and the multiplier, 0.04, has two deci-   0.000948   (6 places)
mal places. Therefore, the product must have
six decimal places.

Add three zeros to the left of the product.   0.00095   Ans
Round 0.000948 to 5 places.

APPLICATION

Multiplying Decimal Fractions

1. Multiply these numbers. Where necessary, round the answers to 4 decimal places.

   a. 3.876 X 0.012_____   b. 2.2 X 1.5_____   c. 40 X 0.15_____   d. 8.93 X 0.32_____

2. A section of a spur gear is shown. Given the circular pitches for various gear sizes, determine the working depths, clearances, and tooth thicknesses. Round the answers to 4 decimal places.

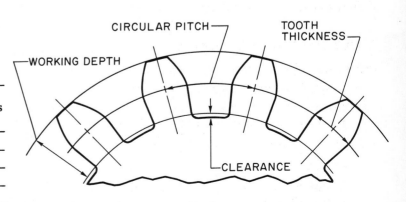

	Circular Pitch (inches)	Working Depth (inches)	Clearance (inches)	Tooth Thickness (inches)
a.	0.3925			
b.	0.1582			
c.	0.8759			
d.	1.2378			
e.	1.5931			

Working depth = 0.6366 × Circular Pitch
Clearance = 0.05 × Circular Pitch
Tooth thickness = 0.5 × Circular Pitch

3. Determine diameters A, B, C, D, and E of this shaft. All dimensions are in millimeters.

   A = _____

   B = _____

   C = _____

   D = _____

   E = _____

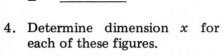

4. Determine dimension x for each of these figures.

   a. All dimensions are in inches.

   _____

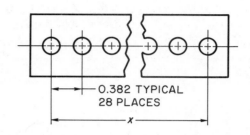

   b. All dimensions are in millimeters.

   _____

   c. Round the answer to 3 decimal places. All dimensions are in inches.

   _____

   d. Round the answer to 3 decimal places. All dimensions are in inches.

   _____

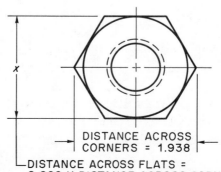

UNIT 11 *DIVISION OF DECIMAL FRACTIONS*

OBJECTIVES

After studying this unit you should be able to

- Divide decimal fractions.
- Divide decimal fractions with whole numbers.
- Divide decimal fractions with mixed decimals.

Division with decimal fractions is used for computing the manufacturing cost and time per piece after total production costs and times have been determined. Division with decimal fractions is also required in order to compute thread pitches, gear tooth thicknesses and depths, cutting speeds, and depths of cut.

DIVIDING DECIMAL FRACTIONS

Moving a decimal point to the right is equivalent to multiplying the decimal by a power of 10.

$$0.237 \times 10 = 2.37 \qquad\qquad 0.237 \times 1000 = 237.$$

$$0.237 \times 100 = 23.7 \qquad\qquad 0.237 \times 10,000 = 2370.$$

When dividing decimal fractions, the value of the answer (quotient) is not changed if the decimal points of both the divisor and the dividend are moved the same number of places to the right. It is the same as multiplying both divisor and dividend by the same number.

$$0.9375 \div 0.612 = (0.9375 \times 1000) \div (0.612 \times 1000) = 937.5 \div 612.$$

$$14.203 \div 6.87 = (14.203 \times 100) \div (6.87 \times 100) = 1420.3 \div 687.$$

Procedure: To divide decimal fractions

- Move the decimal point of the divisor as many places to the right as are necessary to make the divisor a whole number.
- Move the decimal point of the dividend the same number of places as were moved in the divisor.
- Place the decimal point in the quotient directly above the decimal point in the dividend.
- Add zeros to the dividend if necessary.
- Divide as with whole numbers.

Example 1: Divide 0.643 by 0.28. Round the answer to 3 decimal places.

To make the divisor a whole number move the decimal point 2 places, 28.

The decimal point in the dividend is also moved 2 places, 64.3

Add 3 zeros to the dividend. One extra place is necessary in order to round the answer to 3 decimal places.

Place the decimal point of the quotient directly above the decimal point of the dividend.

Divide as with whole numbers.

$$
\begin{array}{r}
2.2964 \approx 2.296 \quad \text{Ans} \\
28\overline{)\,64.3000} \\
\underline{56} \\
8\,3 \\
\underline{5\,6} \\
2\,70 \\
\underline{2\,52} \\
180 \\
\underline{168} \\
120 \\
\underline{112} \\
8
\end{array}
$$

39

Example 2: 3.19 ÷ 0.072 (Round the answer to 2 decimal places.)

Move the decimal point 3 places in the divisor, and 3 places in the dividend.

Add 3 zeros to the dividend.

Place the decimal point of the quotient directly above the decimal point of the dividend.

Divide.

$$
\begin{array}{r}
44.305 \approx 44.31 \quad \text{Ans} \\
72\overline{)\ 3190.000} \\
\underline{288} \\
310 \\
\underline{288} \\
22\ 0 \\
\underline{21\ 6} \\
400 \\
\underline{360} \\
40
\end{array}
$$

When dividing a decimal fraction or a mixed decimal by a whole number, it is not necessary to move the decimal point of either the divisor or the dividend. Add zeros to the right of the dividend, if necessary, to obtain the desired number of decimal places in the answer.

Examples:

1. Divide 0.63 by 12 to 4 decimal places.

$$
\begin{array}{r}
0.0525 \quad \text{Ans} \\
12\overline{)\ 0.6300}
\end{array}
$$

2. Divide 33.97 by 5 to 3 decimal places.

$$
\begin{array}{r}
6.794 \quad \text{Ans} \\
5\overline{)\ 33.970}
\end{array}
$$

APPLICATION

Dividing Decimal Fractions

1. Divide the following numbers. Express the answers to the indicated number of decimal places.

    a. 0.72 ÷ 0.432 (3 places) _____

    b. 0.92 ÷ 0.36 (2 places) _____

    c. 0.001 ÷ 0.1 (4 places) _____

    d. 10 ÷ 0.001 (3 places) _____

    e. 1.017 ÷ 0.07 (3 places) _____

    f. $\dfrac{16.3}{3.8}$ (2 places) _____

    g. $\dfrac{37}{0.273}$ (2 places) _____

    h. $\dfrac{0.003}{0.78}$ (4 places) _____

2. Rack sizes are given according to diametral pitch. Given 4 different diametral pitches, find the linear pitch and the whole depth of each rack to 4 decimal places. All dimensions are in inches.

$$
\text{Linear Pitch} = \frac{3.1416}{\text{Diametral Pitch}} \qquad \text{Whole Depth} = \frac{2.157}{\text{Diametral Pitch}}
$$

	Diametral Pitch	Linear Pitch	Whole Depth
a.	6.75		
b.	2.5		
c.	7.25		
d.	16.125		

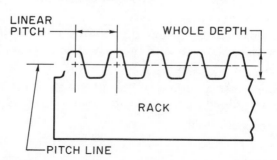

3. Four sets of equally spaced holes are shown in this machined plate. Determine dimensions A, B, C, and D to 2 decimal places. All dimensions are in millimeters.

A = _____
B = _____
C = _____
D = _____

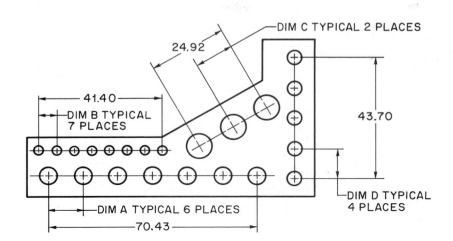

4. A cross-sectional view of a bevel gear is shown. Given the diametral pitch and the number of gear teeth, determine the pitch diameter, the addendum, and the dedendum. Round the answers to 4 decimal places.

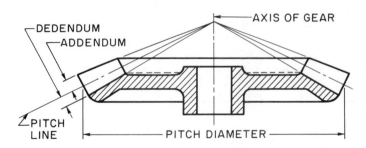

	Diametral Pitch	Number of Teeth	Pitch Diameter (inches)	Addendum (inches)	Dedendum (inches)
a.	4	45			
b.	6	75			
c.	8	44			
d.	3	54			

$$\text{Pitch Diameter} = \frac{\text{Number of Teeth}}{\text{Diametral Pitch}}$$

$$\text{Dedendum} = \frac{1.157}{\text{Diametral Pitch}}$$

$$\text{Addendum} = \frac{1}{\text{Diametral Pitch}}$$

5. How many complete bushings each 15.92 millimeters long can be cut from a bar of bronze which is 468.75 millimeters long? Allow 3.12 millimeters waste for each piece.

_____

6. A shaft is being cut in a lathe. The tool feeds (advances) 0.015 inch each time the shaft turns once (1 revolution). How many revolutions will the shaft turn when the tool advances 3.120 inches? Round the answer to 2 decimal places.

_____

7. How much stock per stroke is removed by the wheel of a surface grinder if a depth of 4.725 millimeters is reached after 75 strokes? Round the answer to 3 decimal places.

_____

8. An automatic screw machine is capable of producing one piece in 0.025 minute. How many pieces can be produced in 1.375 hours?

_____

9. This bolt has 7.7 threads. Determine the pitch to 3 decimal places. All dimensions are in inches.

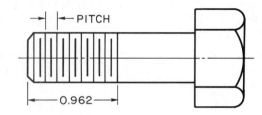

10. This block has a threaded hole with a 0.0625-inch pitch. Determine the number of threads for the given depth to 2 decimal places. All dimensions are in inches.

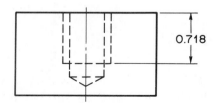

11. The length of a side of a square equals the distance from point A to point B divided by 1.4142. Determine the length of a side of this square plate to 2 decimal places. All dimensions are in millimeters.

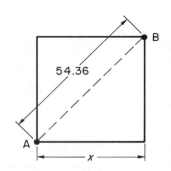

⎺ UNIT 12 *POWERS*

⎺OBJECTIVES

After studying this unit you should be able to

- Raise numbers to indicated powers.
- Solve problems which involve combinations of powers with other basic operations.

Powers of numbers are used to compute areas of square plates and circular sections and to compute volumes of cubes, cylinders, and cones. Use of powers is particularly helpful in determining distances in problems which require applications of geometry and trigonometry.

⎺DESCRIPTION OF POWERS

Two or more numbers multiplied to produce a given number are *factors* of the given number. Two factors of 8 are 2 and 4. The factors of 15 are 3 and 5. A *power* is the product of two or more equal factors. The third power of 5 is 5 × 5 × 5 or 125. An *exponent* shows how many times a number is taken as a factor. It is written smaller than the number, above the number, and to the right of the number. The expression 3^2 means 3 × 3. The exponent 2 shows that 3 is taken as a factor twice. It is read as 3 to the second power or 3 squared.

Examples: Find the indicated powers.
1. 2^5 Two to the fifth power means 2 × 2 × 2 × 2 × 2 or 32.   Ans
2. 3^3 Three cubed means 3 × 3 × 3 or 27.   Ans
3. 0.72^2 0.72 squared means 0.72 × 0.72 or 0.5184.   Ans

A = s² is called a *formula*. A formula is a short method of expressing an arithmetic relationship by the use of symbols. Known values may be substituted for the symbols and other values can be found.

Example 1: Determine the area of the square shown. The area of a square equals the length of a side squared. The answer is given in square units. All dimensions are in inches.

A = s²

A = $\left(\frac{7}{8}\text{ in}\right)^2$

A = $\frac{7}{8}$ in × $\frac{7}{8}$ in

A = $\frac{49}{64}$ sq in   Ans

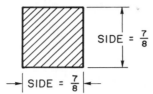

SIDE = $\frac{7}{8}$

SIDE = $\frac{7}{8}$

Example 2: Find the volume of the cube shown. The volume of a cube equals the length of a side cubed. The answer is given in cubic units. All dimensions are in millimeters. Round answer to 1 decimal place.

V = s³

V = (1.6 mm)³

V = 1.6 mm × 1.6 mm × 1.6 mm

V = 4.096 mm³ or 4.1 mm³   Ans

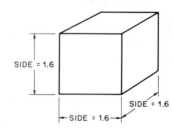

SIDE = 1.6

SIDE = 1.6

SIDE = 1.6

USE OF PARENTHESES

In this example, only the numerator is squared. $\frac{2^2}{3} = \frac{2\times2}{3} = \frac{4}{3} = 1\frac{1}{3}$

In this example, only the denominator is squared. $\frac{2}{3^2} = \frac{2}{3\times3} = \frac{2}{9}$

Parentheses are used as grouping symbols. Parentheses indicate that both the numerator and the denominator of a fraction are raised to the given power.

$$\left(\frac{2}{3}\right)^2 = \frac{2^2}{3^2} = \frac{2\times2}{3\times3} = \frac{4}{9}$$

Procedure: To solve problems which involve operations within parentheses

- Perform the operations within the parentheses.
- Raise to the indicated power.

Examples:

1. $(1.2 \times 0.6)^2 = 0.72^2 = 0.72 \times 0.72 = 0.5184$ Ans
2. $(0.5 + 2.4)^2 = 2.9^2 = 2.9 \times 2.9 = 8.41$ Ans
3. $(0.75 - 0.32)^2 = 0.43^2 = 0.43 \times 0.43 = 0.1849$ Ans
4. $\left(\frac{14.4}{3.2}\right)^2 = 4.5^2 = 4.5 \times 4.5 = 20.25$ Ans

When solving power problems which also require addition, subtraction, multiplication, or division perform the power operation first.

Examples:

1. $5 \times 3^2 - 12 = 5 \times 9 - 12 = 45 - 12 = 33$ Ans
2. $33.5 - 5.5^2 + 8.7 = 33.5 - 30.25 + 8.7 = 11.95$ Ans
3. $\frac{2.2^3 - 5.608}{1.4} = \frac{10.648 - 5.608}{1.4} = \frac{5.040}{1.4} = 3.6$ Ans

The symbol π (pi) represents a constant value used in mathematical relationships involving circles. Depending upon the specific problem to be solved, generally, the value of pi used is $3\frac{1}{7}$, 3.14, or 3.1416.

Example: Compute the volume of the cylinder shown to 2 decimal places. The answer is given in cubic units. All dimensions are in inches.

$V = \pi \times r^2 \times H$
$V = 3.14 \times (0.85 \text{ in})^2 \times 1.25 \text{ in}$
$V = 3.14 \times 0.7225 \text{ sq in} \times 1.25 \text{ in}$
$V = 2.8358 \text{ cu in or } 2.84 \text{ cu in}$ Ans

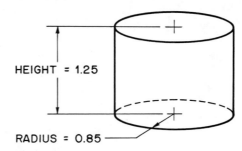

HEIGHT = 1.25

RADIUS = 0.85

Many problems require the application of the same formula more than once or the application of 2 different formulas in the solutions.

Example: Find the metal area of this square plate. Round the answer to 2 decimal places. All dimensions are in inches. $A = s^2$

The metal area equals the area of the large square minus the area of the removed square.

$A_1 = (5.25 \text{ in})^2$
$A_1 = 27.5625 \text{ sq in}$
$A_2 = (2.50 \text{ in})^2$
$A_2 = 6.2500 \text{ sq in}$
$A_3 = 27.5625 \text{ sq in} - 6.2500 \text{ sq in} = 21.31 \text{ sq in}$ Ans

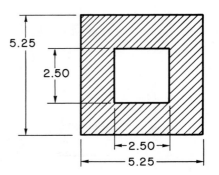

5.25

2.50

2.50
5.25

APPLICATION

Raising a Number to a Power

Raise the following numbers to the indicated power.

1. 3.4^3 _____

2. 1^8 _____

3. 100^4 _____

4. $\left(\frac{3}{4}\right)^3$ _____

5. $\frac{3^3}{4}$ _____

6. $\frac{3}{4^3}$ _____

7. $(0.3 \times 7)^2$ _____

8. $(18.6 + 9.5)^2$ _____

9. $\left(\frac{28.8}{7.2}\right)^3$ _____

Related Problems

In the following table the lengths of the sides of squares are given. Determine the areas of the squares. Round the answers to 2 decimal places where necessary.

	Side	Area		Side	Area
10.	1.25 in		15.	1/4 in	
11.	23.07 mm		16.	7/8 in	
12.	0.19 in		17.	3 1/2 in	
13.	10.7 mm		18.	13/16 in	
14.	0.02 in		19.	13 3/4 in	

$A = s^2$ where A = area
s = side

In the following table the lengths of the sides of cubes are given. Determine the volumes of the cubes. Round answers to 2 decimal places where necessary.

$V = s^3$ where V = volume
s = side

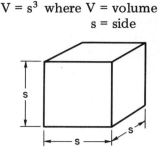

	Side	Volume			Side	Volume
20.	0.37 in		25.		1/3 in	
21.	20.60 mm		26.		7/8 in	
22.	3.93 in		27.		1 1/2 in	
23.	12.00 mm		28.		10 1/4 in	
24.	0.075 in		29.		3/4 in	

In the following table the radii of circles are given. Determine the areas of the circles. Round the answers to 2 decimal places where necessary.

	Radius	Area
30.	16.20 mm	
31.	18.30 mm	
32.	0.07 in	
33.	9.28 in	
34.	12.35 mm	

$A = \pi \times R^2$ where A = area
$\pi = 3.14$
R = radius

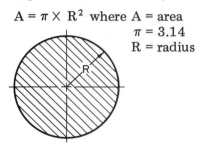

In the following table the diameters of spheres are given. Determine the volumes of the spheres. Round the answers to 2 decimal places where necessary.

	Diameter	Volume
35.	0.45 in	
36.	6.50 mm	
37.	0.75 in	
38.	10.80 mm	
39.	8.08 mm	

$V = \dfrac{\pi \times D^3}{6}$ where V = volume
$\pi = 3.14$
D = diameter

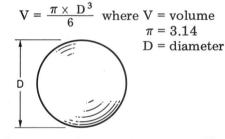

In the following table the radii and heights of cylinders are given. Determine the volumes of the cylinders. Round the answers to 2 decimal places where necessary.

	Radius	Height	Volume
40.	5.00 mm	3.20 mm	
41.	1.50 in	2.30 in	
42.	2.25 in	4.00 in	
43.	0.70 in	6.70 in	
44.	7.81 mm	6.72 mm	

$V = \pi \times R^2 \times H$ where V = volume
$\pi = 3.14$
R = radius
H = height

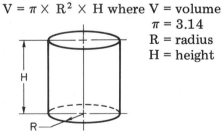

In the following table the diameters and heights of cones are given. Find the volumes of the cones. Round the answers to 2 decimal places where necessary.

	Diameter	Height	Volume
45.	3.20 in	4.00 in	
46.	3.00 in	5.00 in	
47.	11.70 mm	13.10 mm	
48.	9.90 mm	6.20 mm	
49.	0.17 in	0.88 in	

$V = 0.2618 \times D^2 \times H$ where V = volume
D = diameter
H = height

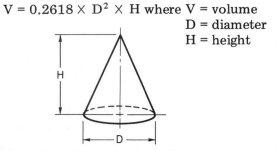

Solve the following problems.   Use $\pi = 3.14$. Round answers to 2 decimal places.

50.   Find the metal area of this washer. All dimensions are in millimeters.   _____

$A = \pi \times R^2$

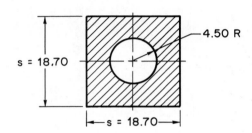

51.   Find the metal area of this spacer. All dimensions are in millimeters.   _____

Area of Square = s^2
Area of Circle = $\pi \times R^2$

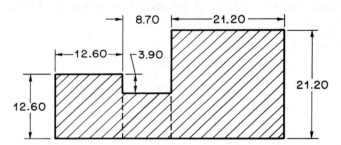

52.   Find the area of this plate. All dimensions are in millimeters. Hint: The
broken lines indicate one method of solution.   _____

$A = s^2$

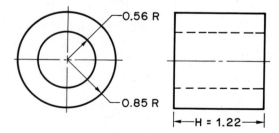

53.   Find the metal volume of this bushing. All dimensions are in inches.   _____

$V = \pi \times R^2 \times H$

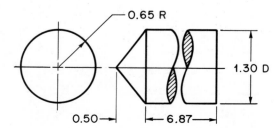

54.   Find the volume of this pin. All dimensions are in inches.   _____

Volume of cylinder = $\pi \times R^2 \times H$
Volume of cone = $0.2618 \times D^2 \times H$

‾UNIT 13 *ROOTS* _____

‾OBJECTIVES _____

After studying this unit you should be able to

- Extract whole number roots.
- Determine square roots to any indicated number of decimal places.
- Solve problems which involve combinations of roots with other basic arithmetic operations.

The operation of extracting roots of numbers is used to determine lengths of sides and heights of squares and cubes and radii of circular sections when areas and volumes are known. The machinist uses roots in computing distances between various parts of machined pieces from given blueprint dimensions.

‾DESCRIPTION OF ROOTS _____

The *root* of a number is a quantity which is taken two or more times as an equal factor of the number. Determining a root is the opposite operation of determining a power. The *radical symbol* ($\sqrt{}$) is used to indicate a root of a number. The *index* is written smaller than the number, to the left and above the radical symbol. The index indicates the number of times that a root is to be taken as an equal factor to produce the given number. The index 2 is omitted for an indicated square root. For example, the square root of 9 is written $\sqrt{9}$. The expression $\sqrt{9}$ means to find the number which can be multiplied by itself and equal 9. Since $3 \times 3 = 9$, 3 is the square root of 9.

Examples: Find the indicated roots.

1. $\sqrt{36}$ Since $6 \times 6 = 36$, the square root of 36 is 6.   Ans
2. $\sqrt{144}$ Since $12 \times 12 = 144$, the square root of 144 is 12.   Ans
3. $\sqrt[3]{8}$ Since $2 \times 2 \times 2 = 8$, the cube root of 8 is 2.   Ans
4. $\sqrt[3]{125}$ Since $5 \times 5 \times 5 = 125$, the cube root of 125 is 5.   Ans
5. $\sqrt[4]{81}$ Since $3 \times 3 \times 3 \times 3 = 81$, the fourth root of 81 is 3.   Ans

Roots must be extracted in determining unknown dimensions represented in certain formulas.

Example 1: Compute the length of the side of the square shown. This square has an area of 25 square inches.

Since $A = s^2$, the length of a side of the square equals the square root of the area.

$s = \sqrt{A}$
$s = \sqrt{25 \text{ sq in}}$
$s = \sqrt{5 \text{ in} \times 5 \text{ in}}$
$s = 5 \text{ inches}$ Ans

Example 2: Compute the length of the side of the cube shown. The volume of this cube equals 64 cubic inches.

$s = \sqrt[3]{V}$
$s = \sqrt[3]{64 \text{ cu in}}$
$s = \sqrt[3]{4 \text{ in} \times 4 \text{ in} \times 4 \text{ in}}$
$s = 4 \text{ inches}$ Ans

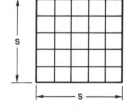

ROOTS OF FRACTIONS

In this example, only the root of the numerator is taken.

$$\frac{\sqrt{16}}{25} = \frac{\sqrt{4 \times 4}}{25} = \frac{4}{25}$$

In this example, only the root of the denominator is taken.

$$\frac{16}{\sqrt{25}} = \frac{16}{\sqrt{5 \times 5}} = \frac{16}{5} = 3\frac{1}{5}$$

A radical sign which encloses a fraction indicates that the roots of both the numerator and denominator are to be taken. The same answer is obtained by extracting both roots first and dividing second as by dividing first and extracting the root second.

Example: Find $\sqrt{\dfrac{36}{9}}$.

METHOD 1: Extract both roots then divide.

$$\sqrt{\frac{36}{9}} = \frac{\sqrt{36}}{\sqrt{9}} = \frac{6}{3} = 2 \quad \text{Ans}$$

METHOD 2: Divide then extract the root.

$$\sqrt{\frac{36}{9}} = \sqrt{4} = 2 \quad \text{Ans}$$

EXPRESSIONS ENCLOSED WITHIN THE RADICAL SYMBOL

The radical symbol is a grouping symbol. An expression consisting of operations within the radical symbol is done using the order of operations.

Procedure: To solve problems which involve operations within the radical symbol

- Perform the operations within the radical symbol first using the order of operations.
- Then find the root.

Examples: Find the indicated roots.

1. $\sqrt{3 \times 12} = \sqrt{36} = \sqrt{6 \times 6} = 6$ Ans
2. $\sqrt{5 + 59} = \sqrt{64} = \sqrt{8 \times 8} = 8$ Ans
3. $\sqrt{128 - 7} = \sqrt{121} = \sqrt{11 \times 11} = 11$ Ans

Problems involving formulas may involve operations within a radical symbol.

Example: Compute the length of the chord of the circular segment shown. All dimensions are in inches.

$C = 2 \times \sqrt{H \times (2 \times R - H)}$
$C = 2 \times \sqrt{1.5 \times (2 \times 3.75 - 1.5)}$
$C = 2 \times \sqrt{1.5 \times 6}$
$C = 2 \times \sqrt{9}$
$C = 2 \times 3$
$C = 6$
Length of chord = 6 inches   Ans

$C = 2 \times \sqrt{H \times (2 \times R - H)}$ where C = length of chord
H = height of segment
R = radius of circle

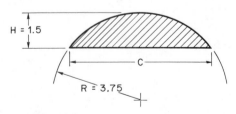

GENERAL METHOD OF COMPUTING SQUARE ROOTS

The square root examples shown have all consisted of perfect squares. *Perfect squares* are numbers which have whole number square roots. These roots are relatively easy to determine by observation. Most numbers do not have whole number square roots; therefore, a definite procedure must be used in computing square roots of most numbers.

The following examples illustrate the procedure used in determining square roots.

Example 1: Find the square root of 5410.218 to 1 decimal place.

Beginning at a decimal point, group the digits in pairs to the left and right of the decimal point.

Annex a zero to the 8 in order to form a pair of digits.

Place the decimal point directly above the decimal point of the number.

$$\sqrt{54\,10\,.\,21\,80}$$

Find the largest perfect square that can be subtracted from the first digit or pair of digits. The largest perfect square is 49. Write the square root of this perfect square above the first digit or group of digits. The square root of 49 is 7.

$$\begin{array}{r} 7 \quad . \\ \sqrt{54\,10\,.\,21\,80} \\ \underline{49} \\ 5 \end{array}$$

Bring down the next pair of digits (10) and place by the remainder (5).

$$\begin{array}{r} 7 \quad . \\ \sqrt{54\,10\,.\,21\,80} \\ \underline{49} \\ 5\,10 \end{array}$$

Double the partial root and use this number as a trial divisor.

$$2 \times 7 = 14$$

$$\begin{array}{r} 7 \quad . \\ \sqrt{54\,10\,.\,21\,80} \\ \underline{49} \\ 14\,\overline{)\ 5\,10} \end{array}$$

Divide the remainder by this trial divisor, disregarding the last digit of the remainder.

$$51 \div 14 = 3$$

Annex the quotient as the next figure in the square root. Also annex the same digit to the trial divisor.

Annex 3 to the root and to the trial divisor.

Multiply the complete divisor by the digit which was annexed to the root, and subtract.

$$\begin{array}{r} 7 \quad 3\,. \\ \sqrt{54\,10\,.\,21\,80} \\ \underline{49} \\ 143\,\overline{)\ 5\,10} \end{array}$$

Multiply: $3 \times 143 = 429$
Subtract: $510 - 429 = 81$

Repeat the process until the desired number of decimal places is obtained.

$$\begin{array}{r} 7 \quad 3\,. \\ \sqrt{54\,10\,.\,21\,80} \\ \underline{49} \\ 143\,\overline{)\ 5\,10} \\ \underline{4\,29} \\ 81 \end{array}$$

Bring down 21.

Double 73: $2 \times 73 = 146$

Divide: $812 \div 146 = 5$

Annex 5 to the root and to the trial divisor.

Multiply: $5 \times 1465 = 7325$

Subtract: $8121 - 7325 = 796$

Bring down 80.

Double 735: $2 \times 735 = 1470$

Divide: $7968 \div 1470 = 5$

Annex 5 to the root and to the trial divisor.

Multiply: $5 \times 14{,}705 = 73{,}525$

Subtract: $79{,}680 - 73{,}525 = 6155$

$$\begin{array}{r} 7 \quad 3\,.\ 5\ \ 5 \approx 73.6 \quad \text{Ans} \\ \sqrt{54\,10\,.\,21\,80} \\ \underline{49} \\ 143\,\overline{)\ 5\,10} \\ \underline{4\,29} \\ 1465\,\overline{)\ 81\,21} \\ \underline{73\,25} \\ 14{,}705\,\overline{)\ 7\,9680} \\ \underline{7\,3525} \\ 6155 \end{array}$$

Round 73.55 to 1 decimal place.

Example 2: Determine the square root of 923.7 to 2 decimal places.

$$3\ 0\ .\ 3\ 9\ 2 \approx 30.39 \quad \text{Ans}$$
$$\sqrt{9\ 23\ .\ 70\ 00\ 00}$$
$$\underline{9}$$
$$60\ \overline{)\ 0\ 23}$$
$$\underline{00}$$
$$603\ \overline{)\ 23\ 70}$$
$$\underline{18\ 09}$$
$$6069\ \overline{)\ 5\ 61\ 00}$$
$$\underline{5\ 46\ 21}$$
$$60{,}782\ \overline{)\ 14\ 79\ 00}$$
$$\underline{12\ 15\ 64}$$
$$2\ 63\ 36$$

Round 30.392 to 2 decimal places.

Example 3: Determine the square root of 0.0039 to 3 decimal places.

$$0.\ 0\ 6\ 2\ 4 \approx 0.062 \quad \text{Ans}$$
$$\sqrt{0.00\ 39\ 00\ 00}$$
$$\underline{36}$$
$$122\ \overline{)\quad 300}$$
$$\underline{2\ 44}$$
$$1244\ \overline{)\quad 56\ 00}$$
$$\underline{49\ 76}$$
$$6\ 24$$

Round 0.0624 to 3 decimal places.

APPLICATION

Radicals That Are Whole Numbers

The following problems have either whole number roots or numerators and denominators which have whole number roots. Determine these roots.

1. $\sqrt[3]{216}$ _____

2. $\sqrt{\dfrac{9}{16}}$ _____

3. $\dfrac{\sqrt{4}}{9}$ _____

4. $\dfrac{25}{\sqrt{36}}$ _____

5. $\sqrt{\dfrac{3}{4} \times \dfrac{3}{4}}$ _____

6. $\sqrt{0.5 \times 8}$ _____

7. $\sqrt{56.7 + 87.3}$ _____

8. $\sqrt{16.4 - 7.4}$ _____

9. $\sqrt[3]{\dfrac{428.8}{6.7}}$ _____

The following problems have whole number square roots. Solve for the missing values in the tables.

10. The areas of squares are given in the following table. Determine the lengths of the sides.

	Area (A)	Side (s)
a.	225 mm²	
b.	121 mm²	
c.	100 mm²	
d.	81 sq in	
e.	1 sq in	

$s = \sqrt{A}$

11. The volumes of cubes are given in the following table. Determine the lengths of the sides.

	Volume (V)	Side (s)
a.	216 mm³	
b.	64 cu in	
c.	343 cu in	
d.	1000 mm³	
e.	1 cu in	

$s = \sqrt[3]{V}$

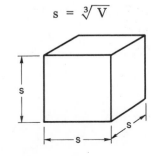

12. The areas of circles are given in this table. Determine the lengths of the radii. Use π = 3.14.

	Area (A)	Radius (R)
a.	50.24 sq in	
b.	12.56 sq in	
c.	314 mm²	
d.	28.26 sq in	
e.	153.86 mm²	

$$R = \sqrt{\frac{A}{\pi}}$$

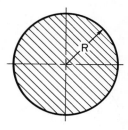

13. The volumes of spheres are given in this table. Determine the lengths of the diameters.

	Volume (V)	Diameter (D)
a.	14.1372 cu in	
b.	113.0976 mm³	
c.	4.1888 cu in	
d.	0.5236 cu in	
e.	523.6 mm³	

$$D = \sqrt[3]{\frac{V}{0.5236}}$$

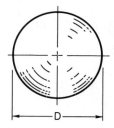

Radicals That Are Not Whole Numbers

The following problems have square roots that are not whole numbers. Compute these roots to the indicated number of decimal places.

14. $\sqrt{12.54}$ (3 places) _____   18. $\sqrt{0.07 \times 32}$ (2 places) _____

15. $\sqrt{391}$ (2 places) _____   19. $\sqrt{15.82 + 3.71}$ (2 places) _____

16. $\sqrt{\frac{4}{5}}$ (3 places) _____   20. $\sqrt{178.5 - 163.7}$ (3 places) _____

17. $\sqrt{3\frac{1}{2}}$ (3 places) _____   21. $\sqrt{\frac{0.441}{70}}$ (4 places) _____

The following problems have roots that are not whole numbers. Solve for the missing values in the tables.

22. The volumes of cylinders and their heights are given in the following table. Find the lengths of the radii to 2 decimal places. Use π = 3.14.

	Volume (V)	Height (H)	Radius (R)
a.	249.896 mm³	5 mm	
b.	132.634 mm³	12 mm	
c.	14 cu in	29 in	
d.	10 cu in	28 in	

$$R = \sqrt{\frac{V}{\pi \times H}}$$

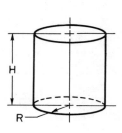

23. The volumes of cones and their heights are given in the following table. Compute the lengths of the diameters to 2 decimal places.

	Volume (V)	Height (H)	Diameter (D)
a.	116.328 mm³	6 mm	
b.	19.388 cu in	2 in	
c.	1257.6 mm³	10 mm	
d.	15 cu in	55 in	

$$D = \sqrt{\frac{V}{0.262 \times H}}$$

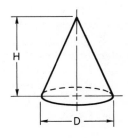

Solve the following problems.

24. The pitch of broach teeth depends upon the length of cut, the depth of cut, and the material being broached.

$$\text{Minimum Pitch} = 3 \times \sqrt{L \times d \times F}$$ where L = length of cut
d = depth of cut
F = a factor related to the type of material being broached

Find the minimum pitch, to 3 decimal places, for broaching cast iron where L = 0.825″, d = 0.005″, and F = 5.   _____

25. The dimensions of keys and keyways are determined in relation to the diameter of the shafts with which they are used.

$$D = \sqrt{\frac{L \times T}{0.3}}$$ where D = shaft diameter
L = key length
T = key thickness

What is the shaft diameter that would be used with a key where L = 2.70″ and T = 0.25″?   _____

UNIT 14 *TABLE OF DECIMAL EQUIVALENTS AND COMBINED OPERATIONS OF DECIMAL FRACTIONS*_____

OBJECTIVES_____

After studying this unit you should be able to

- Write decimal or fraction equivalents using a decimal equivalent table.
- Determine nearer fraction equivalents of decimals by using the decimal equivalent table.
- Solve problems consisting of combinations of operations by applying the order of operations.

Generally, fractional blueprint dimensions are given in multiples of 64ths of an inch. A machinist is often required to express these fractional dimensions as decimal equivalents for machine settings. When laying out parts such as castings that have ample stock allowances, it is sometimes convenient to use a fractional steel scale and to express decimal dimensions to the nearer equivalent fractions. The amount of computation and the chances of error can be reduced by using the decimal equivalent table.

TABLE OF DECIMAL EQUIVALENTS_____

Using a decimal equivalent table saves time and reduces the chance of error. Decimal equivalent tables are widely used in the manufacturing industry. They are posted as large wall charts in work areas and are carried as pocket size cards. Skilled workers memorize many of the equivalents after using decimal equivalent tables.

The decimals listed in the table are given to six places. For actual on-the-job uses, a decimal is rounded to the degree of precision required for a particular application.

DECIMAL EQUIVALENT TABLE

	1/64—0.015 625		33/64–0.515 625
1/32——————0.031 25		17/32——————0.531 25	
	3/64— 0.046 875		35/64–0.546 875
1/16—————— 0.062 5		9/16—————— 0.562 5	
	5/64–0.078 125		37/64–0.578 125
3/32—————— 0.093 75		19/32——————0.593 75	
	7/64— 0.109 375		39/64–0.609 375
1/8——————————— 0.125		5/8——————————— 0.625	
	9/64— 0.140 625		41/64–0.640 625
5/32—————— 0.156 25		21/32—————— 0.656 25	
	11/64–0.171 875		43/64–0.671 875
3/16—————— 0.187 5		11/16——————0.687 5	
	13/64–0.203 125		45/64–0.703 125
7/32—————— 0.218 75		23/32——————0.718 75	
	15/64–0.234 375		47/64–0.734 375
1/4——————————— 0.25		3/4——————————— 0.75	
	17/64–0.265 625		49/64–0.765 625
9/32 —————— 0.281 25		25/32 ————— 0.781 25	
	19/64- 0.296 875		51/64- 0.796 875
5/16—————— 0.312 5		13/16—————— 0.812 5	
	21/64- 0.328 125		53/64–0.828 125
11/32 —————— 0.343 75		27/32 ————— 0.843 75	
	23/64-0.359 375		55/64–0.859 375
3/8——————————— 0.375		7/8——————————— 0.875	
	25/64–0.390 625		57/64–0.890 625
13/32 —————— 0.406 25		29/32 ————— 0.906 25	
	27/64–0.421 875		59/64–0.921 875
7/16 —————— 0.437 5		15/16 ————— 0.937 5	
	29/64–0.453 125		61/64- 0.953 125
15/32 —————— 0.468 75		31/32——————0.968 75	
	31/64-0.484 375		63/64–0.984 375
1/2——————————————— 0.5		1——————————————— 1.	

The following examples illustrate the use of the decimal equivalent table.

Example 1: Find the decimal equivalent of $\frac{23}{32}''$.

The decimal equivalent is shown directly to the right of the common fraction.

$$\frac{23''}{32} = 0.71875'' \quad \text{Ans}$$

Example 2: Find the fractional equivalent of 0.3125''.

The fractional equivalent is shown directly to the left of the decimal fraction.

$$0.3125'' = \frac{5''}{16} \quad \text{Ans}$$

Example 3: Find the nearer fractional equivalents of the decimal dimensions given on the casting shown. All dimensions are in inches.

Compute dimension A. The decimal 0.757 lies between 0.750 and 0.765625. The difference between 0.757 and 0.750 is 0.007. The difference between 0.757 and 0.765625 is 0.008625. Since 0.007 is less than 0.008625, the 0.750 value is closer to 0.757. The nearer fractional equivalent of 0.750'' is 3/4 '' Ans.

Compute dimension B. The decimal 0.978 lies between 0.96875 and 0.984375. The difference between 0.978 and 0.96875 is 0.00925. The difference between 0.978 and 0.984375 is 0.006375. Since 0.006375 is less than 0.00925, the 0.984375 value is closer to 0.978. The fractional equivalent of 0.984375'' is 63/64 '' Ans.

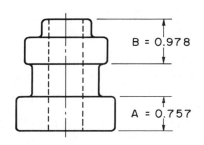

⎯COMBINED OPERATIONS OF DECIMAL FRACTIONS⎯⎯⎯⎯⎯⎯⎯⎯⎯⎯⎯⎯⎯⎯

In the process of completing a job, a machinist must determine stock sizes, cutter sizes, feeds and speeds, and roughing allowances as well as cutting dimensions. Usually most and sometimes all of the fundamental operations of mathematics must be used for computations in the manufacture of a part.

Determination of powers and roots must also be considered in the order of operations. The following procedure incorporates all six fundamental operations.

⎯ORDER OF OPERATIONS⎯⎯⎯⎯⎯⎯⎯⎯⎯⎯⎯⎯⎯⎯⎯⎯⎯⎯⎯⎯⎯⎯

1. Do all operations within the grouping symbol first. Parentheses, the fraction bar and the radical symbol are used to group numbers. If an expression contains parentheses within parentheses or brackets do the work within the innermost parentheses first.
2. Do powers and roots next. The operations are performed in the order in which they occur. If a root consists of two or more operations within the radical symbol, perform all the operations within the radical symbol, then extract the root.
3. Do multiplication and division next in the order in which they occur.
4. Do addition and subtraction last in the order in which they occur.

Example 1: Find the value of $7.875 + 3.2 \times 4.3 - 2.73$.

$7.875 + 3.2 \times 4.3 - 2.73$	Multiply. $3.2 \times 4.3 = 13.76$
$7.875 + 13.76 - 2.73$	Add. $7.875 + 13.76 = 21.635$
$21.635 - 2.73$	Subtract. $21.635 - 2.73 = 18.905$
18.905 Ans	

Example 2: Find the value of $(27.34 - 4.82) \div (2.41 \times 1.78 + 7.89)$. Round the answer to 2 decimal places.

	Perform operations within parentheses.
$(27.34 - 4.82) \div (2.41 \times 1.78 + 7.89)$	Subtract. $27.34 - 4.82 = 22.52$
$22.52 \div (2.41 \times 1.78 + 7.89)$	Multiply. $2.41 \times 1.78 = 4.2898$
$22.52 \div (4.2893 + 7.89)$	Add. $4.2898 + 7.89 = 12.1798$
$22.52 \div 12.1798$	Divide. $22.52 \div 12.1798 = 1.85$
1.85 Ans	

Example 3: Find the value of $\dfrac{13.79 + (27.6 \times 0.3)^2}{\sqrt{23.04} + 0.875 - 3.76}$. Round the answer to 3 decimal places.

Grouping symbol operations are done first. Consider the numerator and the denominator as if each were within parentheses. All of the operations are performed in the numerator and in the denominator before the division is performed.

$$\frac{13.79 + (27.6 \times 0.3)^2}{\sqrt{23.04} + 0.875 - 3.76} =$$

	In the numerator:
$[13.79 + (27.6 \times 0.3)^2] \div (\sqrt{23.04} + 0.875 - 3.76)$	Multiply. $27.6 \times 0.3 = 8.28$
$[13.79 + (8.28)^2] \div (\sqrt{23.04} + 0.875 - 3.76)$	Square. $8.28^2 = 68.5584$
$(13.79 + 68.5584) \div (\sqrt{23.04} + 0.875 - 3.76)$	Add. $13.79 + 68.5584 = 82.3484$
	In the denominator:
$82.3484 \div (\sqrt{23.04} + 0.875 - 3.76)$	Extract the square root. $\sqrt{23.04} = 4.8$
$82.3484 \div (4.8 + 0.875 - 3.76)$	Add. $4.8 + 0.875 = 5.675$
$82.3484 \div (5.675 - 3.76)$	Subtract. $5.675 - 3.76 = 1.915$
$82.3484 \div 1.915$	Divide. $82.3484 \div 1.915 = 43.002$
43.002 Ans	

Example 4: Blanks in the shape of regular pentagons (5-sided figures) are punched from strip stock as shown. Determine the width of strip stock required, using the given dimensions and the formula for dimension R. Round the answer to 3 decimal places. All dimensions are in inches.

$$\text{Width} = R + 0.980 + 2 \times 0.125 \text{ where } R = \sqrt{r^2 + s^2 \div 4}$$

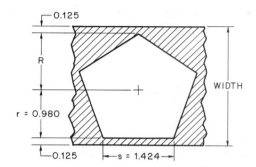

$$\sqrt{0.980^2 + 1.424^2 \div 4} + 0.980 + 2 \times 0.125$$

Substitute the given values.

Compute the operations under the radical sign.

Square. $0.980^2 = 0.9604$

Square. $1.424^2 = 2.027776$

$$\sqrt{0.9604 + 2.027776 \div 4} + 0.980 + 2 \times 0.125$$
$$\sqrt{0.9604 + 0.506944} + 0.980 + 2 \times 0.125$$
$$\sqrt{1.467344} + 0.980 + 2 \times 0.125$$
$$1.211 + 0.980 + 2 \times 0.125$$
$$1.211 + 0.980 + 0.250$$
$$2.441$$

Divide. $2.027776 \div 4 = 0.506944$

Add. $\sqrt{0.9604 + 0.506944} = \sqrt{1.467344}$

Extract the square root. $\sqrt{1.467344} = 1.211$

Multiply: $2 \times 0.125 = 0.250$

Add: $1.211 + 0.980 + 0.250 = 2.441$

Width = 2.441 inches Ans

Note: In solving expressions which consist of numerous multiplication and power operations, it is often necessary to carry out the work to two or three more decimal places than the number of decimal places required in the answer.

APPLICATION

Using the Decimal Equivalent Table

Find the fraction or decimal equivalents of these numbers using the decimal equivalent table.

1. $\frac{25}{32}$ _____

2. $\frac{9}{32}$ _____

3. $\frac{11}{32}$ _____

4. $\frac{15}{16}$ _____

5. $\frac{5}{64}$ _____

6. 0.671875 _____

7. 0.3125 _____

8. 0.28125 _____

9. 0.203125 _____

Find the nearer fraction equivalents of these decimals using the decimal equivalent table.

10. 0.541 _____

11. 0.762 _____

12. 0.459 _____

13. 0.498 _____

14. 0.209 _____

15. 0.805 _____

Combined Operations of Decimal Fractions

Solve these examples of combined operations. Round the answers to 2 decimal places where necessary.

16. $0.5231 + 11.664 \div 4.32 \times 0.521$ _____

17. $81.07 \div 12.1 + 2 \times 3.7$ _____

18. $\dfrac{56.050}{3.8} \times 0.875 - 3.92$ _____

19. $(24.78 - 19.32) \times 4.6$ _____

20. $(13.5 \div 4 - 1.76)^2 \times 4.5$ _____

21. $27.16 \div \sqrt{1.76 + 12.32}$ _____

22. $(\sqrt{3.98} + 0.87 \times 3.9)^2$ _____

23. $(3.29 \times 1.7)^2 \div (3.82 - 0.37)$ _____

24. $0.25 \times \left(\dfrac{\sqrt{64} \times 3.87}{8.32 \times 5.13} \right) + 18.3^2$ _____

25. $18.32 - \sqrt{\dfrac{7.86 \times 13.5}{3.5^2 - 0.52}} \times 0.7$ _____

Solve the following problems which require combined operations.

26. The figure shows the three-wire method of checking screw threads. With proper diameter wires and a micrometer, very accurate pitch diameter measurements can be made. Using the formula given, determine the micrometer dimension over wires of the American (National) Standard threads in the following table. Round the answer to 4 decimal places.

$$M = D - (1.5155 \times P) + (3 \times W)$$

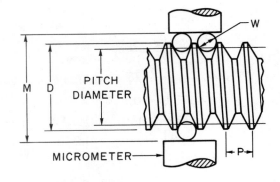

	Major Diameter D (inches)	Pitch P (inches)	Wire Diameter W (inches)	Dimension Over Wires M (inches)
a.	0.8750	0.1250	0.0900	
b.	0.2500	0.0500	0.0350	
c.	0.6250	0.1000	0.0700	
d.	1.3750	0.16667	0.1500	
e.	2.5000	0.2500	0.1500	

27. A bronze bushing with a diameter of 22.225 millimeters is to be pressed into a mounting plate. The assembly print calls for a bored hole in the plate to be 0.038 millimeter less in diameter than the bushing diameter. The hole diameter in the plate checks 22.103 millimeters. How much must the diameter of the plate hole be increased in order to meet the print specification? _____

28. A stamped sheet steel plate is shown. Compute dimensions A–F to 3 decimal places. All dimensions are in inches.

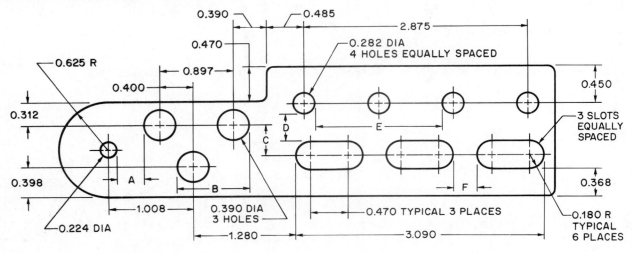

A = _____   B = _____   C = _____   D = _____   E = _____   F = _____

29. A flat is to be milled in three pieces of round stock each of a different diameter. The length of the flat is determined by the diameter of the stock and the depth of cut. The table gives the required length of flat and the stock diameter for each piece. Determine the depth of cut for each piece to 2 decimal places using this formula.

$$C = \frac{D}{2} - 0.5 \times \sqrt{4 \times \left(\frac{D}{2}\right)^2 - F^2}$$

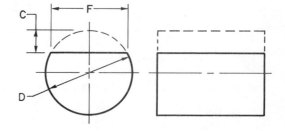

	Diameter D	Length of Flat F	Depth of Cut C
a.	35.60 mm	30.50 mm	
b.	55.90 mm	40.60 mm	
c.	91.40 mm	45.70 mm	

30. A groove is machined in a circular plate with a 41.28-millimeter diameter. Two milling cuts, one 6.30 millimeters deep and the second 3.15 millimeters are made. A grinding operation then removes 0.40 millimeter. What is the distance from the center of the plate to the bottom of the groove? All dimensions are in millimeters.

_____

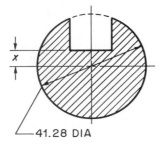

41.28 DIA

31. A 60° groove has been machined in a fixture. The groove is checked by placing a pin in the groove and indicating the distance between the top of the fixture and the top of the pin as shown. Compute distance H to 3 decimal places by using this formula. All dimensions are in inches.

_____

$$H = 1.5 \times D - 0.866 \times W$$

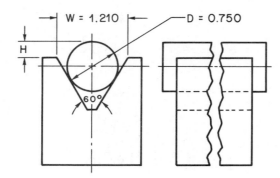

UNIT 15 *ACHIEVEMENT REVIEW —*
SECTION 1 _____

─OBJECTIVE_____

You should be able to solve the exercises and problems in this Achievement Review by applying the principles and methods covered in units 1–14.

1. Express each of the following fractions as equivalent fractions as indicated.

    a. $\frac{3}{8} = \frac{?}{32}$ _____   b. $\frac{3}{10} = \frac{?}{100}$ _____   c. $\frac{1}{4} = \frac{?}{64}$ _____   d. $\frac{9}{16} = \frac{?}{128}$ _____

2. Express each of the following mixed numbers as improper fractions.

    a. $3\frac{1}{5}$ _____   b. $2\frac{9}{10}$ _____   c. $5\frac{3}{4}$ _____   d. $13\frac{3}{8}$ _____   e. $7\frac{7}{32}$ _____

3. Express each of the following improper fractions as mixed numbers.

    a. $\frac{5}{2}$ _____   b. $\frac{23}{5}$ _____   c. $\frac{75}{4}$ _____   d. $\frac{115}{32}$ _____   e. $\frac{329}{64}$ _____

4. Express each of the following fractions as a fraction in lowest terms.

    a. $\frac{8}{16}$ _____   b. $\frac{12}{100}$ _____   c. $\frac{14}{16}$ _____   d. $\frac{18}{64}$ _____   e. $\frac{32}{128}$ _____

5. Express the fractions in each of the following sets as equivalent fractions having the least common denominator.

    a. $\frac{1}{4}$, $\frac{3}{16}$, $\frac{9}{32}$ _____   b. $\frac{7}{16}$, $\frac{1}{32}$, $\frac{9}{64}$ _____   c. $\frac{7}{10}$, $\frac{3}{4}$, $\frac{9}{25}$, $\frac{13}{20}$ _____

6. Add or subtract each of the following values. Express the answers in lowest terms.

    a. $\frac{1}{8} + \frac{5}{8}$ _____   f. $\frac{11}{16} - \frac{5}{16}$ _____

    b. $\frac{7}{16} + \frac{15}{16}$ _____   g. $\frac{17}{20} - \frac{3}{5}$ _____

    c. $\frac{3}{4} + \frac{13}{32}$ _____   h. $\frac{49}{64} - \frac{3}{8}$ _____

    d. $3\frac{7}{10} + \frac{49}{100}$ _____   i. $6 - \frac{13}{16}$ _____

    e. $\frac{9}{32} + \frac{1}{4} + \frac{21}{64}$ _____   j. $13\frac{1}{8} - 9\frac{7}{32}$ _____

7. Multiply or divide each of the following values. Express the answers in lowest terms.

    a. $\frac{1}{2} \times \frac{5}{8}$ _____   f. $\frac{3}{10} \div \frac{2}{5}$ _____

    b. $\frac{3}{4} \times \frac{4}{5} \times \frac{1}{3}$ _____   g. $\frac{14}{15} \div \frac{7}{25}$ _____

    c. $5\frac{7}{32} \times \frac{3}{8}$ _____   h. $16 \div \frac{2}{3}$ _____

    d. $3\frac{1}{10} \times 8\frac{1}{4}$ _____   i. $2\frac{17}{32} \div \frac{9}{24}$ _____

    e. $\frac{3}{16} \times 20 \times 3\frac{1}{2}$ _____   j. $2\frac{29}{32} \div 8\frac{3}{4}$ _____

8. Perform each of the indicated combined operations.

    a. $\frac{3}{4} + \frac{5}{16} - \frac{3}{8}$ _____

    b. $20\frac{1}{2} + 3\frac{3}{8} \times \frac{1}{4}$ _____

    c. $\frac{3}{5} \times \frac{7}{8} + 2\frac{1}{4}$ _____

    d. $\left(18 - 5\frac{3}{4}\right) \div \frac{1}{2} + 3\frac{7}{8}$ _____

    e. $\dfrac{16 - 4\frac{1}{2}}{\frac{1}{2} + 2\frac{1}{8}}$ _____

    f. $\dfrac{5\frac{1}{4} \times \frac{1}{2}}{6 \div 3\frac{3}{4}}$ _____

9. How many complete pieces can be blanked from a strip of aluminum 72 inches long if each stamping requires 1 3/8 inches of material plus an allowance of 3/4 inch at one end of the strip? _____

10. How many inches of bar stock are needed to make 50 spacers each 1 3/16 inches long? Allow 1/8 inch waste for each spacer. _____

11. A shaft is turned at 200 revolutions per minute with a tool feed of 1/32 inch per revolution. How many minutes does it take to cut a distance of 50 inches along the shaft? _____

12. A shop order calls for 1500 steel pins each 1 5/8 inches long. If 3/16 inch is allowed for cutting off and facing each pin, how many complete 10-foot lengths of stock are needed for the order? (12 inches = 1 foot) _____

13. Compute dimensions A, B, C, D, and E of the support bracket shown. All dimensions are given in inches.

    A = _____
    B = _____
    C = _____
    D = _____
    E = _____

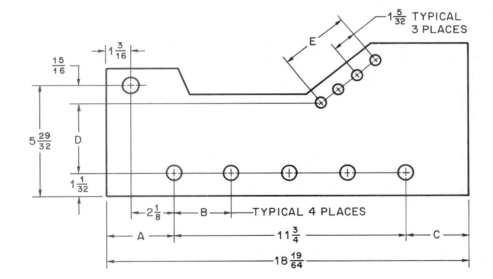

14. Write each of the following numbers as words.

    a. 0.6 _____

    b. 0.52 _____

    c. 0.147 _____

    d. 0.0086 _____

    e. 5.306 _____

    f. 16.0419 _____

15. Write each of the following words as decimal fractions or mixed decimals.

    a. three tenths _____

    b. twenty-six thousandths _____

    c. nine and thirty-four thousandths _____

    d. five and eighty-one ten-thousandths _____

16. Round each of the following numbers to the indicated number of decimal places.

    a. 0.386 (2 places) _____   c. 0.80729 (4 places) _____

    b. 5.0463 (3 places) _____   d. 7.0005 (3 places) _____

17. Express each of the following common fractions as decimal fractions. Where necessary, round the answers to 3 decimal places.

    a. $\frac{3}{4}$ _____ b. $\frac{7}{8}$ _____ c. $\frac{1}{3}$ _____ d. $\frac{2}{25}$ _____ e. $\frac{17}{20}$ _____

18. Express each of the following decimal fractions as common fractions in lowest terms.

    a. 0.7 _____ b. 0.525 _____ c. 0.002 _____ d. 0.915 _____ e. 0.0075 _____

19. Add or subtract each of the following values.

    a. 0.875 + 0.328 _____   f. 0.789 − 0.523 _____

    b. 5.004 + 0.92 + 0.5034 _____   g. 0.1863 − 0.0419 _____

    c. 0.006 + 12.3 + 0.0009 _____   h. 5.400 − 5.399 _____

    d. 2.99 + 6.015 + 0.1003 _____   i. 0.009 − 0.0068 _____

    e. 23 + 0.0007 + 0.007 + 0.9 _____   j. 14.001 − 13.999 _____

20. Multiply or divide each of the following values. Round the answer to 4 decimal places where necessary.

    a. 0.923 × 0.6 _____   f. 0.85 ÷ 0.48 _____

    b. 3.63 × 1.40 _____   g. 0.100 ÷ 0.01 _____

    c. 4.81 × 0.07 _____   h. 4.016 ÷ 0.03 _____

    d. 0.005 × 0.180 _____   i. 123 ÷ 0.873 _____

    e. 12.123 × 0.001 _____   j. 0.0098 ÷ 5.036 _____

21. Raise each of the following values to the indicated powers.

    a. 1.8^2 _____ b. 0.50^3 _____ c. 0.006^2 _____ d. $\left(\frac{2}{5}\right)^2$ _____ e. $\left(\frac{20.8}{6.5}\right)^3$ _____

22. Determine the whole number roots of each of the following values as indicated.

    a. $\sqrt{49}$ _____   c. $\sqrt{\frac{25}{36}}$ _____   e. $\sqrt{39.2 \times 1.25}$ _____

    b. $\sqrt[3]{64}$ _____   d. $\sqrt{46.83 + 17.17}$ _____

23. Determine the square roots of each of the following values to the indicated number of decimal places.

    a. $\sqrt{278}$ (2 places) _____   c. $\sqrt{\frac{3}{5}}$ (3 places) _____

    b. $\sqrt{0.8736}$ (3 places) _____   d. $\sqrt{93.876 - 47.904}$ (3 places) _____

24. Find the decimal or fraction equivalents of each of the following numbers using the decimal equivalent table.

    a. $\frac{3}{8}$ _____ b. $\frac{17}{32}$ _____ c. $\frac{21}{64}$ _____ d. 0.937 5 _____ e. 0.671875 _____

25. Determine the nearer fractional equivalents of each of the following decimals using the decimal equivalent table.

    a. 0.465 _____   b. 0.780 _____   c. 0.038 _____   d. 0.961 _____

26. Solve each of the following combined operations expressions. Round answers to 2 decimal places.

    a. 0.4321 + 10.870 ÷ 3.43 × 0.87 _____   c. $35.98 \div \sqrt{6.35 - 4.81}$ _____

    b. $(12.60 \div 3 - 0.98)^2 \times 3.60$ _____   d. $6 \times \left(\frac{\sqrt{81} \times 4.03}{3.30 \times 2.75}\right) - 1.7^2$ _____

27. The basic form of an ISO Metric Thread is shown. Given a thread pitch of 1.5 millimeters, compute thread dimensions A, B, C, D, E, and F to 3 decimal places.

A = _____

B = _____

C = _____

D = _____

E = _____

F = _____

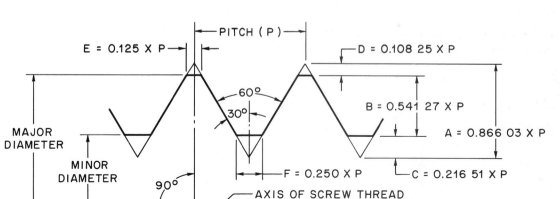

28. A combination of three gage blocks is selected to provide a total thickness of 0.4683 inch. Two blocks 0.250 inch and 0.118 inch thick are selected. What is the required thickness of the third block?

_____

29. A piece of round stock is being turned to a 17.86-millimeter diameter. A machinist measures the diameter of the piece as 18.10 millimeters. What depth of cut should be made to turn the piece to the required diameter?

_____

30. A plate 57.20 millimeters thick is to be machined to a thickness of 44.10 millimeters. The plate is to be rough cut with the last cut a finish cut 0.30 millimeter deep. If each rough cut is 3.20 millimeters deep, how many rough cuts are required?

_____

31. A shaft is turned in a lathe at 120 revolutions per minute. The cutting tool advances 0.030 inch per revolution. How long is the length of cut along the shaft at the end of 5 minutes?

_____

SECTION 2 LINEAR MEASUREMENT: ENGLISH AND METRIC

UNIT 16 *ENGLISH AND METRIC UNITS OF MEASURE*

OBJECTIVES

After studying this unit you should be able to

- Express English lengths as larger or smaller English linear units.
- Express metric lengths as larger or smaller metric linear units.
- Express metric length units as English length units.
- Express English length units as metric length units.

The ability to measure with tools and instruments and to compute measurements is a basic requirement in the machine trades. The units of measure used in the United States are established and maintained by the Bureau of Standards. These units of measure are based on international standards. Industrial standards of measure are determined by the American National Standards Institute (ANSI). This institute, with the cooperation of other similar organizations throughout the world, establishes and maintains industrial standards of measure, specifications and practice.

In 1960, a modernized system of metrics called the International System of Units (SI) was established by international agreement. All but a few countries are now converting from nonmetric systems or are revising their version of the metric system to SI standards. Presently, the United States uses both the English and the SI metric systems of measure. The use of the metric system in this country is continually increasing. It is important that the machine craftsperson be able to compute and measure with metric units as well as with English units.

Both the English and metric systems include all types of units of measure, such as length, area, volume, and capacity. In the machine trades, linear or length measure is used most often.

MEASUREMENT DEFINITIONS

Measurement is the comparison of a quantity with a standard unit. A *linear measurement* is a means of expressing the distance between two points; it is the measurement of lengths. A linear measurement has two parts: a unit of length and a multiplier.

The measurements 3.872 inches and 27.18 millimeters are examples of denominate numbers. A *denominate number* is a number that specifies a unit of measure.

ENGLISH UNITS OF LINEAR MEASURE

The yard is the standard unit of linear measure in the English system. From the yard, other units such as the inch and foot are established. The smallest unit is the inch. Common English units of length with their symbols are shown in this table.

English Units of Linear Measure	
1 yard (yd)	= 3 feet (ft)
1 yard (yd)	= 36 inches (in)
1 foot (ft)	= 12 inches (in)
1 mile (mi)	= 1760 yards (yd)
1 mile (mi)	= 5280 feet (ft)

In the machine trades, English linear units other than the inch are seldom used. English measure dimensions on engineering drawings are given in inches. Although English linear units other than the inch are rarely required for on-the-job applications, you should be able to use any units in the system.

⌐EXPRESSING LARGER ENGLISH UNITS OF LINEAR MEASURE AS SMALLER UNITS

Procedure: To express a larger unit of length as a smaller unit of length

- Multiply the given length by the number of smaller units contained in one of the larger units.

Example 1: Express $2\frac{1}{2}$ feet as inches.

Since 12 inches equal 1 foot, multiply $2\frac{1}{2}$ by 12.

$$2\frac{1}{2} \times 12 = 30$$
$$2\frac{1}{2} \text{ feet} = 30 \text{ inches}\quad \text{Ans}$$

Example 2: How many inches are in 0.25 yard?

Since 36 inches equal 1 yard, multiply 0.25 by 36.

$$0.25 \times 36 = 9$$
$$0.25 \text{ yard} = 9 \text{ inches}\quad \text{Ans}$$

⌐EXPRESSING SMALLER ENGLISH UNITS OF LINEAR MEASURE AS LARGER UNITS

Procedure: To express a smaller unit of length as a larger unit of length

- Divide the given length by the number of smaller units contained in one of the larger units.

Example 1: Express 67.2 inches as feet.

Since 12 inches equal 1 foot, divide 67.2 by 12

$$67 \div 12 = 5.6$$
$$67.2 \text{ inches} = 5.6 \text{ feet}\quad \text{Ans}$$

Example 2: How many yards are in 122.4 inches?

Since 36 inches equal 1 yard, divide 122.4 by 36

$$122.4 \div 36 = 3.4$$
$$122.4 \text{ inches} = 3.4 \text{ yards}\quad \text{Ans}$$

⌐METRIC UNITS OF LINEAR MEASURE

An advantage of the metric system is that it allows easy and fast computations. Since metric units are based on powers of ten, computations are simplified. To express a metric unit as a smaller or larger unit, all that is required is to move the decimal point a certain number of places to the left or right.

The metric system does not require difficult conversions as with the English system. For example, it is easier to remember that 1000 meters equal 1 kilometer than to remember that 1720 yards equal 1 mile. The meter is the standard unit of linear measure in the metric system. Other linear metric units are based on the meter. The smallest unit is the millimeter.

You will observe that in various technical publications and materials the spelling of metric units end in *er* or *re*. For example, units are expressed as *meter* or *metre* and *millimeter* or *millimetre*. Generally, in the manufacturing industry, the *er* spelling is used. Therefore, throughout this book *er* spellings of metric units are used.

Metric measure dimensions on engineering drawings are given in millimeters. In the machine trades, metric linear units other than the millimeter are seldom used. However, you should be able to use any units in the system. Metric units of length with their symbols are shown in this table. Observe that each unit is ten times greater than the unit directly above it.

Metric Units of Linear Measure			
1 millimeter (mm)	= 0.001 meter (m)	1000 millimeters (mm)	= 1 meter (m)
1 centimeter (cm)	= 0.01 meter (m)	100 centimeters (cm)	= 1 meter (m)
1 decimeter (dm)	= 0.1 meter (m)	10 decimeters (dm)	= 1 meter (m)
1 meter (m)	= 1 meter (m)	1 meter (m)	= 1 meter (m)
1 dekameter (dam)	= 10 meters (m)	0.1 dekameter (dam)	= 1 meter (m)
1 hectometer (hm)	= 100 meters (m)	0.01 hectometer (hm)	= 1 meter (m)
1 kilometer (km)	= 1000 meters (m)	0.001 kilometer (km)	= 1 meter (m)

The following metric power of ten prefixes are based on the meter.

milli means one thousandth (0.001)   *deka* means ten (10)
centi means one hundredth (0.01)   *hecto* means hundred (100)
deci means one tenth (0.1)   *kilo* means thousand (1000)

The most frequently used metric units of length are the kilometer (km), meter (m), centimeter (cm), and millimeter (mm). In actual applications, the dekameter (dam) and hectometer (hm) are not used. The decimeter (dm) is seldom used.

WRITING METRIC QUANTITIES

Periods are *not* used after the unit symbols. For example, write 1 mm, *not* 1 m.m. or 1 mm., when expressing the millimeter as a symbol. A comma is *not* used to separate digits in groups of three. Many countries use commas for decimal markers. To avoid confusion, a space is left between groups of three digits counting from the decimal point. If there are only four digits to the left or right of the decimal point, the space is optional.

Examples: 1. Write 11 240 mm, *not* 11,240 mm.
                    2. Write 0.869 54 mm, *not* 0.86954 mm.

EXPRESSING EQUIVALENT UNITS WITHIN THE METRIC SYSTEM

To express a given unit of length as a larger unit, move the decimal point a certain number of places to the left. To express a given unit of length as a smaller unit, move the decimal point a certain number of places to the right. The procedure of moving decimal points is shown in the following examples. Refer to the table of metric units of linear measure.

Example 1: Express 72 millimeters (mm) as centimeters (cm).

Since a centimeter is the next larger unit to a millimeter, move the decimal point 1 place to the left. (In moving the decimal point 1 place to the left, you are actually dividing by 10.)

7 2.

72 mm = 7.2 cm   Ans

Example 2: Express 0.96 centimeter (cm) as millimeters (mm).

Since a millimeter is the next smaller unit to a centimeter, move the decimal point 1 place to the right. (In moving the decimal point 1 place to the right, you are actually multiplying by 10.)

0 . 9 6

0.96 cm = 9.6 mm   Ans

Example 3: Express 0.245 meter (m) as millimeters (mm).

Since a millimeter is three smaller units from a meter, move the decimal point 3 places to the right. (In moving the decimal point 3 places to the right, you are actually multiplying by 10^3 or 1000.)

0 . 2 4 5

0.245 m = 245 mm   Ans

Example 4: Add. 0.3 meter (m) + 12.6 centimeters (cm) + 76 millimeters (mm)
                    Express the answer in millimeters.

Express each value in millimeters.

Add.

0.3 m	=	300 mm
12.6 cm	=	126 mm
+ 76 mm	=	76 mm
		502 mm Ans

METRIC-ENGLISH LINEAR EQUIVALENTS (CONVERSION FACTORS)

Since both the English and metric systems are used in this country, it is sometimes necessary to express equivalents between systems. Dimensioning an engineering drawing with both English and metric dimensions is called dual dimensioning. Since dual dimensioning tends to clutter a drawing and introduces additional opportunities for error, many companies do not use the system. Instead, some companies use metric dimensions only, with an inch-millimeter conversion table on or attached to the print. However, certain companies use dual dimensioning; it can be a practical method for industries which have plants in foreign countries. Examples of two types of dual dimensioning are shown.

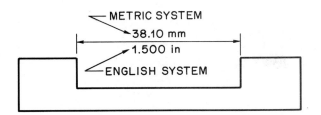
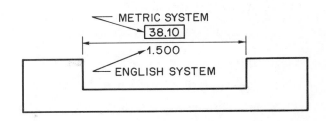

The commonly used equivalent factors of linear measure are shown in this table. Equivalent factors are commonly called conversion factors.

Metric-English Linear Equivalents (Conversion Factors)	
Metric to English Units	English to Metric Units
1 millimeter (mm) = 0.03937 inch (in)	1 inch (in) = 25.4 millimeters (mm)
1 centimeter (cm) = 0.3937 inch (in)	1 inch (in) = 2.54 centimeters (cm)
1 meter (m) = 39.37 inches (in)	1 foot (ft) = 0.3048 meter (m)
1 meter (m) = 3.2808 feet (ft)	1 yard (yd) = 0.9144 meter (m)
1 kilometer (km) = 0.6214 mile (mi)	1 mile (mi) = 1.609 kilometers (km)

Metric-English linear equivalents other than millimeter-inch equivalents are seldom used in the machine trades. However, you should be able to express any unit in one measuring system as a unit in the other system. The relationship between English decimal inch units and metric millimeter units is shown by comparing these scales.

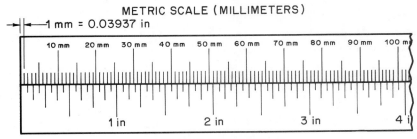

DECIMAL INCH SCALE

Procedure: To express a unit in one system as an equivalent unit in the other system

- Multiply the given measurement by the appropriate conversion factor in the Metric-English Linear Equivalent Table.

Examples:

1. Express 12.700 inches as millimeters.
   Since 1 in = 25.4 mm, 12.700 × 25.4 mm = 322.58 mm   Ans.

2. Express 6.78 centimeters as inches. Round the answer to 3 decimal places.
   Since 1 cm = 0.393 7 in, 6.78 × 0.393 7 in = 2.669 in   Ans.

3. This template is dimensioned in millimeters. Determine, in inches, the total length of the template. Round the answer to 3 decimal places.

   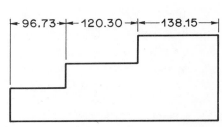

   Add the dimensions in millimeters as they are given and express the sum in inches.

   96.73 mm + 120.30 mm + 138.15 mm = 355.18 mm
   Since 1 mm = 0.03937 in, 355.18 × 0.03937 in = 13.983 in   Ans.

─APPLICATION_____

English Units of Linear Measure

1. Express each of the following lengths as indicated.

    a. 84 inches as feet   _____   k. 42 feet as yards   _____

    b. 123 inches as feet   _____   l. $\frac{1}{3}$ yard as inches   _____

    c. $3\frac{1}{2}$ feet as inches   _____   m. 258 inches as feet   _____

    d. 0.3 yard as inches   _____   n. $7\frac{2}{3}$ feet as inches   _____

    e. $1\frac{1}{4}$ yards as inches   _____   o. 0.25 yard as inches   _____

    f. 144 inches as yards   _____   p. 140.25 feet as yards   _____

    g. 75 inches as feet   _____   q. 333 inches as yards   _____

    h. 8 yards as feet   _____   r. 177 inches as feet   _____

    i. 3.6 yards as feet   _____   s. $20\frac{2}{3}$ yards as feet   _____

    j. 27 feet as yards   _____   t. 9.25 feet as inches   _____

2. A 3 1/2-inch diameter milling cutter revolving at 120 revolutions per minute has a cutting speed of 110 feet per minute. What is the cutting speed in inches per minute?   _____

3. How many complete 6-foot lengths of round stock should be ordered to make 230 pieces each 1.300 inches long? Allow 1 1/2 lengths of stock for cutoff and scrap.   _____

4. Pieces each 3.25 inches long are to be cut from lengths of bar stock. Allowing 0.10 inch for cutoff per piece, how many complete pieces can be cut from twelve 8-foot lengths of stock?   _____

Metric Units of Linear Measure

5. Express each of the following lengths as indicated.

    a. 2.9 centimeters as millimeters   _____   k. 0.0098 meter as millimeters   _____

    b. 14.68 centimeters as millimeters   _____   l. 1.046 meters as centimeters   _____

    c. 219.75 millimeters as centimeters   _____   m. 30.03 centimeters as millimeters   _____

    d. 97.83 millimeters as centimeters   _____   n. 979.76 millimeters as centimeters   _____

    e. 0.93 meter as centimeters   _____   o. 2039 millimeters as meters   _____

    f. 0.17 meter as millimeters   _____   p. 3.47 centimeters as meters   _____

    g. 153 millimeters as meters   _____   q. 0.056 meter as millimeters   _____

    h. 784 centimeters as meters   _____   r. 7.321 meters as centimeters   _____

    i. 0.93 millimeter as centimeters   _____   s. 6.377 centimeters as millimeters   _____

    j. 0.08 centimeter as millimeters   _____   t. 0.898 meter as millimeters   _____

6. Perform the indicated operations. Express the answer in the indicated unit.

    a. 25.73 mm + 7.6 cm = ? mm   _____   f. 0.793 m − 523.8 mm = ? mm   _____

    b. 2.4 m + 98 cm = ? m   _____   g. 214 mm + 87.6 cm + 0.9 m = ? m   _____

    c. 59.6 cm − 63.7 mm = ? cm   _____   h. 0.063 m + 4.93 cm + 57.3 mm = ? mm   _____

    d. 184.8 mm − 12.3 cm = ? mm   _____   i. 54.4 mm + 5.05 cm + 204.3 mm = ? mm   _____

    e. 1.06 m − 46.3 cm = ? cm   _____   j. 3.927 m − 812 mm = ? m   _____

7. An aluminum slab 0.076 meter thick is machined with three equal cuts, each cut is 10 millimeters deep. Determine the finished thickness of the slab in millimeters.   _____

8. A piece of sheet metal is 1.12 meters wide. Strips each 3.4 centimeters wide
    are cut. Allow 3 millimeters for cutting each strip.
    a. Determine the number of complete strips cut.   _____
    b. Determine the width of the waste strip in millimeters.   _____

Metric-English Linear Equivalents

9. Express each of the following English units of length as the indicated metric
    unit of length. Where necessary round the answer to 3 decimal places.
    a. 34 millimeters as inches   _____   h. 10.2 meters as feet   _____
    b. 126.8 millimeters as inches   _____   i. 736 millimeters as inches   _____
    c. 17.3 centimeters as inches   _____   j. 30.09 millimeters as inches   _____
    d. 0.97 centimeter as inches   _____   k. 56.3 centimeters as inches   _____
    e. 2.4 meters as inches   _____   l. 2 meters as yards   _____
    f. 0.09 meter as inches   _____   m. 50 centimeters as feet   _____
    g. 6 meters as feet   _____   n. 780 millimeters as feet   _____

10. Express each of the following metric units of length as the indicated English
    unit of length. Where necessary, round the answer to 2 decimal places.
    a. 3 inches as millimeters   _____   h. 1.3 yards as meters   _____
    b. 0.360 inch as millimeters   _____   i. 2.368 inches as millimeters   _____
    c. 34 inches as centimeters   _____   j. 0.67 inch as centimeters   _____
    d. 19.83 inches as centimeters   _____   k. 216 inches as meters   _____
    e. 6 feet as meters   _____   l. $\frac{1}{2}$ inch as millimeters   _____
    f. 0.75 foot as meters   _____   m. $4\frac{1}{4}$ inches as centimeters   _____
    g. 4 yards as meters   _____   n. $75\frac{3}{8}$ inches as meters   _____

11. Determine the total length of stock in inches required to make 35 bushings,
    each 38.1 mm long. Allow 3/16″ waste for each bushing.   _____

12. The part shown is to be made in a machine shop using decimal-inch machinery
    and tools. All dimensions are in millimeters. Express each of the dimensions,
    A–L, in inches to 3 decimal places.

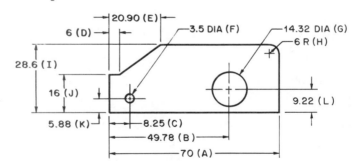

A = _____   B = _____   C = _____   D = _____   E = _____   F = _____
G = _____   H = _____   I = _____   J = _____   K = _____   L = _____

13. The shaft shown is dimensioned in inches. Express each dimension, A–J,
    in millimeters and round each dimension to 2 decimal places.

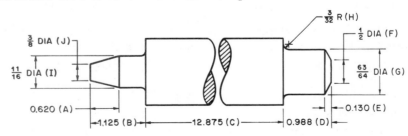

A = _____   B = _____   C = _____   D = _____   E = _____   F = _____
G = _____   H = _____   I = _____   J = _____

UNIT 17 DEGREE OF PRECISION AND GREATEST POSSIBLE ERROR

OBJECTIVES

After studying this unit you should be able to
- Determine the degree of precision of any given number.
- Compute the greatest possible error of English and metric length units.

DEGREE OF PRECISION

The cost of producing a part increases with the degree of precision called for; therefore, no greater degree of precision than is actually required should be specified on a drawing. The degree of precision specified for a particular machining operation dictates the type of machine, the machine setup, and the measuring instrument used for that operation.

The *exact* length of an object cannot be measured. All measurements are approximations. By increasing the number of graduations on a measuring instrument, the degree of precision is increased. Increasing the number of graduations enables the user to get closer to the *exact* length. The precision of a measurement depends on the measuring instrument used. The degree of precision of a measuring instrument depends on the smallest graduated unit of the instrument.

Machinists often work to 0.001-inch or 0.02-millimeter precision. In the manufacture of certain products, very precise measurements to 0.00001 inch or 0.0003 millimeter and 0.000001 inch or 0.000 03 millimeter are sometimes required.

Various measuring instruments have different limitations on the degree of precision possible. The accuracy achieved in measurement does not only depend on the limitations of the measuring instrument. Accuracy can also be affected by errors of measurement. Errors can be caused by defects in the measuring instruments and by environmental changes such as differences in temperature. Perhaps the greatest cause of error is the inaccuracy of the person using the measuring instrument.

LIMITATIONS OF MEASURING INSTRUMENTS

Following are the limitations on the degree of precision possible of some commonly used manufacturing measuring instruments.

Steel rules: 1/64" (fractional-inch); 0.01" (decimal-inch); 0.5 mm (metric).

Micrometers: 0.001" (decimal-inch) and 0.0001" (with vernier scale); 0.01 mm (metric) and 0.002 mm (with vernier scale).

Vernier and dial calipers: 0.001" (decimal-inch); 0.02 mm (metric).

Dial indicators (comparison measurement): graduations as small as 0.00005" (decimal-inch); 0.002 mm (metric).

Precision gage blocks (comparison measurement): Accurate to 0.000002" (decimal-inch); 0.000 06 mm (metric). The degree of precision of measurement is only as precise as the measuring instrument that is used with the blocks.

High amplification comparators (mechanical, optical, pneumatic, electronic): Graduations as small as 0.000001".

DEGREE OF PRECISION OF NUMBERS

The degree of precision of a number depends upon the unit of measurement. The degree of precision of a number increases as the number of decimal places increases.

Example 1: The degree of precision of 2" is to the nearer inch as shown in A. The range of values includes all numbers equal to or greater than 1.5" or less than 2.5".

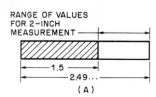

RANGE OF VALUES FOR 2-INCH MEASUREMENT

1.5
2.49...
(A)

Example 2: The degree of precision of 2.0″ is to the nearer 10th of an inch as shown in B. The range of values includes all numbers equal to or greater than 1.95″ and less than 2.05″.

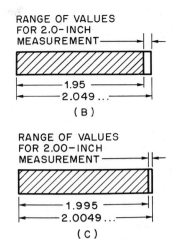

RANGE OF VALUES
FOR 2.0-INCH
MEASUREMENT

1.95
2.049...
(B)

Example 3: The degree of precision of 2.00″ is to the nearer 100th of an inch as shown in C. The range of values includes all numbers equal to or greater than 1.995″ and less than 2.005″.

RANGE OF VALUES
FOR 2.00-INCH
MEASUREMENT

1.995
2.0049...
(C)

Example 4: The degree of precision of 2.000″ is to the nearer 1000th of an inch. The range of values includes all numbers equal to or greater than 1.9995″ and less than 2.0005″.

GREATEST POSSIBLE ERROR

The *greatest possible error* of a measurement is one-half the smallest graduated unit of the measurement used to make the measurement. Therefore, the greatest possible error is equal to 1/2 or 0.5 of the precision.

Examples:

1. A machinist reads a measurement of 36 millimeters on a steel rule. The smallest graduation on the rule used is 1 millimeter, therefore, the precision is 1 millimeter. Since the greatest possible error is one-half of the smallest graduated unit, the greatest possible error is 0.5 × 1 mm or 0.5 mm. The actual length measured is between 36 mm – 0.5 mm and 36 mm + 0.5 mm or between 35.5 mm and 36.5 mm.

2. A tool and die maker reads a measurement of 0.4754 inch on a vernier scale micrometer. The smallest graduation on the micrometer is 0.0001 inch, therefore the precision is 0.0001 inch. The greatest possible error is 0.5 × 0.0001″ or 0.00005″. The actual length measured is between 0.4754″ – 0.00005″ and 0.4754″ + 0.00005″ or between 0.47535″ and 0.47545″.

APPLICATION

Degree of Precision

For each measurement find

    a. the degree of precision.
    b. the value which is equal to or less than the range of values.
    c. the value which is greater than the range of values.

1. 3.6″   a. _____   b. _____   c. _____
2. 1.62″   a. _____   b. _____   c. _____
3. 4.3″   a. _____   b. _____   c. _____
4. 7.08″   a. _____   b. _____   c. _____
5. 15.885″   a. _____   b. _____   c. _____
6. 9.1837″   a. _____   b. _____   c. _____
7. 12.002″   a. _____   b. _____   c. _____
8. 36.0″   a. _____   b. _____   c. _____
9. 7.01″   a. _____   b. _____   c. _____

10. 23.00″   a. _____   b. _____   c. _____

11. 9.1″   a. _____   b. _____   c. _____

12. 14.01070″   a. _____   b. _____   c. _____

13. 26.87 mm   a. _____   b. _____   c. _____

14. 15.4 mm   a. _____   b. _____   c. _____

15. 123.08 mm   a. _____   b. _____   c. _____

16. 0.976 mm   a. _____   b. _____   c. _____

17. 48.01 mm   a. _____   b. _____   c. _____

18. 203.899 mm   a. _____   b. _____   c. _____

19. 7.00 mm   a. _____   b. _____   c. _____

20. 34.082 5 mm   a. _____   b. _____   c. _____

21. 9.001 mm   a. _____   b. _____   c. _____

22. 14.0000 mm   a. _____   b. _____   c. _____

Greatest Possible Error

For each of the exercises in the following tables, the measurement made and the smallest graduation of the measuring instrument are given. Determine the greatest possible error and the smallest and largest possible actual length for each.

English System

	Measurement Made (inches)	Smallest Graduation of Measuring Instrument Used (inches)	Greatest Possible Error (inches)	Actual Length	
				Smallest Possible (inches)	Largest Possible (inches)
23.	8.30	0.05 (steel rule)			
24.	15.68	0.02 (steel rule)			
25.	0.753	0.001 (vernier caliper)			
26.	0.376	0.001 (micrometer)			
27.	0.9369	0.0001 (vernier micrometer)			
28.	3 5/8	1/64 (steel rule)			

Metric System

	Measurement Made (millimeters)	Smallest Graduation of Measuring Instrument Used (millimeters)	Greatest Possible Error (millimeters)	Actual Length	
				Smallest Possible (millimeters)	Largest Possible (millimeters)
29.	64	1 (steel rule)			
30.	105	0.5 (steel rule)			
31.	98.5	0.5 (steel rule)			
32.	58.26	0.02 (vernier caliper)			
33.	13.37	0.01 (micrometer)			
34.	12.778	0.002 (vernier micrometer)			

UNIT 18 *TOLERANCE, CLEARANCE, AND INTERFERENCE*

OBJECTIVES

After studying this unit you should be able to
- Compute total tolerances and maximum and minimum limits of dimensions.
- Compute maximum and minimum clearances of mating parts.
- Compute maximum and minimum interferences of mating parts.
- Express unilateral tolerances as bilateral tolerances.

TOLERANCE

Tolerance is the amount of variation permitted on the dimensions or surfaces of manufactured parts. *Limits* are the extreme permissible dimensions of a part. Tolerance is equal to the difference between the maximum and minimum limits of any specified dimension of a part.

Example: The maximum limit of a hole diameter is 0.878 inch and the minimum limit is 0.872 inch. Find the tolerance.

The tolerance is 0.878″ − 0.872″ = 0.006″   Ans

A *basic dimension* is the standard size from which the maximum and minimum limits are determined. Usually tolerances are given in such a way as to show the amount of variation and in which direction from the basic dimension these variations can occur. *Unilateral tolerance* means that the total tolerance is taken in one direction from the basic dimension. *Bilateral tolerance* means that the tolerance is divided partly plus (+) or above and partly minus (−) or below the basic dimension. A *mean dimension* is a value which is midway between the maximum and minimum limits. Where bilateral tolerances are used with equal plus and minus tolerances, the mean dimension is equal to the basic dimension.

Example 1: The part shown is dimensioned with a unilateral tolerance. The dimensions are given in inches. The basic dimension is 3.7500″. The total tolerance is a minus (−) tolerance. Find the maximum permissible dimension (maximum limit) and minimum permissible dimension (minimum limit).

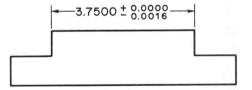

Maximum Limit: 3.7500″ + 0.0000″ = 3.7500″   Ans
Minimum Limit: 3.7500″ − 0.0016″ = 3.7484″   Ans

Example 2: The part shown is dimensioned with a bilateral tolerance. The dimensions are given in millimeters. The basic dimension is 62.79 mm. The tolerance is given in two directions, plus (+) and minus (−). Find the maximum limit and the minimum limit.

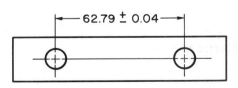

Maximum Limit: 62.79 mm + 0.04 mm = 62.83 mm   Ans
Minimum Limit: 62.79 mm − 0.04 mm = 62.75 mm   Ans

Example 3: What is the mean dimension of a part if the maximum dimension (maximum limit) is 46.35 millimeters and the minimum dimension (minimum limit) is 46.27 millimeters?

Subtract.   46.35 mm − 46.27 mm = 0.08 mm
Divide.   0.08 mm ÷ 2 = 0.04 mm
Subtract.   46.35 mm − 0.04 mm = 46.31 mm   Ans
Note: The mean dimension is midway between 46.35 mm and 46.27 mm.

$$46.31 \text{ mm} \begin{array}{l} + \\ - \end{array} \begin{array}{l} 0.04 \text{ mm} = 46.35 \text{ mm (Max. limit)} \\ 0.04 \text{ mm} = 46.27 \text{ mm (Min. limit)} \end{array}$$

‾EXPRESSING UNILATERAL TOLERANCE AS BILATERAL TOLERANCE_____

In the actual processing of parts, given unilateral tolerances are sometimes changed to bilateral tolerances. A machinist may prefer to work to a mean dimension and take equal plus and minus tolerances while machining a part. The following example shows the procedure for expressing a unilateral tolerance as a bilateral tolerance.

Example: The part shown is dimensioned with unilateral tolerances. Express the unilateral tolerance as a bilateral tolerance. Dimensions are given in inches.

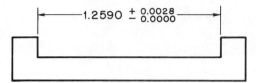

Divide the total tolerance by 2.
0.0028″ ÷ 2 = 0.0014″

Determine the mean dimension.
1.2590″ + 0.0014″ = 1.2604″

Show as a bilateral tolerance.
1.2604″ ± 0.0014″   Ans

‾FITS OF MATING PARTS_____

Fits between mating parts, such as between shafts and holes, have wide application in the manufacturing industry. The tolerances applied to each of the mating parts determines the relative looseness or tightness of fit between parts.

When one part is to move within another there is a *clearance* between the parts. A shaft made to turn in a bushing is an example of a clearance fit. The shaft diameter is less than the bushing hole diameter. When one part is made to be forced into the other there is *interference* between parts. A pin pressed into a hole is an example of an interference fit. The pin diameter is greater than the hole diameter.

Allowance is the intentional difference in the dimensions of mating parts which provides for different classes of fits. *Allowance* is the minimum clearance or the maximum interference which is intended between mating parts. Allowance represents the condition of the tightest permissible fit.

Example 1: A mating shaft and hole with a clearance fit dimensioned with bilateral tolerances is shown. All dimensions are in inches. Determine the following:

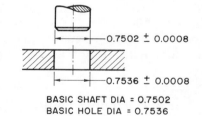

a. Maximum shaft diameter
   0.7502″ + 0.0008″ = 0.7510″   Ans

b. Minimum shaft diameter
   0.7502″ − 0.0008″ = 0.7494″   Ans

c. Maximum hole diameter
   0.7536″ + 0.0008″ = 0.7544″   Ans

d. Minimum hole diameter
   0.7536″ − 0.0008″ = 0.7528″   Ans

e. Maximum clearance equals maximum hole diameter minus minimum shaft diameter
   0.7544″ − 0.7494″ = 0.0050″   Ans

f. Minimum clearance equals minimum hole diameter minus maximum shaft diameter
   0.7528″ − 0.7510″ = 0.0018″   Ans
   Since allowance is defined as the minimum clearance, the allowance = 0.0018″   Ans

Example 2: A pin which is to be pressed into a hole is shown. This is an example of an interference fit dimensioned with unilateral tolerances. Dimensions are in millimeters. Determine the following:

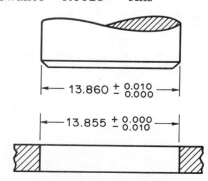

a. Maximum pin diameter
   13.860 mm + 0.010 mm = 13.870 mm   Ans

b. Minimum pin diameter
   13.860 mm − 0.000 mm = 13.860 mm   Ans

c. Maximum hole diameter
   13.855 mm + 0.000 mm = 13.855 mm   Ans

d. Minimum hole diameter
   13.855 mm − 0.010 mm = 13.845 mm   Ans

e. Minimum interference equals minimum pin diameter minus maximum hole diameter
   13.860 mm − 13.855 mm = 0.005 mm   Ans

f. Maximum interference equals maximum pin diameter minus minimum hole diameter
   13.870 mm − 13.845 mm = 0.025 mm   Ans
   Since allowance is defined as the maximum interference, the allowance = 0.025 mm   Ans

APPLICATION

Tolerance, Maximum and Minimum Limits

Refer to the following tables and determine the tolerance, maximum limit, or minimum limit as required for each problem.

1. English System

	Tolerance	Maximum Limit	Minimum Limit
a.		5 7/16″	5 13/32″
b.		7′–9 1/16″	7′–8 15/16″
c.	0.02″	16.76″	
d.	0.007″		0.904″
e.		1.7001″	1.6998″
f.	0.005″		10.999″

2. Metric System

	Tolerance	Maximum Limit	Minimum Limit
a.		50.7 mm	49.8 mm
b.		26.8 cm	26.6 cm
c.	0.04 mm		258.03 mm
d.	0.12 mm	80.09 mm	
e.	0.006 cm		12.731 cm
f.		4.01 mm	3.98 mm

Unilateral and Bilateral Tolerance

3. Refer to this figure. Dimension A with its tolerance is given in each of the following problems. Determine the maximum dimension (maximum limit) and the minimum dimension (minimum limit) for each.

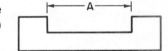

a. Dimension A = 3.750″ $^{+0.003″}_{-0.000″}$
   maximum _____ minimum _____

b. Dimension A = 5.927″ $^{+0.005″}_{-0.000″}$
   maximum _____ minimum _____

c. Dimension A = 2.004″ $^{+0.000″}_{-0.004″}$
   maximum _____ minimum _____

d. Dimension A = 4.8739″ $^{+0.0000″}_{-0.0012″}$
   maximum _____ minimum _____

e. Dimension A = 1.0875″ $^{+0.0009″}_{-0.0000″}$
   maximum _____ minimum _____

f. Dimension A = 28.16 mm $^{+0.00\ mm}_{-0.06\ mm}$
   maximum _____ minimum _____

g. Dimension A = 43.94 mm $^{+0.04\ mm}_{-0.00\ mm}$
   maximum _____ minimum _____

h. Dimension A = 120.88 mm $^{+0.07\ mm}_{-0.00\ mm}$
   maximum _____ minimum _____

i. Dimension A = 73.398 mm $^{+0.000\ mm}_{-0.012\ mm}$
   maximum _____ minimum _____

j. Dimension A = 45.106 mm $^{+0.009\ mm}_{-0.000\ mm}$
   maximum _____ minimum _____

4. The following dimensions are given with bilateral tolerances. For each value determine the maximum dimension (maximum limit) and the minimum dimension (minimum limit).

a. 2.812″ ± 0.006″
   maximum _____ minimum _____

b. 5.003″ ± 0.004″
   maximum _____ minimum _____

c. 3.971″ ± 0.010″
   maximum _____ minimum _____

d. 4.0562″ ± 0.0012″
   maximum _____ minimum _____

e. 1.3799″ ± 0.0009″
maximum _____ minimum _____

f. 2.0000″ ± 0.0007″
maximum _____ minimum _____

g. 54.36 mm ± 0.05 mm
maximum _____ minimum _____

h. 107.07 mm ± 0.08 mm
maximum _____ minimum _____

i. 62.04 mm ± 0.10 mm
maximum _____ minimum _____

j. 10.203 mm ± 0.024 mm
maximum _____ minimum _____

k. 295.005 mm ± 0.007 mm
maximum _____ minimum _____

l. 66.761 mm ± 0.015 mm
maximum _____ minimum _____

5. Express each of the following unilateral tolerances as bilateral tolerances having equal plus and minus values.

a. $0.938'' \,^{+0.010''}_{-0.000''}$ _____

b. $1.734'' \,^{+0.002''}_{-0.000''}$ _____

c. $3.000'' \,^{+0.000''}_{-0.004''}$ _____

d. $0.073'' \,^{+0.000''}_{-0.008''}$ _____

e. $4.1873'' \,^{+0.0014''}_{-0.0000''}$ _____

f. $1.0012'' \,^{+0.0000''}_{-0.0074''}$ _____

g. $0.0010'' \,^{+0.0000''}_{-0.0008''}$ _____

h. $8.4649'' \,^{+0.0022''}_{-0.0000''}$ _____

i. $38.60 \text{ mm} \,^{+0.02 \text{ mm}}_{-0.00 \text{ mm}}$ _____

j. $10.06 \text{ mm} \,^{+0.00 \text{ mm}}_{-0.08 \text{ mm}}$ _____

k. $64.89 \text{ mm} \,^{+0.06 \text{ mm}}_{-0.00 \text{ mm}}$ _____

l. $37.988 \text{ mm} \,^{+0.055 \text{ mm}}_{-0.000 \text{ mm}}$ _____

m. $250.000 \text{ mm} \,^{+0.000 \text{ mm}}_{-0.017 \text{ mm}}$ _____

n. $43.091 \text{ mm} \,^{+0.000 \text{ mm}}_{-0.026 \text{ mm}}$ _____

o. $79.979 \text{ mm} \,^{+0.009 \text{ mm}}_{-0.000 \text{ mm}}$ _____

Fits of Mating Parts

The following problems require computations with both clearance fits and interference fits between mating parts. Find the missing values in the following tables.

6. Refer to the figure to determine the values in the table. The answer to the first problem is given. Allowance is equal to the minimum clearance. All dimensions are in inches.

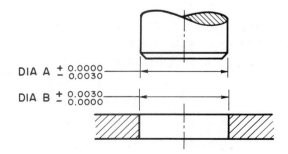

		Basic Dimension	Maximum Diameter (Max. Limit)	Minimum Diameter (Min. Limit)	Maximum Clearance	Minimum Clearance (Allowance)
a.	DIA A	1.4580	1.4580	1.4550	0.0090	0.0030
	DIA B	1.4610	1.4640	1.4610		
b.	DIA A	0.9345				
	DIA B	0.9365				
c.	DIA A	2.1053				
	DIA B	2.1078				

7. Refer to the figure to determine the values in the table. Allowance is equal to the maximum interference. All dimensions are in millimeters.

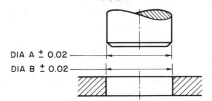

DIA A ± 0.02
DIA B ± 0.02

		Basic Dimension	Maximum Diameter (Max. Limit)	Minimum Diameter (Min. Limit)	Maximum Interference (Allowance)	Minimum Interference
a.	DIA A	20.73				
	DIA B	20.68				
b.	DIA A	32.07				
	DIA B	32.01				
c.	DIA A	10.82				
	DIA B	10.75				

8. Refer to the figure to determine the values in the table. Allowance is equal to minimum clearance. All dimensions are in inches.

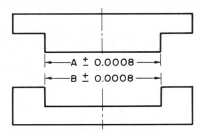

A ± 0.0008
B ± 0.0008

		Basic Dimension	Maximum Dimension (Max. Limit)	Minimum Dimension (Min. Limit)	Maximum Clearance	Minimum Clearance (Allowance)
a.	DIM A	0.9995				
	DIM B	1.0020				
b.	DIM A	2.0554				
	DIM B	2.0580				
c.	DIM A	1.4392				
	DIM B	1.4412				

9. Refer to the figure to determine the values in the table. Allowance is equal to the maximum interference. All dimensions are in millimeters.

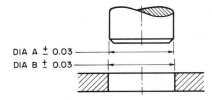

DIA A ± 0.03
DIA B ± 0.03

		Basic Dimension	Maximum Diameter (Max. Limit)	Minimum Diameter (Min. Limit)	Maximum Interference (Allowance)	Minimum Interference
a.	DIA A	78.78				
	DIA B	78.70				
b.	DIA A	9.94				
	DIA B	9.85				
c.	DIA A	130.03				
	DIA B	129.96				

Related Problems

10. Spacers are manufactured to the mean dimension and tolerance shown in this figure. An inspector measures 10 spacers and records the following thicknesses:

0.372″	0.379″	0.370″	0.377″	0.373″
0.376″	0.375″	0.373″	0.378″	0.380″

Which spacers are defective (above the maximum limit or below the minimum limit)? All dimensions are in inches.

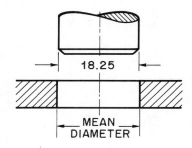

_____

11. A tool-and-die maker grinds a pin to an 18.25-millimeter diameter as shown. The pin is to be pressed (an interference fit) in a hole. The minimum interference allowed is 0.03 millimeter. The maximum interference allowed is 0.07 millimeter. Determine the mean diameter of the hole. All dimensions are in millimeters.

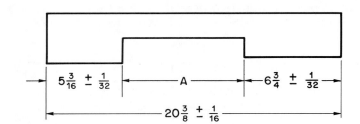

_____

12. A piece is to be cut to the dimensions and tolerances shown. Determine the maximum permissible value of length A. All dimensions are in inches.

_____

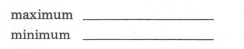

13. Determine the maximum and minimum permissible wall thickness of the steel sleeve shown. All dimensions are in millimeters.

maximum _____

minimum _____

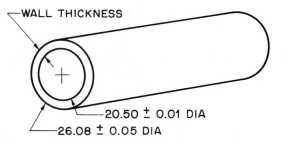

14. Mating parts are shown. The pins in the top piece fit into the holes in the bottom piece. All dimensions are in inches. Determine the following:

   a. The mean pin diameters.

   b. The mean hole diameters.

   c. The maximum dimension A.

   d. The minimum dimension A.

   e. The maximum dimension B.

   f. The minimum dimension B.

   g. The maximum total clearance between dimension C and dimension D.

   h. The minimum total clearance between dimension C and dimension D.

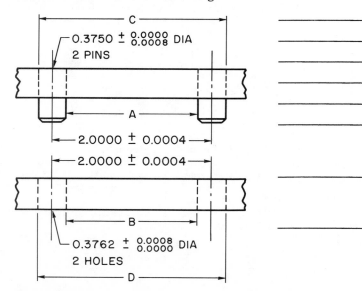

_____
_____
_____
_____
_____
_____

_____

_____

UNIT 19 ENGLISH AND METRIC STEEL RULES

OBJECTIVES

After studying this unit you should be able to

- Read measurements on fractional-inch and decimal-inch steel rules.
- Measure lengths using fractional-inch and decimal-inch scales.
- Read measurements on metric steel rules.
- Measure lengths using metric scales.

Steel rules are widely used for machine shop applications which do not require a high degree of precision. The steel rule is often the most practical measuring instrument to use for checking dimensions where stock allowances for finishing are provided. Steel rules are also used for locating roughing cuts on machined pieces and for determining the approximate locations of parts for machine setups. Steel rules used in the machine shop are generally six inches long, although rules anywhere from a fraction of an inch to several inches in length are also used.

CORRECT PROCEDURE IN THE USE OF STEEL RULES

The end of a rule receives more wear than the rest of the rule. Therefore, the end should not be used as a reference point unless it is used with a knee (a straight block).

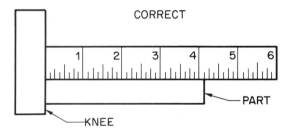

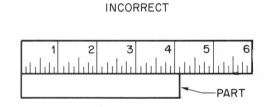

If a knee is not used, the 1-inch graduation of English measure rules should be used as a reference point as shown. The 1 inch must be subtracted from the measurement obtained. For metric measure rules, use the 10-millimeter graduation as the reference point. The 10 millimeters must be subtracted from the measurement obtained.

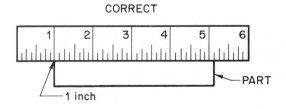

The scale edge of the rule should be put on the part to be measured. Following the correct procedure eliminates parallax error (error caused by the scale and the part being in different planes).

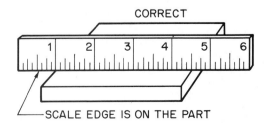

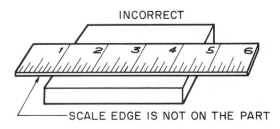

READING FRACTIONAL-INCH RULES

The smallest division of fractional rules is 1/64 inch. An enlarged fractional-inch rule is shown. The top scale is graduated in 64ths of an inch and the bottom scale in 32nds of an inch. The staggered graduations are halves, quarters, eighths, sixteenths, and thirty-seconds.

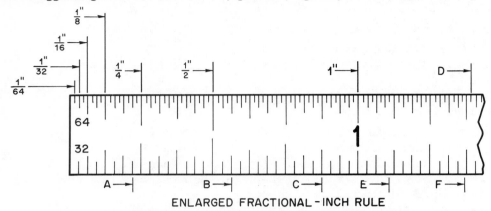

ENLARGED FRACTIONAL - INCH RULE

Measurements can be read by noting the last complete inch unit and counting the number of fractional units past the inch unit. Generally, a short-cut method of reading measurements is used as illustrated by the following examples.

Examples: Read the following measurements on the enlarged fractional-inch rule shown.

1. Length A: Subtract one $\frac{1}{32}''$ graduation from $\frac{1}{4}''$. $\frac{1}{4}'' - \frac{1}{32}'' = \frac{8''}{32} - \frac{1''}{32} = \frac{7''}{32}$ Ans

2. Length B: Add one $\frac{1}{16}''$ graduation to $\frac{1}{2}''$. $\frac{1}{2}'' + \frac{1}{16}'' = \frac{8''}{16} + \frac{1''}{16} = \frac{9''}{16}$ Ans

3. Length C: Subtract one $\frac{1}{8}''$ graduation from $1''$. $1'' - \frac{1}{8}'' = \frac{8''}{8} - \frac{1''}{8} = \frac{7''}{8}$ Ans

4. Length D: Add one $\frac{1}{64}''$ graduation to $1\frac{3}{8}''$. $1\frac{3}{8}'' + \frac{1}{64}'' = 1\frac{24''}{64} + \frac{1''}{64} = 1\frac{25''}{64}$ Ans

Often the edge of an object being measured does not fall exactly on a rule graduation. In these cases, read the measurement to the nearer rule graduation.

Examples: Read the following measurements, to the nearer graduation, on the enlarged fractional-inch rule shown.

1. Length E: Since the measurement is nearer to $1\frac{3}{32}''$ than $1\frac{1}{8}''$, Length E is read as $1\frac{3}{32}''$. Ans

2. Length F: Since the measurement is nearer to $1\frac{3}{8}''$ than $1\frac{11}{32}''$, Length F is read $1\frac{3}{8}''$.   Ans

READING DECIMAL-INCH RULES

An enlarged decimal-inch rule is shown. The top scale is graduated in hundredths of an inch (0.01″). The bottom scale is graduated in fiftieths of an inch (0.02″). The staggered graduations are halves, tenths, and fiftieths.

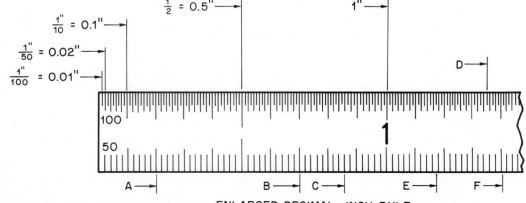

ENLARGED DECIMAL - INCH RULE

Examples: Read the following measurements on the enlarged decimal-inch rule shown.

1. Length A: Count two 0.1″ graduations. 2 × 0.1″ = 0.2″   Ans

2. Length B: Add two 0.1″ graduations to the 0.5″. 0.5″ + 0.2″ = 0.7″   Ans

3. Length C: Add three 0.02″ graduations to 0.8″. 0.8″ + 0.06″ = 0.86″   Ans

4. Length D: Add 1″, plus three 0.1″ graduations, plus five 0.01″ graduations. 1″ + 0.3″ + 0.05″ = 1.35″   Ans

5. Length E: Since the measurement is nearer to 1.18″ than 1.16″, Length E is read as 1.18″.   Ans

6. Length F: Since the measurement is nearer 1.40″ than 1.42″, Length F is read as 1.40″.   Ans

READING A METRIC RULE

An enlarged metric rule is shown. The top scale is graduated in half millimeters (0.5 mm). The bottom scale is graduated in millimeters (1 mm).

The following examples show the method of reading measurements with a metric rule with 0.5 mm and 1 mm scales.

Examples: Read the following measurements on the enlarged metric rule shown.

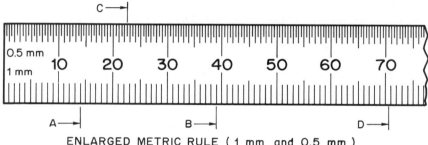

ENLARGED METRIC RULE (1 mm and 0.5 mm)

1. Length A: Add four 1-mm graduations to 10 mm. 10 mm + 4 mm = 14 mm   Ans

2. Length B: Subtract one 1-mm graduation from 40 mm. 40 mm – 1 mm = 39 mm   Ans

3. Length C: Add 20 mm, plus two 1-mm graduations, plus one 0.5-mm graduation. 20 mm + 2 mm + 0.5 mm = 22.5 mm   Ans

4. Length D: Since the measurement is nearer 71 mm than 70 mm, Length D is read as 71 mm.   Ans

APPLICATION

Fractional-Inch Steel Rules

1. Read measurements a–p on the enlarged fractional-inch rule shown.

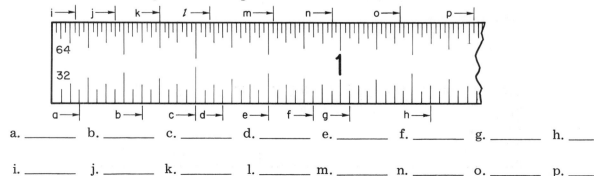

a. _____   b. _____   c. _____   d. _____   e. _____   f. _____   g. _____   h. _____

i. _____   j. _____   k. _____   l. _____   m. _____   n. _____   o. _____   p. _____

2. Measure the length of each of the following line segments to the nearer $\frac{1''}{16}$.

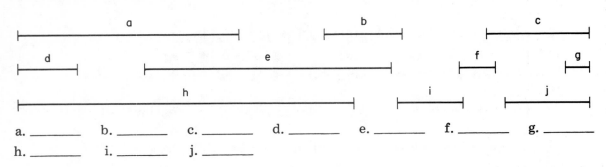

a. _____   b. _____   c. _____   d. _____   e. _____   f. _____   g. _____

h. _____   i. _____   j. _____

3. Measure the lengths of dimensions a–n of the template shown to the nearer $\frac{1''}{32}$.

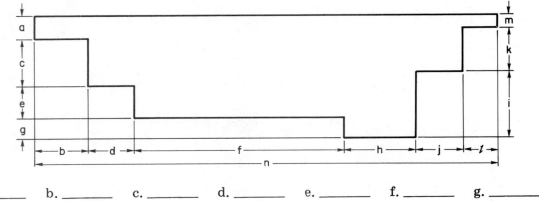

a. _____   b. _____   c. _____   d. _____   e. _____   f. _____   g. _____

h. _____   i. _____   j. _____   k. _____   l. _____   m. _____   n. _____

4. Measure the length of each of the following line segments to the nearer $\frac{1''}{64}$.

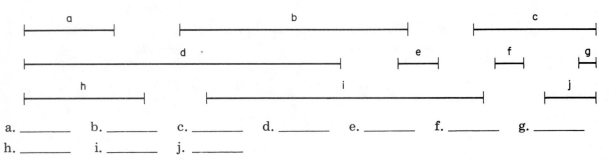

a. _____   b. _____   c. _____   d. _____   e. _____   f. _____   g. _____

h. _____   i. _____   j. _____

Decimal-Inch Steel Rules

5. Read measurements a–p on the enlarged decimal-inch rule shown.

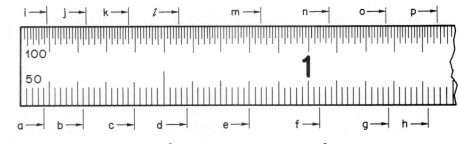

a. _____   b. _____   c. _____   d. _____   e. _____   f. _____   g. _____   h. _____

i. _____   j. _____   k. _____   l. _____   m. _____   n. _____   o. _____   p. _____

6. Measure the length of each of the following line segments to the nearer fiftieth of an inch (0.02″).

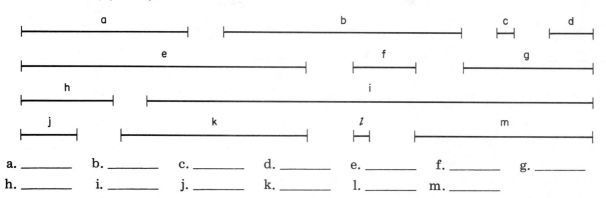

a. _____ b. _____ c. _____ d. _____ e. _____ f. _____ g. _____

h. _____ i. _____ j. _____ k. _____ l. _____ m. _____

7. Measure the diameters of the holes in the plate shown to the nearer fiftieth of an inch (0.02″).

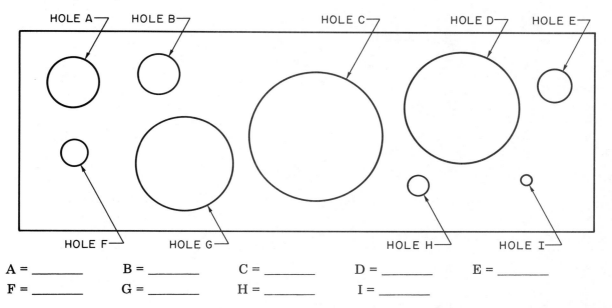

A = _____ B = _____ C = _____ D = _____ E = _____

F = _____ G = _____ H = _____ I = _____

Metric Steel Rules

8. Read measurements a–p on the enlarged metric rule with 1-millimeter and 0.5-millimeter graduations shown.

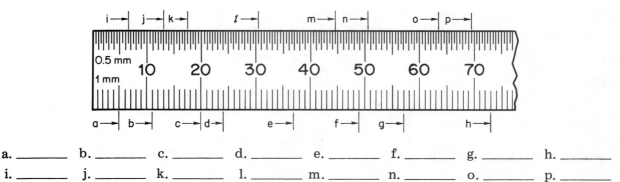

a. _____ b. _____ c. _____ d. _____ e. _____ f. _____ g. _____ h. _____

i. _____ j. _____ k. _____ l. _____ m. _____ n. _____ o. _____ p. _____

9. Measure the length of each of the following line segments to the nearer whole millimeter.

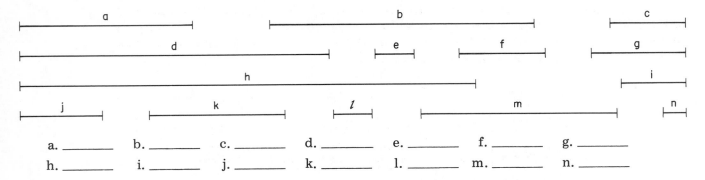

a. _____ b. _____ c. _____ d. _____ e. _____ f. _____ g. _____
h. _____ i. _____ j. _____ k. _____ l. _____ m. _____ n. _____

10. Measure dimensions a–k on the pattern shown to the nearer whole millimeter.

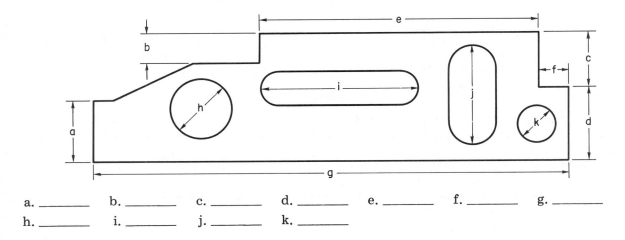

a. _____ b. _____ c. _____ d. _____ e. _____ f. _____ g. _____
h. _____ i. _____ j. _____ k. _____

UNIT 20 *ENGLISH VERNIER CALIPERS AND HEIGHT GAGES*

OBJECTIVES

After studying this unit you should be able to
- Read measurements set on a decimal-inch vernier caliper.
- Set given measurements on a decimal-inch vernier caliper.
- Read measurements set on a decimal-inch vernier height gage.
- Set given measurements on a decimal-inch vernier height gage.

Decimal-inch vernier calipers are used in machine shop applications when the degree of precision to thousandths of an inch is adequate. They are used for measuring lengths of parts, distances between holes, and both inside and outside diameters of cylinders.

Vernier height gages are widely used on surface plates and on machine tables. The height gage with an indicator attachment is used for checking locations of surfaces and holes. The height gage with a scriber attachment is used to mark reference lines, locations, and stock allowances on castings and forgings.

DECIMAL-INCH VERNIER CALIPER

The basic parts of a vernier caliper are a main scale which is similar to a steel rule with a fixed jaw and a sliding jaw with a vernier scale. The vernier scale slides parallel to the main scale and provides a degree of precision to 0.001″. Calipers are available in a wide range of lengths with different types of jaws and scale graduations. A vernier caliper which is commonly used in machine shops is shown.

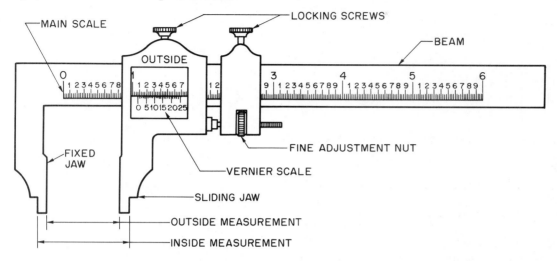

The main scale is divided into inches and the inches are divided into 10 divisions each equal to 0.1″. The 0.1″ divisions are divided into 4 parts each equal to 0.025″. The vernier scale consists of 25 divisions.

A vernier scale is shown. The vernier scale has 25 divisions in a length equal to a length on the main scale that has 24 divisions. The difference between a main scale division and a vernier division is 1/25 of 0.025″ or 0.001″.

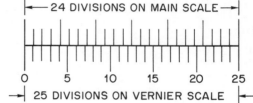

READING AND SETTING A MEASUREMENT ON A DECIMAL-INCH VERNIER CALIPER

A measurement is read by adding the thousandths reading on the vernier scale to the reading from the main scale.

Procedure: To read a measurement on a decimal-inch vernier caliper

- Read the number of 1″ graduations, 0.1″ graduations, and 0.025″ graduations on the main scale that are left of the zero graduation on the vernier scale.

- On the vernier scale, find the graduation that most closely coincides with a graduation on the main scale. Add this vernier reading which indicates the number of 0.001″ graduations to the main scale reading.

Setting a given measurement is the reverse procedure of reading a measurement on the vernier caliper.

Example 1: Read the measurement set on this vernier caliper.

In reference to the zero division on the vernier scale read two 1″ divisions, seven 0.1″ divisions, and three 0.025″ divisions on the main scale.
(2″ + 0.7″ + 0.075″ = 2.775″)

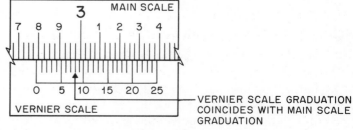

Observe which vernier scale graduation most closely coincides with a main scale graduation. The eight vernier scale graduation coincides; therefore, 0.008″ is added to 2.775″.

Measurement: 2.775″ + 0.008″ = 2.783″   Ans

Example 2: Set 1.237″ on a vernier caliper.

Move the vernier zero graduation to 1″ + 0.2″ + 0.025″ on the main scale.

An additional 0.012″ (1.237″ − 1.225″) is set by adjusting the sliding jaw until the 12 graduation on the vernier scale coincides with a graduation on the main scale.

The 1.237-inch setting is shown.

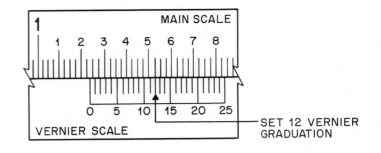

The accuracy of measurement obtainable with a vernier caliper depends on the user's ability to align the caliper with the part which is being measured and the user's "feel" when measuring. The line of measurement must be parallel to the beam of the caliper and lie in the same plane as the caliper. Care must be used to prevent too loose or too tight a caliper setting.

DECIMAL-INCH VERNIER HEIGHT GAGE

The vernier height gage and vernier caliper are similar in operation. The height gage also has a sliding jaw; the fixed jaw is the surface plate with which the height gage is usually used. The gage can be used with a scriber, a depth gage attachment, or an indicator. The indicator is the most widely used and, generally, the most accurate attachment. The parts of a vernier height gage are shown.

Measurements on the vernier height gage are read and set using the same procedure as with the vernier caliper.

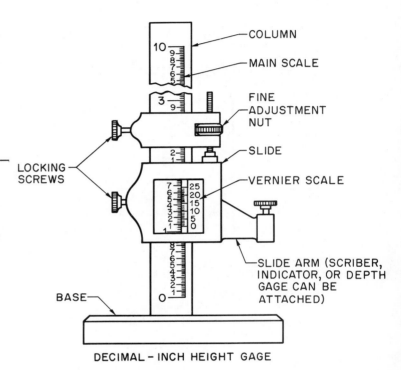

DECIMAL – INCH HEIGHT GAGE

Example 1: Read the measurement set on this vernier height gage.

In reference to the zero division on the vernier scale read 5″, four 0.1″ divisions, and two 0.025″ divisions on the main scale. (5″ + 0.4″ + 0.050″ = 5.450″)

Observe which vernier scale graduation most closely coincides with the main scale graduation. The twenty-first vernier scale graduation coincides; therefore, 0.021″ is added to 5.450″.

Measurement = 5.450″ + 0.021″ = 5.471″   Ans

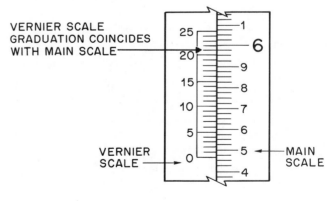

Example 2: Set 8.398″ on a vernier height gage.

Move the vernier zero graduation to 8″ + 0.3″ + 0.075″ = 8.375″.

An additional .023″ (8.398″ − 8.375″) is set by turning the fine adjustment screw until the 23 graduation on the vernier scale coincides with a graduation on the main scale.

The 8.398-inch setting is shown.

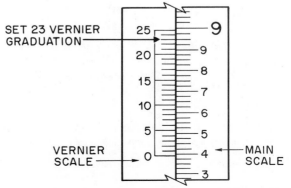

───APPLICATION──────────────────────────────────────

Decimal-Inch Vernier Caliper

1. Read the decimal-inch vernier caliper measurements a–h for the following settings.

a. _____

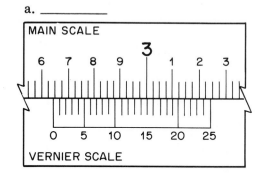

e. _____

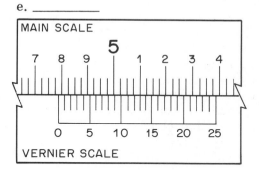

b. _____

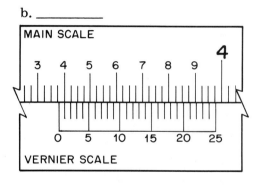

f. _____

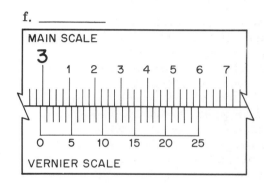

c. _____

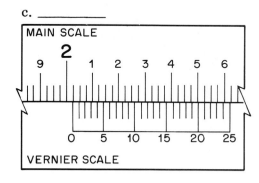

g. _____

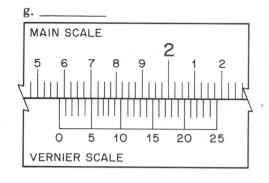

d. _____

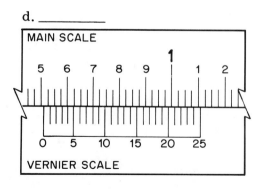

h. _____

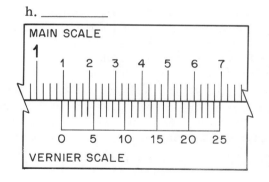

2. The following tables give the position of the zero graduation on the vernier scale in reference to the main scale and the vernier scale graduation that coincides with a main scale graduation. Determine the vernier caliper settings. The answer to the first problem is given.

	Zero Vernier Graduation Lies Between These Main Scale Graduations (inches)	Vernier Graduation That Coincides With A Main Scale Graduation	Vernier Caliper Setting (inches)		Zero Vernier Graduation Lies Between These Main Scale Graduations (inches)	Vernier Graduation That Coincides With A Main Scale Graduation	Vernier Caliper Setting (inches)
a.	1.875–1.900	19	1.894	m.	0.000–0.025	5	
b.	3.025–3.050	21		n.	0.825–0.850	15	
c.	0.050–0.075	3		o.	3.550–3.575	23	
d.	5.775–5.800	11		p.	5.075–5.100	20	
e.	1.225–1.250	7		q.	3.325–3.350	17	
f.	0.075–0.100	16		r.	2.075–2.100	6	
g.	4.000–4.025	4		s.	4.400–4.425	10	
h.	2.650–2.675	9		t.	1.025–1.050	14	
i.	1.000–1.025	13		u.	0.675–0.700	18	
j.	5.975–6.000	24		v.	0.050–0.075	2	
k.	2.825–2.850	8		w.	3.000–3.025	19	
l.	4.950–4.975	1		x.	2.925–2.950	22	

3. Refer to the following sentence and to the following given vernier caliper settings to find values A, B, and C. "The zero vernier scale graduation lies between A and B on the main scale and the vernier graduation C coincides with the main scale graduation." The answer to the first problem is given.

	Vernier Caliper Setting (inches)	A (inches)	B (inches)	C		Vernier Caliper Setting (inches)	A (inches)	B (inches)	C
a.	3.242	3.225	3.250	17	h.	1.646			
b.	2.877				i.	6.024			
c.	5.939				j.	0.022			
d.	0.611				k.	3.333			
e.	4.369				l.	5.999			
f.	0.094				m.	0.388			
g.	7.857				n.	0.965			

4. The distance between the centers of two holes can be checked with a vernier caliper. The position of the caliper in measuring the inside distance between two holes is shown. To determine the setting on the caliper, subtract the radius of each hole (one-half the diameter) from the center distance. The following problems give the hole diameters and the distances between centers. For each determine (1) the main scale setting and (2) the vernier scale setting. All dimensions are in inches.

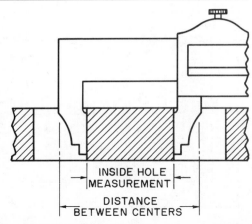

INSIDE HOLE MEASUREMENT

DISTANCE BETWEEN CENTERS

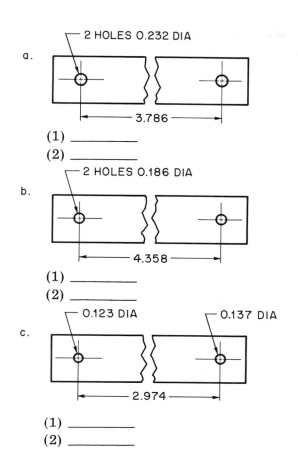

a. 2 HOLES 0.232 DIA
3.786
(1) _____
(2) _____

b. 2 HOLES 0.186 DIA
4.358
(1) _____
(2) _____

c. 0.123 DIA   0.137 DIA
2.974
(1) _____
(2) _____

d. NOTE: HOLE TOLERANCES ARE SHOWN. MAXIMUM AND MINIMUM VERNIER SCALE SETTINGS ARE REQUIRED.

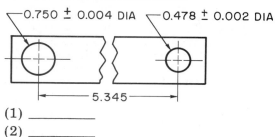

0.750 ± 0.004 DIA   0.478 ± 0.002 DIA
5.345
(1) _____
(2) _____

e. NOTE: HOLE TOLERANCES AND CENTER DISTANCE TOLERANCES ARE SHOWN. MAXIMUM AND MINIMUM VERNIER SCALE SETTINGS ARE REQUIRED.

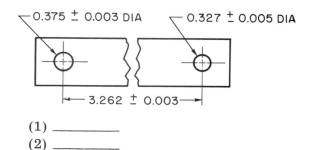

0.375 ± 0.003 DIA   0.327 ± 0.005 DIA
3.262 ± 0.003
(1) _____
(2) _____

Decimal-Inch Height Gage

5. Read height gage measurements a–h for the following settings.

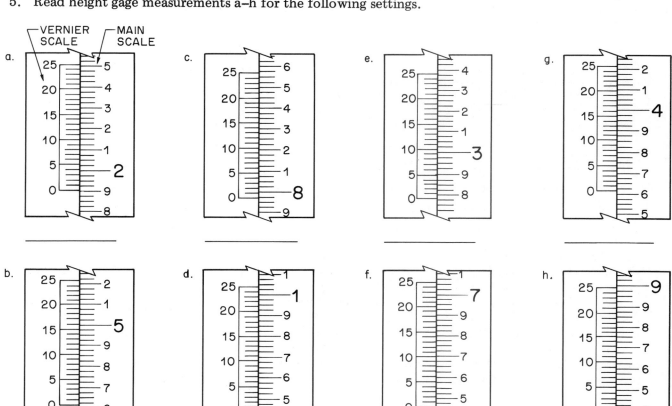

6. The hole locations of this block are checked by placing the block on a surface plate and indicating the bottom of each hole using a height gage with an indicator attachment. Determine the height gage settings from the bottom of the part to the bottom of the holes. Assume that the actual hole diameters and locations are the same as the given dimensions. The setting for the first problem is given.

Hole Number	Hole Diameter (inches)	Given Locations to Centers of Holes (inches)	Height Gage Settings	
			Main Scale Setting (inches)	Vernier Scale Setting
1	0.376	A = 0.640	0.450–0.475	2
2	0.250	B = 1.008		
3	0.188	C = 0.514		
4	0.496	D = 0.312		
5	0.132	E = 0.810		

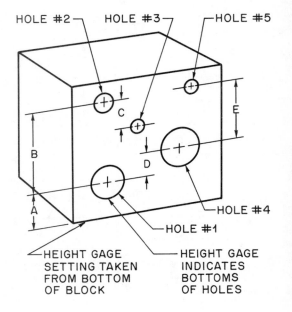

UNIT 21 *METRIC VERNIER CALIPERS AND HEIGHT GAGES*

OBJECTIVES

After studying this unit you should be able to

- Read measurements set on a metric vernier caliper.
- Set given measurements on a metric vernier caliper.
- Read measurements set on a metric vernier height gage.
- Set given measurements on a metric vernier height gage.

Metric vernier calipers and height gages are used in machine shop applications when the degree of precision to 0.02 millimeter is adequate.

METRIC VERNIER CALIPER

The same principles are used in reading and setting metric measure vernier calipers as those in decimal-inch measure. The main scale is divided in millimeter divisions. Each millimeter division is divided in half or 0.5-millimeter divisions. Every tenth millimeter graduation is numbered in sequence as 10 mm, 20 mm, 30 mm, etc. The vernier scale has 25 divisions; each division is 1/25 of 0.5 millimeter or 0.02 millimeter.

READING AND SETTING MEASUREMENTS ON A METRIC VERNIER CALIPER

A measurement is read by adding the 0.02-millimeter reading on the vernier scale to the reading from the main scale.

Procedure: To read a measurement on a metric vernier caliper

- Read the number of millimeter divisions and 0.5 millimeter divisions on the main scale that are to the left of the zero graduation on the vernier scale.

- On the vernier scale, find the graduation that most closely coincides with a graduation on the main scale. Multiply the graduation by 0.02 millimeter and add the value obtained to the main scale reading.

Setting a given measurement is the reverse procedure of reading a measurement on the vernier caliper.

Example 1:   Read the measurement set on the metric scales shown.

To the left of the zero division on the vernier scale read 21 millimeter divisions and one 0.5-millimeter division on the main scale. (21 mm + 0.5 mm = 21.5 mm)

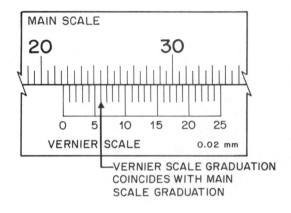

Observe which vernier scale graduation most closely coincides with a main scale graduation. The sixth vernier scale graduation coincides. Since each vernier scale graduation represents 0.02 mm, multiply 6 times 0.02 mm = 0.12 mm. Add 0.12 mm to 21.5 mm.

Measurement: 0.12 mm + 21.5 mm = 21.62 mm   Ans

Example 2:   Set 50.96 millimeters on a vernier caliper.

Move the vernier scale zero graduation to 50 millimeters plus 0.5 millimeter on the main scale. (50 mm + 0.5 mm = 50.5 mm)

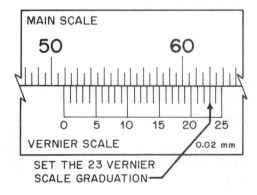

An additional 0.46 millimeter (50.96 mm – 50.5 mm) is set by adjusting the sliding jaw. Since each vernier scale graduation represents 0.02 mm, divide 0.46 mm by 0.02 mm; 0.46 mm ÷ 0.02 mm = 23. Adjust the sliding jaw until the 23 graduation on the vernier scale coincides with a graduation on the main scale.

The 50.96-millimeter setting is shown.

METRIC VERNIER HEIGHT GAGE

As with the English vernier height gage and caliper, the metric vernier height gage and metric vernier caliper are similar in operation.

READING AND SETTING MEASUREMENTS ON A METRIC VERNIER HEIGHT GAGE

Measurements on the metric vernier height gage are read and set using the same procedure as with the metric vernier caliper.

Example 1: Read the measurement set on the metric vernier height gage scales shown.

Below the zero division on the vernier scale read 70 millimeter divisions and one 0.5 millimeter division on the main scale. (70 mm + 0.5 mm = 70.5 mm)

Observe which vernier scale graduation most closely coincides with a main scale graduation. The eighth vernier scale graduation coincides. Since each vernier scale graduation represents 0.02 mm, multiply 8 times 0.02 mm = 0.16 mm. Add 0.16 mm to 70.5 mm.

Measurement: 0.16 mm + 70.5 mm = 70.66 mm   Ans

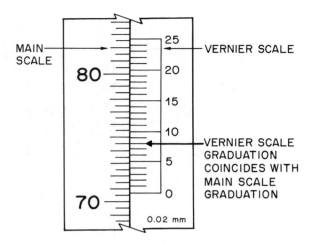

Example 2: Set 42.74 millimeters on a vernier height gage.

Move the vernier scale zero graduation to 42 millimeters plus 0.5 millimeter on the main scale. (42 mm + 0.5 mm = 42.5 mm)

An additional 0.24 millimeter (42.74 mm − 42.5 mm) is set by carefully adjusting the sliding jaw. Since each vernier scale graduation represents 0.02 mm, divide 0.24 mm by 0.02 mm; 0.24 mm ÷ 0.02 mm = 12. Adjust the sliding jaw until the 12 graduation on the vernier scale coincides with a graduation on the main scale.

The 42.74-millimeter setting is shown.

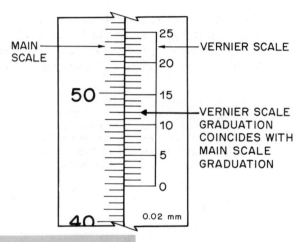

APPLICATION

Metric Vernier Caliper

1. Read the metric vernier caliper measurements for the following settings.

a. _____

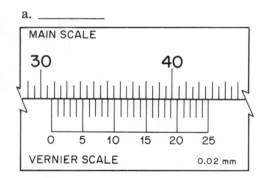

b. _____

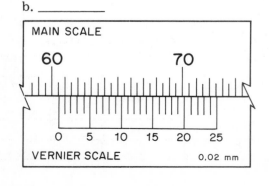

c. _____

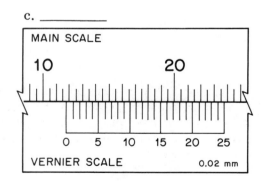

d. _____

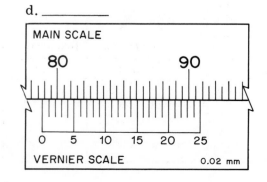

e. _____

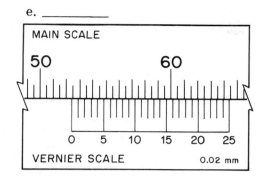

f. _____

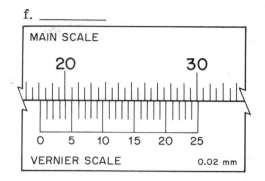

2. The following table gives the position of the zero graduation on the vernier scale in reference to the main scale of a metric vernier caliper. Also listed is the vernier scale graduation that coincides with a main scale graduation. Determine each vernier caliper setting. The answer to the first exercise is given.

	Zero Vernier Graduation Lies Between These Main Scale Graduations (millimeters)	Vernier Graduation That Coincides With A Main Scale Graduation	Vernier Caliper Setting (millimeters)		Zero Vernier Graduation Lies Between These Main Scale Graduations (millimeters)	Vernier Graduation That Coincides With A Main Scale Graduation	Vernier Caliper Setting (millimeters)
a.	52.5–53.0	14	52.78	g.	48.0–48.5	24	
b.	14.5–15.0	2		h.	77.5–78.0	21	
c.	86.0–86.5	19		i.	16.5–17.0	1	
d.	70.5–71.0	21		j.	98.0–98.5	9	
e.	26.0–26.5	8		k.	41.0–41.5	17	
f.	39.5–40.0	13		l.	56.5–57.0	20	

3. Refer to the following sentence and to the given vernier caliper settings in the following table to determine the values A, B, and C. "The zero vernier scale graduation lies between A and B on the main scale, and the vernier graduation C coincides with a main scale graduation." The answer to the first problem is given.

	Vernier Caliper Setting (millimeters)	A (millimeters)	B (millimeters)	C		Vernier Caliper Setting (millimeters)	A (millimeters)	B (millimeters)	C
a.	37.68	37.5	38.0	9	f.	20.28			
b.	19.76				g.	43.06			
c.	42.04				h.	77.40			
d.	88.82				i.	81.22			
e.	63.74				j.	96.98			

4. The distance between the two holes in the object shown is checked with
   vernier calipers. The position of the caliper in measuring the outside distance
   between the two holes in the part is shown. To determine the high and low
   limits for the setting on the caliper, add the radius of each hole (one-half the
   diameter) to the maximum or minimum center distance.

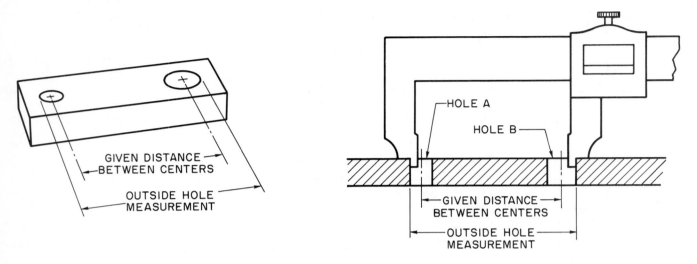

Refer to the data in the following table. Determine the vernier caliper scale
settings for each problem. The answer to the first problem is given. All dimensions are in millimeters.

	Actual Hole Diameters			Vernier Caliper Scale Settings		
	Hole A	Hole B	Given Distance Between Hole Centers	Main Scale Setting	High Limit Vernier Scale Setting	Low Limit Vernier Scale Setting
a.	10.52	12.86	56.92 ± 0.07	68.5–69.0	9	2
b.	14.10	17.18	72.08 ± 0.08			
c.	17.34	19.06	95.36 ± 0.04			
d.	9.98	14.80	44.41 ± 0.06			
e.	8.40	11.66	67.33 ± 0.10			
f.	19.36	21.82	86.57 ± 0.12			

Metric Height Gage

5. Read the metric vernier height gage measurements for the following settings.

a. _____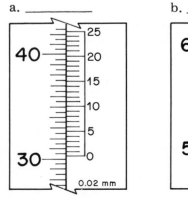

b. _____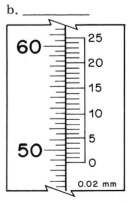

c. _____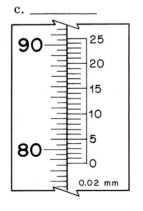

d. _____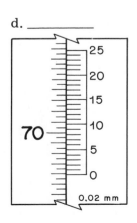

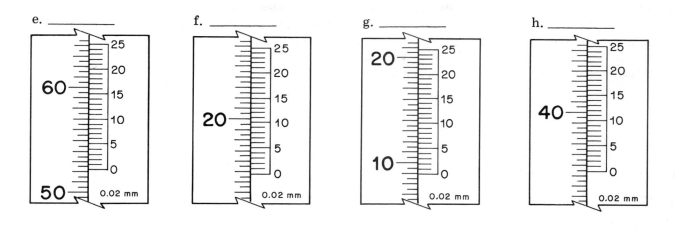

e. _____   f. _____   g. _____   h. _____

6. The hole locations of this block are checked by placing the block on a surface plate and indicating the bottom of each hole using a height gage with an indicator attachment. Determine the height gage settings from the bottom of the block to the bottom of the holes. Assume that the actual hole diameters and locations are the same as the given dimensions. The setting for the first hole is given.

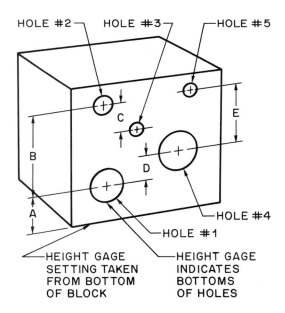

HOLE #2   HOLE #3   HOLE #5

E

B

C

D

A

HOLE #4

HOLE #1

HEIGHT GAGE SETTING TAKEN FROM BOTTOM OF BLOCK

HEIGHT GAGE INDICATES BOTTOMS OF HOLES

Hole Number	Hole Diameter (millimeters)	Given Locations To Centers of Holes (millimeters)	Height Gage Settings Main Scale Setting (millimeters)	Vernier Scale Setting
1	12.32	A = 15.78	9.5–10.0	6
2	6.38	B = 25.75		
3	4.50	C = 13.26		
4	14.76	D = 8.04		
5	5.84	E = 21.44		

UNIT 22 *ENGLISH MICROMETERS*

OBJECTIVES

After studying this unit you should be able to

- Read settings from the barrel and thimble scales of a 0.001-inch micrometer.

- Set given dimensions on the scales of 0.001-inch and 0.0001-inch micrometers.

- Read settings from the barrel, thimble, and vernier scales of 0.0001-inch micrometers.

Micrometers are basic measuring instruments used by machinists in the processing and checking of parts. Micrometers are available in a wide range of sizes and types. Outside micrometers are used to measure dimensions between parallel surfaces of parts and outside diameters of cylinders. Other types, such as depth micrometers, screw thread micrometers, disc and blade micrometers, bench micrometers, and inside micrometers, also have wide application in the machine shop. A few of the many types of micrometers are shown.

Anvil Micrometer
(The L.S. Starrett Company)

Bow Micrometer
(The L.S. Starrett Company)

Inside Micrometer
(The L.S. Starrett Company)

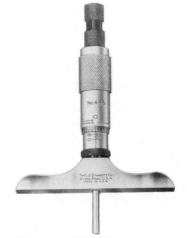

Micrometer Depth Gage
(The L.S. Starrett Company)

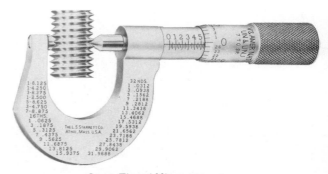

Screw Thread Micrometer
(The L.S. Starrett Company)

94

THE 0.001-INCH MICROMETER

A 0.001-inch outside micrometer is shown with its principal parts labeled.

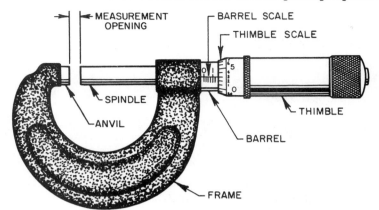

The part to be measured is placed between the anvil and the spindle. The barrel of a micrometer consists of a scale which is one inch long. The one-inch length is divided into ten divisions each equal to 0.100 inch. The 0.100-inch divisions are further divided in four divisions each equal to 0.025 inch.

The thimble has a scale which is divided into twenty-five parts. One revolution of the thimble moves 0.025 inch on the barrel scale. Therefore, a movement of one graduation on the thimble equals 1/25 of 0.025 inch or 0.001 inch along the barrel.

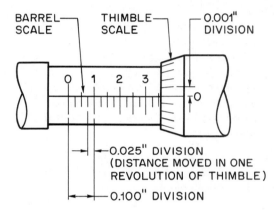

READING AND SETTING A 0.001-INCH MICROMETER

A micrometer is read by observing the position of the bevel edge of the thimble in reference to the scale on the barrel. Observe the greatest 0.100-inch division and the number of 0.025-inch divisions on the barrel scale. To this barrel reading, add the number of the 0.001-inch divisions on the thimble that coincide with the horizontal line (reading line) on the barrel scale.

Procedure: To read a 0.001-inch micrometer

- Observe the greatest 0.100-inch division on the barrel scale.
- Observe the number of 0.025-inch divisions on the barrel scale.
- Add the thimble scale reading (0.001-inch division) that coincides with the horizontal line on the barrel scale.

Example 1: Read the micrometer setting shown.

Observe the greatest 0.100-inch division on the barrel scale. (three 0.100″ = 0.300″)

Observe the number of 0.025-inch divisions between the 0.300-inch mark and the thimble. (two 0.025″ = 0.050″)

Add the thimble scale reading that coincides with the horizontal line on the barrel scale. (eight 0.001″ = 0.008″)

Micrometer reading: 0.300″ + 0.050″ + 0.008″ = 0.358″   Ans

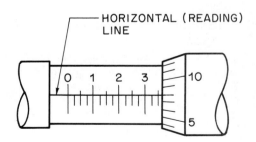

Example 2: Read the micrometer setting shown.

On the barrel scale, two 0.100″ = 0.200″.

On the barrel scale, zero 0.025″ = 0″.

On the thimble scale, twenty-three 0.001″ = 0.023″.

Micrometer reading: 0.200″ + 0.023″ = 0.223″   Ans

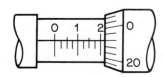

Procedure: To set a 0.001-inch micrometer to a given dimension

- Turn the thimble until the barrel scale indicates the required number of 0.100-inch divisions plus the necessary number of 0.025-inch divisions.
- Turn the thimble until the thimble scale indicates the required additional 0.001-inch divisions.

Example 1:   Set 0.949 inch on a micrometer.

Turn the thimble to nine 0.100-inch divisions plus one 0.025-inch division on the barrel scale. (9 × 0.100″ + 0.025″ = 0.925″)

Turn the thimble an additional twenty-four 0.001-inch thimble scale divisions. (0.949″ − 0.925″ = 0.024″)

The 0.949-inch setting is shown.

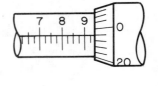

Example 2:   Set 0.520 inch on a micrometer.

Turn the thimble to five 0.100-inch divisions on the barrel scale. (5 × 0.100″ = 0.500″)

Turn the thimble an additional twenty 0.001-inch divisions. (0.520″ − 0.500″ = 0.020″)

The 0.520-inch setting is shown.

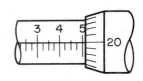

THE VERNIER (0.0001-INCH) MICROMETER

The addition of a vernier scale on the barrel of a 0.001-inch micrometer increases the degree of precision of the instrument to 0.0001 inch.   The barrel scale and thimble scale of a vernier micrometer are identical to that of a 0.001-inch micrometer.   The figure shows the relative positions of the barrel scale, thimble scale, and vernier scale of a 0.0001-inch vernier micrometer.

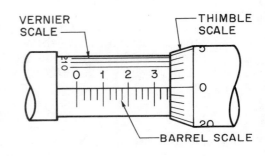

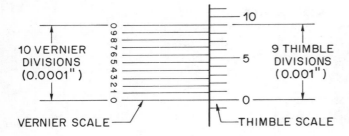

The vernier scale consists of ten divisions.   Ten vernier divisions on the circumference of the barrel are equal in length to nine divisions of the thimble scale.   The difference between one vernier division and one thimble division is 0.0001-inch.   A flattened view of a vernier and a thimble scale is shown.

READING AND SETTING THE VERNIER (0.0001-INCH) MICROMETER

Reading a vernier micrometer is the same as reading a 0.001-inch micrometer except for the addition of reading the vernier scale.   A particular vernier graduation coincides with a thimble scale graduation.   This vernier graduation gives the number of 0.0001-inch divisions that are added to the barrel and thimble scale readings.

Example 1:   Read the vernier micrometer setting shown in this flattened view.

Read the barrel scale reading.   Three 0.100″ divisions plus three 0.025″ divisions = 0.375″

Read the thimble scale.   The reading is between the 0.009″ and 0.010″ divisions, therefore, the thimble reading is 0.009″.

Read the vernier scale.   The 0.0004″ division of the vernier scale coincides with a thimble division.

Vernier micrometer reading: 0.375″ + 0.009″ + 0.0004″ = 0.3844″   Ans

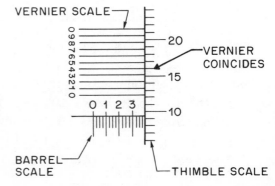

Example 2: Read the vernier micrometer setting shown in this flattened view.

On the barrel scale read 0.200″

On the thimble scale read 0.020″

On the vernier scale read 0.0008″

Vernier micrometer reading:
0.200″ + 0.020″ + 0.0008″ = 0.2208″   Ans

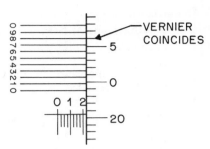

Setting a vernier (0.0001-inch) micrometer is the same as setting a 0.001-inch micrometer except for the addition of setting the vernier scale.

Example: Set 0.2336 inch on a vernier micrometer.

Turn the thimble to two 0.100-inch divisions plus one 0.025-inch division on the barrel scale. (2 × 0.100″ + 0.025″ = 0.225″)

Turn the thimble an additional eight 0.001-inch divisions.
(0.2336″ − 0.225″ = 0.0086″)

Turn the thimble carefully until a graduation on the thimble scale coincides with the 0.0006-inch division on the vernier scale. (0.2336″ − 0.233″ = 0.0006″)

The 0.2336-inch setting is shown.

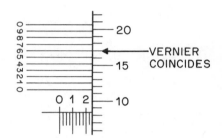

APPLICATION

0.001-Inch Micrometer

Read the settings on the following 0.001-inch micrometer scales.

1. _____

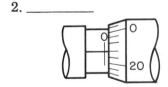

4. _____

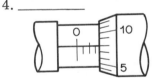

7. _____

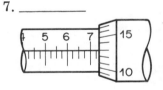

10. _____

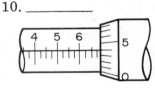

2. _____

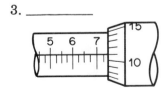

5. _____

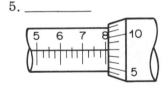

8. _____

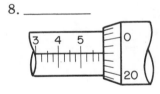

11. _____

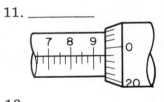

3. _____

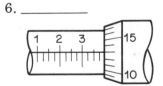

6. _____

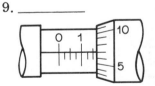

9. _____

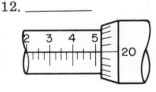

12. _____

Given the following barrel scale and thimble scale settings of a 0.001-inch micrometer, determine the readings in the tables. The answer to the first problem is given.

	Barrel Scale Setting is Between: (inches)	Thimble Scale Setting (inches)	Micrometer Reading (inches)
13.	0.425–0.450	0.016	0.441
14.	0.075–0.100	0.007	
15.	0.150–0.175	0.003	
16.	0.875–0.900	0.012	
17.	0.300–0.325	0.024	

	Barrel Scale Setting is Between: (inches)	Thimble Scale Setting (inches)	Micrometer Reading (inches)
18.	0.000–0.025	0.021	
19.	0.025–0.050	0.013	
20.	0.750–0.775	0.017	
21.	0.975–1.000	0.006	
22.	0.625–0.650	0.016	

Given the following 0.001-inch micrometer readings, determine the barrel scale and thimble scale settings. The answer to the first problem is given.

	Micrometer Reading (inches)	Barrel Scale Setting is Between: (inches)	Thimble Scale Setting (inches)		Micrometer Reading (inches)	Barrel Scale Setting is Between: (inches)	Thimble Scale Setting (inches)
23.	0.387	0.375–0.400	0.012	28.	0.998		
24.	0.839			29.	0.036		
25.	0.973			30.	0.281		
26.	0.002			31.	0.517		
27.	0.059			32.	0.666		

The Vernier (0.0001-Inch) Micrometer

Read the settings on the following 0.0001-inch micrometer scales. The vernier, thimble, and barrel scales are shown in flattened views.

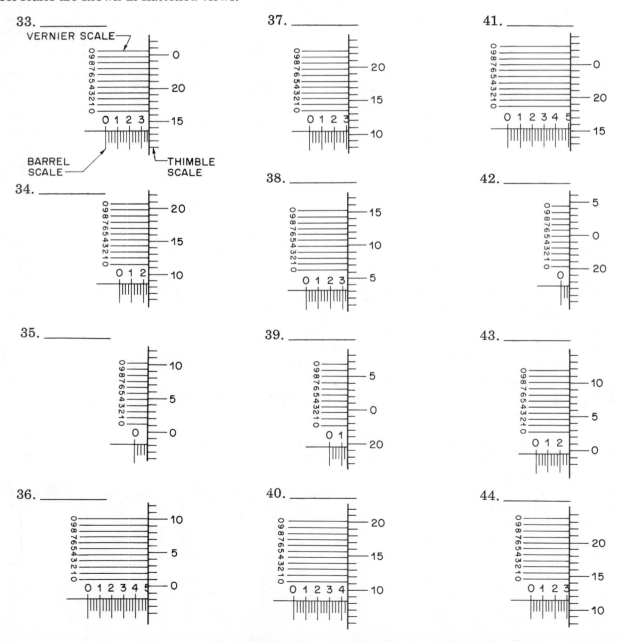

33. _____

VERNIER SCALE

BARREL SCALE   THIMBLE SCALE

34. _____

35. _____

36. _____

37. _____

38. _____

39. _____

40. _____

41. _____

42. _____

43. _____

44. _____

Given the following barrel scale, thimble scale, and vernier scale settings of a 0.0001-inch micrometer, determine the micrometer readings in these tables. The answer to the first problem is given.

	Barrel Scale Setting is Between: (inches)	Thimble Scale Setting is Between: (inches)	Vernier Scale Setting (inches)	Micrometer Reading (inches)		Barrel Scale Setting is Between: (inches)	Thimble Scale Setting is Between: (inches)	Vernier Scale Setting (inches)	Micrometer Reading (inches)
45.	0.375–0.400	0.017–0.018	0.0008	0.3928	50.	0.625–0.650	0.021–0.022	0.0003	
46.	0.125–0.150	0.008–0.009	0.0003		51.	0.000–0.025	0.000–0.001	0.0009	
47.	0.950–0.975	0.021–0.022	0.0009		52.	0.275–0.300	0.020–0.021	0.0007	
48.	0.075–0.100	0.011–0.012	0.0005		53.	0.850–0.875	0.009–0.010	0.0004	
49.	0.200–0.225	0.000–0.001	0.0004		54.	0.125–0.150	0.014–0.015	0.0008	

Given the following 0.0001-inch micrometer readings, determine the barrel scale, thimble scale, and vernier scale settings. The answer to the first problem is given.

	Micrometer Reading (inches)	Barrel Scale Setting is Between: (inches)	Thimble Scale Setting is Between: (inches)	Vernier Scale Setting (inches)		Micrometer Reading (inches)	Barrel Scale Setting is Between: (inches)	Thimble Scale Setting is Between: (inches)	Vernier Scale Setting (inches)
55.	0.7846	0.775–0.800	0.009–0.010	0.0006	60.	0.0008			
56.	0.1035				61.	0.8008			
57.	0.0079				62.	0.3135			
58.	0.9898				63.	0.9894			
59.	0.3001				64.	0.0379			

UNIT 23 *METRIC MICROMETERS*

OBJECTIVES

After studying this unit you should be able to

- Read settings from the barrel and thimble scales of a 0.01-millimeter micrometer.
- Set given dimensions on the scales of 0.01-millimeter and 0.002-millimeter micrometers.
- Read settings from the barrel, thimble, and vernier scales of 0.002-millimeter micrometers.

THE 0.01-MILLIMETER MICROMETER

The construction, parts, and operation of a 0.01-millimeter micrometer are basically the same as a 0.001-inch micrometer. A 0.01-millimeter outside micrometer is shown.

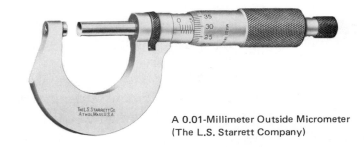

A 0.01-Millimeter Outside Micrometer
(The L.S. Starrett Company)

The barrel of a 0.01-millimeter micrometer consists of a scale which is 25 millimeters long. Refer to the barrel and thimble scales shown. The 25-millimeter barrel scale length is divided into 25 divisions each equal to 1 millimeter. Every fifth millimeter is numbered from 0 to 25 (0, 5, 10, 15, 20, 25). On the lower part of the barrel scale each millimeter is divided in half (0.5 mm).

The thimble has a scale which is divided into 50 parts. One revolution of the thimble moves 0.5 millimeter on the barrel scale. Therefore, a movement of one graduation on the thimble equals 1/50 of 0.5 millimeter or 0.01 millimeter along the barrel.

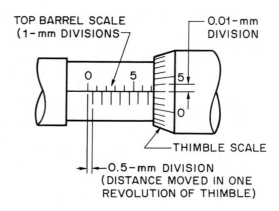

READING AND SETTING A 0.01-MILLIMETER MICROMETER

Procedure: To read a 0.01-millimeter micrometer

- Observe the number of 1-millimeter divisions on the barrel scale.
- Observe the number of 0.5-millimeter divisions (either 0 or 1) on the lower part of the barrel scale.
- Add the thimble scale reading (0.01-millimeter division) that coincides with the horizontal line (reading line) on the barrel side.

Example 1:   Read the micrometer setting shown.

Observe the number of 1-millimeter divisions on the barrel scale. Read as 4 millimeters.

Observe the number of 0.5-millimeter divisions on the lower barrel scale.

Read as zero 0.5-millimeter divisions.

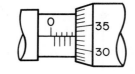

Add the thimble scale reading that coincides with the horizontal line on the barrel scale. Read as thirty-three 0.01-millimeter divisions.

Micrometer reading: 4 mm + 0.33 mm = 4.33 mm   Ans

Example 2:   Read this micrometer setting.

On the barrel scale read 17 millimeters.

On the lower barrel scale read one 0.5 millimeter.

On the thimble scale read twenty-six 0.01 millimeter.

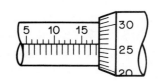

Micrometer reading: 17 mm + 0.5 mm + 0.26 mm = 17.76 mm   Ans

Procedure: To set a 0.01-millimeter micrometer

- Turn the thimble until the scale indicates the required number of 1-millimeter divisions plus the necessary number of 0.5-millimeter divisions.
- Turn the thimble until the thimble scale indicates the required additional 0.01-millimeter divisions.

Example: Set 14.94 millimeters on a micrometer.

Turn the thimble to fourteen 1-millimeter divisions plus one 0.5-millimeter division on the barrel scale.
(14 mm + 0.5 mm = 14.5 mm)

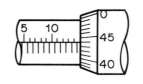

Turn the thimble an additional forty-four 0.01-millimeter thimble scale divisions. (14.94 mm − 14.5 mm = 0.44 mm)

The 14.94-millimeter setting is shown.

THE VERNIER (0.002-MILLIMETER) MICROMETER

The addition of a vernier scale on the barrel of a 0.01-millimeter micrometer increases the degree of precision of the instrument to 0.002 millimeter. The barrel scale and thimble scale of a vernier micrometer are identical to that of a 0.01-millimeter micrometer. The relative positions of the barrel scale, thimble scale and vernier scale of a vernier micrometer are shown.

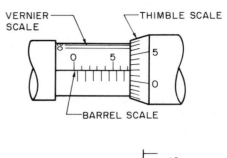

The vernier scale consists of five divisions. Each division equals one-fifth of a thimble division or 1/5 of 0.01 millimeter or 0.002 millimeter. A flattened view of a vernier and thimble scale is shown.

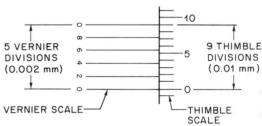

READING AND SETTING A VERNIER (0.002-MILLIMETER) MICROMETER

Reading a vernier (0.002-millimeter) micrometer is the same as reading a 0.01-millimeter micrometer except for the addition of reading the vernier scale. Observe which division on the vernier scale coincides with a division on the thimble scale. If the vernier division which coincides is marked 2, add 0.002 millimeter to the barrel and thimble scale reading. Add 0.004 millimeter for a coinciding vernier division marked 4, add 0.006 millimeter for a division marked 6, and add 0.008 millimeter for a division marked 8.

Example 1: A flattened view of a vernier micrometer is shown. Read this setting.

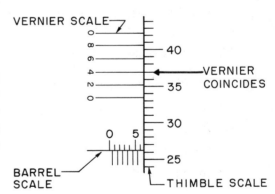

On the barrel scale read 6 millimeters.

On the lower barrel scale read zero 0.5 millimeter.

On the thimble scale read the number of 0.01-millimeter divisions. The reading is between the twenty-six and twenty-seven 0.01-millimeter divisions. Read as 0.26 millimeter.

Read the vernier scale. The 4 on the vernier scale coincides with a thimble scale division. Add 0.004 millimeter to the barrel and thimble scale readings.

The vernier micrometer reading:
6 mm + 0.26 mm + 0.004 mm = 6.264 mm   Ans

Example 2: A flattened view of a vernier micrometer is shown. Read this setting.

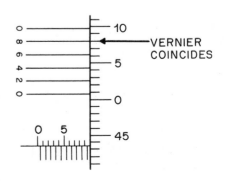

On the barrel scale read 9 millimeters.

On the lower barrel scale read one 0.5 millimeter.

On the thimble scale read the number of 0.01-millimeter divisions. The reading is between the forty-three and forty-four 0.01-millimeter divisions. Read as 0.43 millimeter.

Read the vernier scale. The 8 on the vernier scale coincides with a thimble scale division. Add 0.008 millimeter to the barrel and thimble scale readings.

The vernier micrometer reading:
9 mm + 0.5 mm + 0.43 mm + 0.008 mm = 9.938 mm   Ans

Setting a 0.002-millimeter vernier micrometer is the same as setting a 0.01-millimeter micrometer except for the addition of setting the vernier scale.

Example: Set 1.862 millimeters on a vernier micrometer.

Turn the thimble to one 1 millimeter-division plus one 0.5 millimeter-division on the barrel scale. (1 mm + 0.5 mm = 1.5 mm)

Turn the thimble an additional thirty-six 0.01-millimeter thimble scale divisions. (1.862 mm – 1.5 mm = 0.362 mm)

Turn the thimble carefully until a graduation on the thimble coincides with the 0.002 millimeter division on the vernier scale. (0.362 mm – 0.36 mm = 0.002 mm)

The 1.862-millimeter setting is shown.

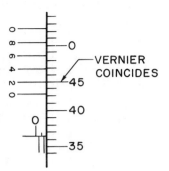

APPLICATION

0.01-Millimeter Micrometer

Read the settings on the following 0.01-millimeter scales.

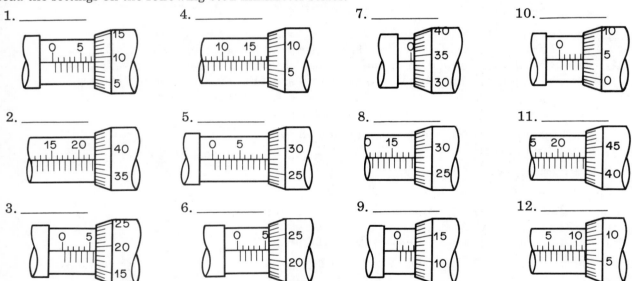

1. _____
2. _____
3. _____
4. _____
5. _____
6. _____
7. _____
8. _____
9. _____
10. _____
11. _____
12. _____

Given the micrometer readings in the following table. Determine the barrel scale and thimble scale settings. The answer to the first problem is given.

	Micrometer Reading (millimeters)	Barrel Scale Setting is Between: (millimeters)	Thimble Scale Setting (millimeters)		Micrometer Reading (millimeters)	Barrel Scale Setting is Between: (millimeters)	Thimble Scale Setting (millimeters)
13.	12.86	12.5–13.0	0.36	20.	6.66		
14.	9.34			21.	8.44		
15.	15.08			22.	19.72		
16.	3.92			23.	23.08		
17.	0.88			24.	5.66		
18.	7.06			25.	21.82		
19.	18.12			26.	13.90		

Vernier (0.002-Millimeter) Micrometer

Read the settings on the following 0.002-millimeter vernier micrometer scales.

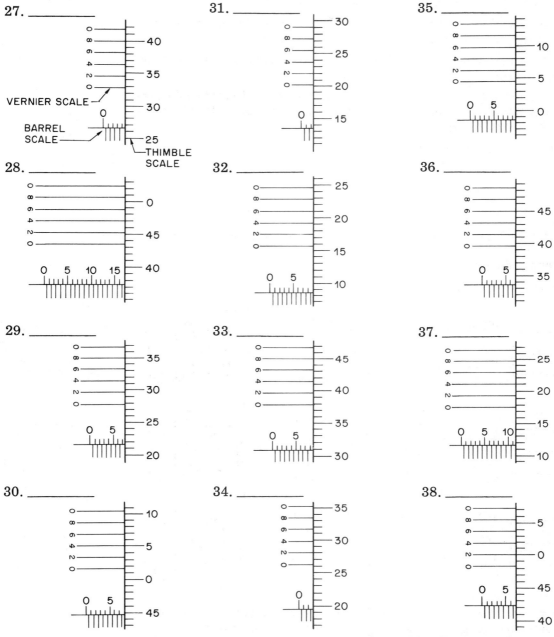

Given the following 0.002-millimeter vernier micrometer readings in the table, determine the barrel scale, thimble scale, and vernier scale settings. The answer to the first problem is given.

	Micrometer Reading (millimeters)	Barrel Scale Setting is Between: (millimeters)	Thimble Scale Setting is Between: (millimeters)	Vernier Scale Setting (millimeters)		Micrometer Reading (millimeters)	Barrel Scale Setting is Between: (millimeters)	Thimble Scale Setting is Between: (millimeters)	Vernier Scale Setting (millimeters)
39.	14.874	14.5–15.0	0.37–0.38	0.004	46.	20.292			
40.	21.168				47.	5.708			
41.	9.238				48.	13.998			
42.	11.862				49.	8.324			
43.	3.046				50.	0.756			
44.	8.768				51.	14.582			
45.	7.004				52.	9.776			

UNIT 24 ENGLISH AND METRIC GAGE BLOCKS

OBJECTIVE

After studying this unit you should be able to

- Determine proper gage block combinations for specified English or metric system dimensions.

Gage blocks are used in machine shops as standards for checking and setting (calibration) of micrometers, calipers, dial indicators, and other measuring instruments. Other applications of gage blocks are for layout, machine setups, and surface plate inspection.

DESCRIPTION OF GAGE BLOCKS

Gage blocks are square or rectangular shaped hardened steel blocks which are manufactured to a high degree of accuracy, flatness, and parallelism. Gage blocks, when properly used, provide millionths of an inch accuracy with millionths of an inch precision.

By *wringing* blocks (slipping blocks one over the other using light pressure), a combination of the proper blocks can be achieved which provides a desired length. Wringing the blocks produces a very thin air gap that is similar to liquid film in holding the blocks together. There are a variety of both English unit and metric gage block sets available. These tables list the thicknesses of blocks of a frequently used English gage block set and the thicknesses of blocks of a commonly used metric gage block set.

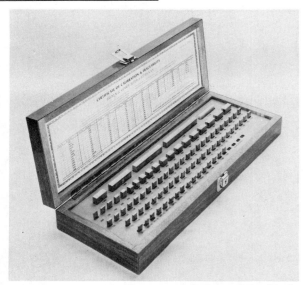

A complete set of gage blocks
(Brown & Sharpe Mfg. Co.)

BLOCK THICKNESSES OF AN ENGLISH GAGE BLOCK SET
NOTE: ALL THICKNESSES ARE IN INCHES

9 BLOCKS 0.000 1" SERIES

0.100 1	0.100 2	0.100 3	0.100 4	0.100 5	0.100 6	0.100 7	0.100 8	0.100 9

49 BLOCKS 0.001" SERIES

0.101	0.102	0.103	0.104	0.105	0.106	0.107	0.108	0.109
0.110	0.111	0.112	0.113	0.114	0.115	0.116	0.117	0.118
0.119	0.120	0.121	0.122	0.123	0.124	0.125	0.126	0.127
0.128	0.129	0.130	0.131	0.132	0.133	0.134	0.135	0.136
0.137	0.138	0.139	0.140	0.141	0.142	0.143	0.144	0.145
0.146	0.147	0.148	0.149					

19 BLOCKS 0.050" SERIES

0.050	0.100	0.150	0.200	0.250	0.300	0.350	0.400	0.450
0.500	0.550	0.600	0.650	0.700	0.750	0.800	0.850	0.900
0.950								

4 BLOCKS 1.000" SERIES

1.000	2.000	3.000	4.000

BLOCK THICKNESSES OF A METRIC GAGE BLOCK SET NOTE: ALL THICKNESSES ARE IN MILLIMETERS								
9 BLOCKS 0.001 mm SERIES								
1.001	1.002	1.003	1.004	1.005	1.006	1.007	1.008	1.009
9 BLOCKS 0.01 mm SERIES								
1.01	1.02	1.03	1.04	1.05	1.06	1.07	1.08	1.09
9 BLOCKS 0.1 mm SERIES								
1.1	1.2	1.3	1.4	1.5	1.6	1.7	1.8	1.9
9 BLOCKS 1 mm SERIES								
1	2	3	4	5	6	7	8	9
9 BLOCKS 10 mm SERIES								
10	20	30	40	50	60	70	80	90

DETERMINING GAGE BLOCK COMBINATIONS

Usually there is more than one combination of blocks which will give a desired length. The most efficient procedure for determining block combinations is to eliminate digits of the desired measurement from right to left. This procedure saves time, minimizes the number of blocks, and reduces the chances of error. The following examples show how to apply the procedure in determining block combinations.

Example 1: Determine a combination of gage blocks for 2.9468 inches. Refer to the gage block sizes given in the Table of Block Thicknesses of an English Gage Block Set. All dimensions are in inches.

Choose the block which eliminates the last digit to the right, the 8. Choose the 0.1008″ block. Subtract. (2.9468″ − 0.1008″ = 2.846″)

Eliminate the last digit, 6, of 2.846″. Choose the 0.146″ block which eliminates the 4 as well as the 6. Subtract. (2.846″ − 0.146″ = 2.700″)

Eliminate the last non-zero digit, 7, of 2.700″. Choose the 0.700″ block. Subtract. (2.700″ − 0.700″ = 2.000″)

The 2.000″ block completes the required dimension as shown.

Check. Add the blocks chosen.
0.1008″ + 0.146″ + 0.700″ + 2.000″ = 2.9468″

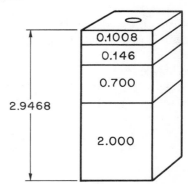

Example 2: Determine a combination of gage blocks for 10.2843 inches. Refer to the gage block sizes given in the Table of Block Thicknesses for an English Gage Block Set. All dimensions are in inches.

Eliminate the 3. Choose the 0.1003″ block. Subtract. (10.2843″ − 0.1003″ = 10.184″)

Eliminate the 4. Choose the 0.134″ block. Subtract. (10.184″ − 0.134″ = 10.050″)

Eliminate the 5. Choose the 0.050″ block. Subtract. (10.050″ − 0.050″ = 10.000″)

The 1.000″, 2.000″, 3.000″ and 4.000″ blocks complete the required dimensions as shown.

Check. (0.1003″ + 0.134″ + 0.050″ + 1.000″ + 2.000″ + 3.000″ + 4.000″ = 10.2843″)

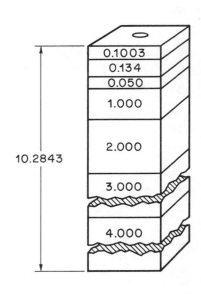

Example 3: Determine a combination of gage blocks for 157.372 millimeters. Refer to the gage block sizes given in the Table of Block Thicknesses for a Metric Gage Block Set. All dimensions are in millimeters.

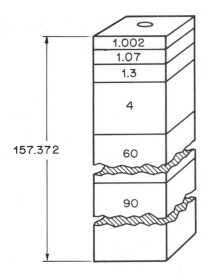

Eliminate the 2. Choose the 1.002 mm block. Subtract. (157.372 mm – 1.002 mm = 156.37 mm)

Eliminate the 7. Choose the 1.07 mm block. Subtract. (156.37 mm – 1.07 mm = 155.3 mm)

Eliminate the 3. Choose the 1.3 mm block. Subtract. (155.3 mm – 1.3 mm = 154 mm)

Eliminate the 4. Choose the 4 block. Subtract. (154 mm – 4 mm = 150 mm)

The 60 and 90 block complete the required dimension as shown.

Check. (1.002 mm + 1.07 mm + 1.3 mm + 4 mm + 60 mm + 90 mm = 157.372 mm)

APPLICATION

English Gage Blocks

Using the Table of Block Thicknesses for an English Gage Block Set, determine a combination of gage blocks for each of the following dimensions. Note: Usually more than one combination of blocks will give the desired dimension.

1. 4.8638″	_____	10. 9.050″	_____	18. 0.8754″	_____
2. 1.8702″	_____	11. 4.8757″	_____	19. 7.7777″	_____
3. 3.1222″	_____	12. 1.0001″	_____	20. 10.0101″	_____
4. 0.6333″	_____	13. 0.2731″	_____	21. 9.4346″	_____
5. 0.3759″	_____	14. 2.7311″	_____	22. 3.9208″	_____
6. 5.8002″	_____	15. 5.090″	_____	23. 6.003″	_____
7. 7.973″	_____	16. 6.0907″	_____	24. 10.0021″	_____
8. 0.9999″	_____	17. 2.9789″	_____	25. 0.7998″	_____
9. 10.375″	_____				

Metric Gage Blocks

Using the Table of Block Thicknesses for a Metric Gage Block Set, determine a combination of gage blocks for each of the following dimensions. Note: Usually more than one combination of blocks will give the desired dimension.

26. 63.385 mm	_____	38. 85.111 mm	_____
27. 14.073 mm	_____	39. 39.099 mm	_____
28. 34.356 mm	_____	40. 122.22 mm	_____
29. 146.09 mm	_____	41. 67.005 mm	_____
30. 213.9 mm	_____	42. 41.87 mm	_____
31. 43.707 mm	_____	43. 2.007 mm	_____
32. 9.999 mm	_____	44. 206.23 mm	_____
33. 83.38 mm	_____	45. 193.03 mm	_____
34. 157.08 mm	_____	46. 73.061 mm	_____
35. 13.86 mm	_____	47. 12.704 mm	_____
36. 38.727 mm	_____	48. 149.007 mm	_____
37. 6.071 mm	_____	49. 55.555 mm	_____

UNIT 25 ACHIEVEMENT REVIEW —
SECTION 2 _____

OBJECTIVE_____

You should be able to solve the exercises and problems in this Achievement Review by applying the principles and methods covered in units 16–24.

1. Express each of the following lengths as indicated.

   a. 90 inches as feet _____

   b. $6\frac{1}{4}$ feet as inches _____

   c. 9.6 yards as feet _____

   d. 3.8 centimeters as millimeters _____

   e. 0.8 meter as millimeters _____

   f. 218 millimeters as centimeters _____

2. Holes are to be drilled in the length of angle iron as shown. What is the distance between 2 consecutive holes?

   _____

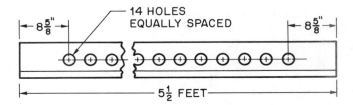

3. How many complete 3-meter lengths of tubing are required to make 220 pieces each 54 millimeters long? Allow a total one-half length of tubing for cutoff and scrap. _____

4. Express each of the following lengths as indicated. When necessary, round the answer to 3 decimal places.

   a. 56 millimeters as inches _____

   b. 5.5 meters as feet _____

   c. 16.8 centimeters as inches _____

   d. 4.75 inches as millimeters _____

   e. 27 inches as centimeters _____

   f. 4.5 feet as meters _____

5. For each of the exercises in the following table, the measurement made and the smallest graduation of the measuring instrument is given. Determine the greatest possible error and the smallest and largest possible actual length measured for each.

	Measurement Made	Smallest Graduation of Measuring Instrument Used	Greatest Possible Error	Actual Length Smallest Possible	Largest Possible
a.	5.18″	0.02″ (steel rule)			
b.	0.8367″	0.0001″ (vernier micrometer)			
c.	46.16 mm	0.02 mm (vernier caliper)			
d.	15.35 mm	0.01 mm (micrometer)			

6. The following dimensions with tolerances are given. Determine the maximum dimension (maximum limit) and the minimum dimension (minimum limit) for each.

   a. 1.714″ ± 0.005″

   maximum _____ minimum _____

   b. 3.0798″ $^{+0.0000''}_{-0.0012''}$

   maximum _____ minimum _____

   c. 5.9047″ $^{+0.0008''}_{-0.0000''}$

   maximum _____ minimum _____

   d. 56.93 mm ± 0.08 mm

   maximum _____ minimum _____

   e. 173.003 mm $^{+0.000\ mm}_{-0.013\ mm}$

   maximum _____ minimum _____

107

7. Express each of the following unilateral tolerances as bilateral tolerances having equal plus and minus values.

a. 0.876″ $^{+0.006″}_{-0.000″}$ _____

b. 5.2619″ $^{+0.0000″}_{-0.0012″}$ _____

c. 43.63 mm $^{+0.00\ mm}_{-0.03\ mm}$ _____

d. 78.909 mm $^{+0.009\ mm}_{-0.000\ mm}$ _____

8. The following problems require computations with both clearance fits and interference fits between mating parts. Determine the clearance or interference values as indicated. All dimensions are given in inches.

a.

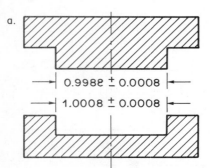

0.9982 ± 0.0008

1.0008 ± 0.0008

(1) FIND THE MAXIMUM CLEARANCE. _____
(2) FIND THE MINIMUM CLEARANCE. _____

c.

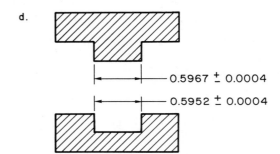

1.6250 $^{+0.0005}_{-0.0000}$

1.6232 $^{+0.0000}_{-0.0005}$

(1) FIND THE MAXIMUM INTERFERENCE (ALLOWANCE). _____
(2) FIND THE MINIMUM INTERFERENCE. _____

b.

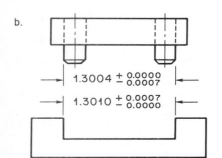

1.3004 $^{+0.0000}_{-0.0007}$

1.3010 $^{+0.0007}_{-0.0000}$

(1) FIND THE MAXIMUM CLEARANCE. _____
(2) FIND THE MINIMUM CLEARANCE. _____

d.

0.5967 ± 0.0004

0.5952 ± 0.0004

(1) FIND THE MAXIMUM INTERFERENCE (ALLOWANCE). _____
(2) FIND THE MINIMUM INTERFERENCE. _____

9. Determine the minimum permissible length of distance A of the part shown. All dimensions are in millimeters.

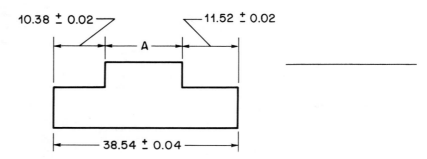

10.38 ± 0.02   11.52 ± 0.02

A

38.54 ± 0.04

_____

10. Read measurements a–p on the enlarged 32nds and 64ths graduated fractional rule shown.

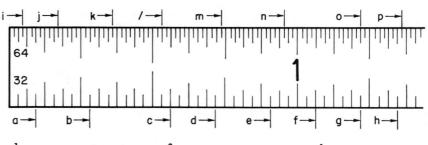

a. _____ b. _____ c. _____ d. _____ e. _____ f. _____ g. _____ h. _____

i. _____ j. _____ k. _____ l. _____ m. _____ n. _____ o. _____ p. _____

11. Read measurements a–p on the enlarged 50th and 100ths graduated decimal-inch rule shown.

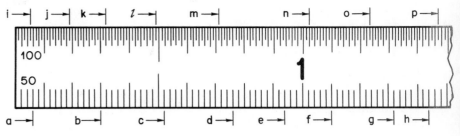

a. _____ b. _____ c. _____ d. _____ e. _____ f. _____ g. _____ h. _____

i. _____ j. _____ k. _____ l. _____ m. _____ n. _____ o. _____ p. _____

12. Read measurements a–p on the enlarged 1 millimeter and 0.5 millimeter graduated metric rule shown.

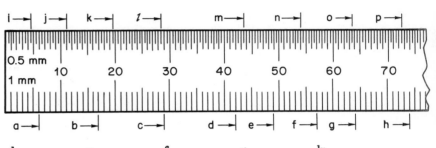

a. _____ b. _____ c. _____ d. _____ e. _____ f. _____ g. _____ h. _____

i. _____ j. _____ k. _____ l. _____ m. _____ n. _____ o. _____ p. _____

13. Read the vernier caliper and height gage measurements for the following settings.

ENGLISH MEASUREMENTS

a. _____

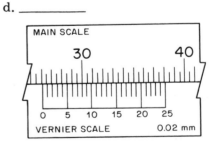

METRIC MEASUREMENTS

d. _____

b. _____

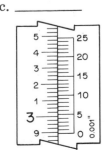

e. _____

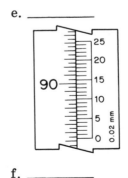

c. _____

f. _____

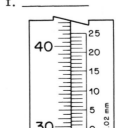

14. Read the settings on the following micrometer scales.

a. 0.001 Decimal-Inch Micrometer

(1) _____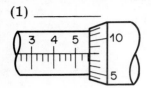

(2) _____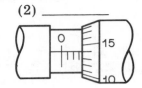

(3) _____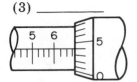

(4) _____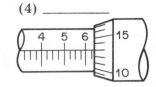

b. 0.0001 Decimal-Inch Vernier Micrometer

(1) _____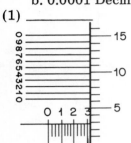

(2) _____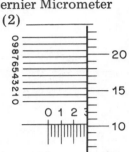

(3) _____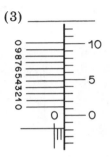

(4) _____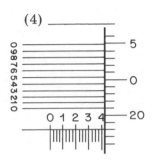

c. 0.01-Millimeter Metric Micrometer

(1) _____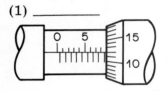

(2) _____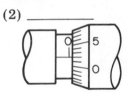

(3) _____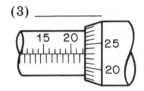

(4) _____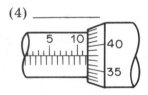

d. 0.002-Millimeter Metric Micrometer

(1) _____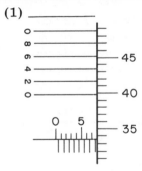

(2) _____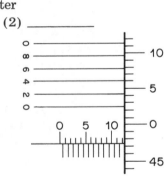

(3) _____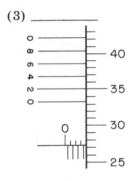

15. Using the Table of Block Thicknesses for an English Gage Block Set found in Unit 24, determine a combination of gage blocks for each of the following dimensions. Note: Usually more than one combination of blocks will give the desired dimension.

a. 0.3784″ _____   d. 5.7573″ _____   g. 7.8895″ _____

b. 2.3486″ _____   e. 3.0901″ _____   h. 9.0016″ _____

c. 1.7062″ _____   f. 0.2009″ _____

16. Using the Table of Block Thicknesses for a Metric Gage Block Set, found in Unit 24 determine a combination of gage blocks for each of the following dimensions. Note: Usually more than one combination of blocks will give the desired dimension.

a. 47.63 mm _____   d. 13.274 mm _____   g. 99.998 mm _____

b. 125.22 mm _____   e. 66.066 mm _____   h. 108.061 mm _____

c. 85.092 mm _____   f. 34.404 mm _____

SECTION 3 FUNDAMENTALS OF ALGEBRA

⌐UNIT 26 *SYMBOLISM* _____

⌐OBJECTIVES _____

After studying this unit you should be able to

- Express word statements as algebraic expressions.

- Express diagram dimensions as algebraic expressions.

- Evaluate algebraic expressions by substituting numbers for symbols.

Algebra is a branch of mathematics in which letters are used to represent numbers. By the use of letters, general rules called *formulas* can be stated mathematically. Algebra is an extension of arithmetic; therefore, the rules and procedures which apply to arithmetic also apply to algebra. Many problems which are difficult or impossible to solve by arithmetic can be solved by algebra.

The basic principles of algebra discussed in this text are intended to provide a practical background for machine shop applications. A knowledge of algebraic fundamentals is essential in the use of trade handbooks and for the solutions of many geometric and trigonometric problems.

⌐SYMBOLISM _____

Symbols are the language of algebra. Both arithmetic numbers and literal numbers are used in algebra. *Arithmetic numbers* are numbers which have definite numerical values, such as 4, 5.17, and $\frac{7}{8}$. *Literal numbers* are letters which represent arithmetic numbers, such as a, x, V, and P. Depending on how it is used, a literal number can represent one particular arithmetic number, a wide range of numerical values, or all numerical values.

Customarily the multiplication sign ($\times$) is not used in algebra because it can be misinterpreted as the letter x. When a literal number is multiplied by a numerical value, or when two or more literal numbers are multiplied, no sign of operation is required.

Examples:

1. 5 times a is written $5a$
2. 17 times c is written $17c$
3. V times P is written VP
4. 6 times a times b times c is written $6abc$

Parentheses ( ) are often used in place of the multiplication sign ($\times$) when numerical values are multiplied; 3×4 is written $3(4)$; $18 \times 3.4 \times 5^2$ is written $18(3.4)(5^2)$.

An *algebraic expression* is a word statement put into mathematical form by using literal numbers, arithmetic numbers, and signs of operation. The following are examples of algebraic expressions.

Example 1: A dimension is increased by 0.5 inch. How long is the increased dimension? All dimensions are in inches.

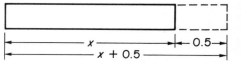

If x is the original dimension, the increased dimension is $x + 0.5''$. Ans

Example 2: The production rate of a new machine is 4 times as great as an old machine. Write an algebraic expression for the production rate of the new machine.

If the old machine produced y parts per hour, the new machine produces $4y$ parts per hour. Ans

Example 3:   A drill rod is cut in 3 equal pieces. How long is each piece? (Disregard waste)

If L is the length of the drill rod, the length of each piece is $\frac{L}{3}$.   Ans

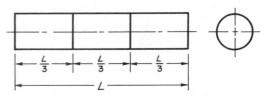

Example 4:   In the step block shown, dimension B equals $\frac{3}{4}$ of dimension A and dimension C is twice dimension A. Find the total height of the block.

If d is the length of dimension A, dimension B is $\frac{3}{4} d$ and dimension C is $2d$. The total height is $d + \frac{3}{4} d + 2d$ or $3\frac{3}{4} d$. Ans

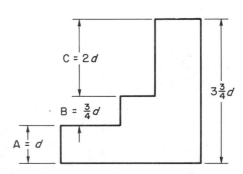

Note: If no arithmetic number appears before a literal number, it is assumed that the value is the same as if a one (1) appeared before the letter, $d = 1d$.

Example 5:   A plate with 8 drilled holes is shown. The distance from the left edge of the plate to hole 1 and the distance from the right edge of the plate to hole 8 are each represented by a. The distances between holes 1 and 2, holes 2 and 3, and holes 3 and 4 are each represented by b. The distances between holes 4 and 5, holes 5 and 6, holes 6 and 7, and holes 7 and 8 are each represented by c. Find the total length of the plate. All dimensions are in millimeters.

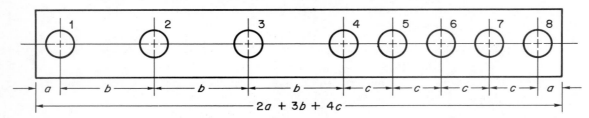

The total length of the plate is $a + b + b + b + c + c + c + c + a$, or $2a + 3b + 4c$.   Ans

Note: Only like literal numbers may be arithmetically added.

─EVALUATION OF ALGEBRAIC EXPRESSIONS

The value of an algebraic expression is found by substituting given numerical values for literal values and solving the expression by following the order of operations as in arithmetic. The order of operations follows:

- Do all operations within the grouping symbol first. Parentheses, the fraction bar, and the radical symbol are used to group numbers. If an expression contains parentheses within parentheses or brackets, do the work within the innermost parentheses first.

- Do powers and roots next. The operations are performed in the order in which they occur. If a root consists of two or more operations within the radical symbol, perform all the operations within the radical symbol, then extract the root.

- Do multiplication and division next in the order which they occur.

- Do addition and subtraction last in the order in which they occur.

Example 1:   The formula for finding the perimeter of a rectangle is given. Find the perimeter of the rectangle shown. All dimensions are in millimeters.

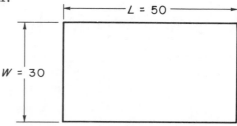

$P = 2L + 2W$
$P = 2(50 \text{ mm}) + 2(30 \text{ mm})$
$P = 100 \text{ mm} + 60 \text{ mm}$
$P = 160 \text{ mm}$ Ans

$P = 2L + 2W$ where
P = perimeter
L = length
W = width

Example 2: The formula for finding the area of a ring is given. Find the area of the ring shown. All dimensions are in inches.

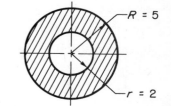

$A = \pi R^2 - \pi r^2$ where A = area
$A = 3.14(5 \text{ in})^2 - 3.14(2 \text{ in})^2$ π = 3.14
$A = 3.14(25 \text{ sq in}) - 3.14(4 \text{ sq in})$ R = outside radius
$A = 78.50 \text{ sq in} - 12.56 \text{ sq in}$ r = inside radius
$A = 65.94 \text{ sq in}$ Ans

Example 3: The formula for the approximate perimeter of an ellipse is given. Find the perimeter of the ellipse shown. All dimensions are in inches.

$P = \pi\sqrt{2(a^2 + b^2)}$ where P = perimeter
$P = 3.14\sqrt{2[(8 \text{ in})^2 + (6 \text{ in})^2]}$ π = 3.14
$P = 3.14\sqrt{2(64 \text{ sq in} + 36 \text{ sq in})}$ a = 0.5 (major axis)
$P = 3.14\sqrt{2(100 \text{ sq in})}$ b = 0.5 (minor axis)
$P = 3.14\sqrt{200 \text{ sq in}}$
$P = 3.14(14.14 \text{ in})$
$P = 44.40 \text{ in}$ Ans

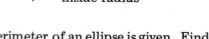

Example 4: Find the value of $\dfrac{3(2b + 3dy)}{4(7d - bd)}$ when $b = 6$, $d = 4$ and $y = 2$.

$$\frac{3[2(6) + 3(4)(2)]}{4[7(4) - 6(4)]} = \frac{3(12 + 24)}{4(28 - 24)} = \frac{3(36)}{4(4)} = \frac{108}{16} = 6.75 \quad \text{Ans}$$

Example 5: Find the value of $3m[4p + 5(x - m) + p]^2$ when $m = 2$, $p = 3$, $x = 8$.
$3(2)[4(3) + 5(8 - 2) + 3]^2 = 6[12 + 5(6) + 3]^2 = 6(45)^2 = 6(2025) = 12{,}150 \quad \text{Ans}$

Example 6: Find the value of $\dfrac{6a}{b} + \dfrac{abc}{20}(a^3 - 12b)$ when $a = 5$, $b = 10$, and $c = 8$.
$\dfrac{6(5)}{10} + \dfrac{5(10)(8)}{20}[5^3 - 12(10)] = \dfrac{30}{10} + \dfrac{400}{20}(125 - 120) = 3 + 20(5) = 3 + 100 = 103 \quad \text{Ans}$

APPLICATION

Algebraic Expressions

Express each of the following problems as an algebraic expression.

1. The product of 6 and x increased by y. _____
2. The sum of a and 12. _____
3. Subtract b from 25. _____
4. Subtract 25 from b. _____
5. Divide r by s. _____
6. Twice L minus one-half P. _____
7. The product of x and y divided by the square of m. _____

8. In the part shown, all dimensions are in inches.
   a. What is the total length of this part? _____
   b. What is the length from point A to point B? _____

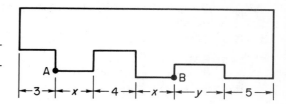

9. Find the distance between the indicated points.
   a. Point A to point B _____
   b. Point F to point C _____
   c. Point B to point C _____
   d. Point D to point E _____

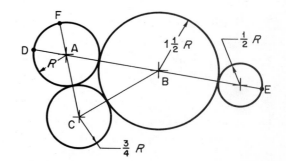

10. What is the length of the following dimensions? All dimensions are in millimeters.

    a. Dimension A   _____

    b. Dimension B   _____

    c. Dimension C   _____

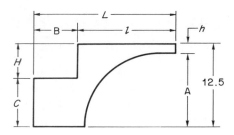

11. Stock is removed from a block in two operations. The original thickness of the block is represented by n. The thickness removed by the milling operation is represented by p and the thickness removed by the grinding operation is represented by t. What is the final thickness of the block?   _____

12. Given: s as the length of a side of a hexagon, r as the radius of the inside circle, and R as the radius of the outside circle.

    a. What is the length of r if r equals the product of 0.866 and the length of a side of the hexagon?   _____

    b. What is the length of R if R equals the product of 1.155 and the radius of the inside circle?   _____

    c. What is the area of the hexagon if the area equals the product of 2.598 and the square of the radius of the outside circle?   _____

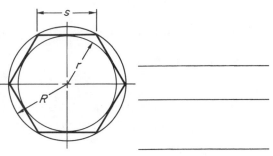

Evaluation of Algebraic Expressions

Substitute the given numbers for letters and find the values of the following expressions.

13. If $a = 4$ and $c = 2$, find

    a. $5a + 3c^2$ _____

    c. $\dfrac{10c}{a}$ _____

    e. $\dfrac{a + 5c}{ac + a}$ _____

    b. $5c + a$ _____

    d. $\dfrac{a + c}{a - c}$ _____

14. If $b = 8$, $d = 4$, and $e = 2$, find

    a. $\dfrac{b}{d} + e - 3$ _____

    d. $3e(b - e) - d\left(\dfrac{b}{2}\right)$ _____

    b. $bd(3 + 4d - b)$ _____

    e. $\dfrac{12d}{e} - [3b - (d + e) + 4]$ _____

    c. $5b - (bd + 3)$ _____

15. If $x = 6$ and $y = 3$, find

    a. $2xy + 7$ _____

    c. $\dfrac{5xy - 2y}{8x - xy}$ _____

    e. $6x - 3y + xy$ _____

    b. $3x - 2y + xy$ _____

    d. $\dfrac{4x - 4y}{3}$ _____

16. If $m = 5$, $p = 4$, and $r = 3$, find

    a. $m + mp^2 - r^3$ _____

    d. $\dfrac{p^3 + 3p - 12}{m^2 + 15}$ _____

    b. $(p + 2)^2 (m - r)^2$ _____

    e. $\dfrac{r^3}{3p - 9} + m^2 (mp - 6r)^2$ _____

    c. $\dfrac{(pr)^2}{2} - pr + m^3$ _____

17. All dimensions are in inches.

    a. Find the area (A) of this square.

    $A = \dfrac{1}{2} d^2$ _____

    b. Find the side (S) of this square.

    $S = 0.7071d$ _____

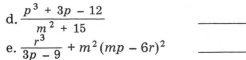

18. All dimensions are in millimeters.

    a. Find the length of this arc (l).
    $$l = \frac{\pi R \alpha}{180°}$$ _____

    b. Find the area of this sector (A).
    $$A = \frac{1}{2} Rl$$ _____

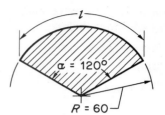

19. All dimensions are in inches. Refer to the triangle shown.

    a. Find S when $S = \frac{1}{2}(a + b + c)$. _____

    b. Find the area (A) when
    $$A = \sqrt{S(S-a)(S-b)(S-c)}$$ _____

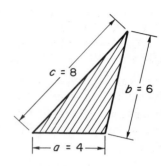

20. All dimensions are in millimeters.

    a. Find the radius of this circle.
    $$r = \frac{c^2 + 4h^2}{8h}$$ _____

    b. Find the length of the arc (l).
    $$l = 0.0175 r\alpha$$ _____

21. All dimensions are in inches.
    Find the shaded area.
    $$\text{Area} = \frac{(H + h)b + ch + aH}{2}$$ _____

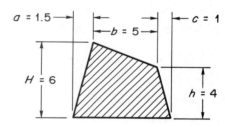

22. All dimensions are in inches. Find
    the length of belt on the pulleys.
    $$\text{Length of belt} = 2C + \frac{11D + 11d}{7} + \frac{(D - d)^2}{4C}$$

    _____

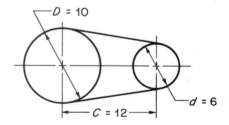

23. All dimensions are in millimeters.
    Find the shaded area. _____
    $$\text{Area} = dt + 2a(s + n)$$

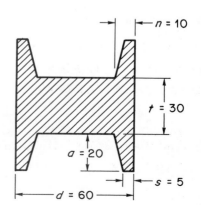

24. All dimensions are in inches.
     Find the shaded area.   _____

     Area $= \pi(ab - cd)$

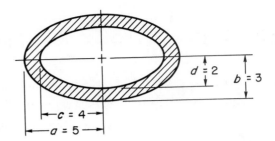

25. All dimensions are in inches.
     Find the shaded area.   _____

     Area $= \dfrac{\pi(R^2 - r^2)}{2}$

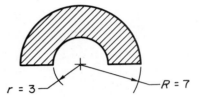

26. All dimensions are in millimeters.
     Find the shaded area.   _____

     Area $= t[b + 2(a - t)]$

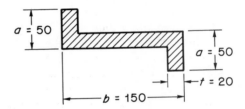

27. All dimensions are in inches.
     a. Find the slant height (S).   _____

        $S = \sqrt{(R - r)^2 + h^2}$

     b. Find the volume.   _____

        Volume $= 1.05h(R^2 + Rr + r^2)$

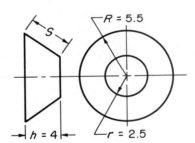

28. All dimensions are in inches.
     Find the volume.   _____

     Volume $= \dfrac{(2a + c)bh}{6}$

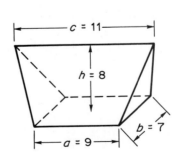

UNIT 27 *SIGNED NUMBERS*

OBJECTIVES

After studying this unit you should be able to

- Compare signed numbers according to size and direction using the number scale.
- Determine absolute values of signed numbers.
- Perform basic operations of addition, subtraction, multiplication, division, powers, and roots using signed numbers.
- Solve expressions which involve combined operations of signed numbers.

Signed numbers are required for solving problems in mechanics and trigonometry. Positive and negative numbers express direction, such as machine table movement from a reference point. Signed numbers are particularly useful in programming machining operations for numerical control.

MEANING OF SIGNED NUMBERS

Plus and minus signs which you have worked with so far in this book have been *signs of operation*. These are signs used in arithmetic, with the plus sign (+) indicating the operation of addition and the minus sign (−) indicating the operation of subtraction.

In algebra, plus and minus signs are used to indicate both operation and direction from a reference point or zero. A *positive number* is indicated either with no sign or with a plus sign (+) preceding the number. For example, +7 or 7 is a positive number which is 7 units greater than zero. A *negative number* is indicated with a minus sign (−) preceding the number. For example, −7 is a negative number which is 7 units less than zero. Positive and negative numbers are called *signed numbers* or directed numbers.

THE NUMBER SCALE

The number scale shows the relationship of positive and negative numbers. It shows both distance and direction between numbers. Considering a number as a starting point and counting to a number to the right represents positive (+) direction with numbers increasing in value. Counting to the left represents negative (−) direction with numbers decreasing in value.

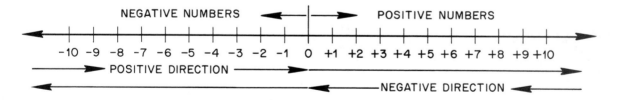

Examples:

1. Starting at 0 and counting to the right to +5 represents 5 units in a positive (+) direction; +5 is 5 units greater than 0.

2. Starting at 0 and counting to the left to −5 represents 5 units in a negative (−) direction; −5 is 5 units less than 0.

3. Starting at −2 and counting to the right to +6 represents 8 units in a positive (+) direction; +6 is 8 units greater than −2.

4. Starting at +6 and counting to the left to −2 represents 8 units in a negative (−) direction; −2 is 8 units less than +6.

5. Starting at –3 and counting to the left to –10 represents 7 units in a (–) direction; –10 is 7 units less than –3.

6. Starting at –9 and counting to the right to 0 represents 9 units in a (+) direction; 0 is 9 units greater than –9.

OPERATIONS USING SIGNED NUMBERS

In order to solve problems in algebra, you must be able to perform basic operations using signed numbers. The following procedures and examples show how to perform operations of addition, subtraction, multiplication, division, powers, and roots with signed numbers. The procedure for performing certain operations of signed numbers are based on an understanding of absolute value.

The *absolute value* of a number is the number without regard to its sign. For example, the absolute value of +4 is 4, the absolute value of –4 is also 4. Therefore, the absolute value of +4 and –4 is the same value, 4.

The absolute value of –20 is 15 greater than the absolute value of +5; 20 is 15 greater than 5.

ADDITION OF SIGNED NUMBERS

Procedure: To add two or more positive numbers

- Add the numbers as in arithmetic.

Examples: Add the following numbers.

1. +3
   +5
   ——
   +8 Ans

2. 15
   7
   ——
   22 Ans

3. $2 + 9 + 13 = 24$ Ans

4. $+12 + (+15) = +27$ Ans

Procedure: To add two or more negative numbers

- Add the absolute values of the numbers.
- Prefix a minus sign to the sum.

Examples: Add the following numbers.

1. –5
   –2
   ——
   –7 Ans

2. –13
   – 4
   –15
   ——
   –32 Ans

3. $–6 + (–5) = –11$ Ans

4. $–8 + (–10) + (–4) + (–3) = –25$ Ans

Procedure: To add a positive and a negative number

- Subtract the smaller absolute value from the larger absolute value.
- Prefix the sign of the number having the larger absolute value to the difference.

Examples: Add the following numbers.

1. +5
   –3
   ——
   +2 Ans

2. –5
   +3
   ——
   –2 Ans

3. –17
   +17
   ——
   0 Ans

4. $+12 + (–8) = +4$ Ans

5. $–12 + (+8) = –4$ Ans

Procedure: To add more than two positive and negative numbers

- Add all the positive numbers.
- Add all the negative numbers.
- Add their sums following the procedure for adding signed numbers.

Examples: Add the following numbers.

1. $–2 + 4 + (–10) + 5 = 9 + (–12) = –3$ Ans
2. $8 + 7 + (–6) + 4 + (–3) + (–5) + 10 = 29 –14 = 15$ Ans
3. $4 + (–6) + 12 + 3 + (–7) + 1 + (–5) + (–2) = 20 – 20 = 0$ Ans

SUBTRACTION OF SIGNED NUMBERS

Procedure: To subtract signed numbers

- Change the sign of the number subtracted (subtrahend) to the opposite sign.
- Follow the procedure for addition of signed numbers.

Note: When the sign of the subtrahend is changed, the problem becomes one in addition. Therefore, subtracting a negative number is the same as adding a positive number. Subtracting a positive number is the same as adding a negative number.

Examples:

1. Subtract 5 from 8. $8 - (+5) = 8 + (-5) = 3$ Ans
2. Subtract 8 from 5. $5 - (+8) = 5 + (-8) = -3$ Ans
3. Subtract -5 from 8. $8 - (-5) = 8 + (+5) = 13$ Ans
4. Subtract -5 from -8. $-8 - (-5) = -8 + (+5) = -3$ Ans
5. $-3 - (+7) = -3 + (-7) = -10$ Ans
6. $0 - (-14) = 0 + (+14) = 14$ Ans
7. $0 - (+14) = 0 + (-14) = -14$ Ans
8. $-14 - (-14) = -14 + (+14) = 0$ Ans

MULTIPLICATION OF SIGNED NUMBERS

Procedure: To multiply two or more signed numbers

- Multiply the absolute values of the numbers.
- Count the number of negative signs.

    If there is an odd number of negative signs, the product is negative.

    If there is an even number of negative signs, the product is positive.

    If all numbers are positive, the product is positive.

It is not necessary to count the number of positive values in an expression consisting of both positive and negative numbers. Count only the number of negative values to determine the sign of the product.

Examples:

1. $4(-3) = -12$ Ans
   (There is one negative sign. Since one is an odd number, the product is negative.)
2. $-4(-3) = +12$ Ans
   (There are two negative signs. Since two is an even number, the product is positive.)
3. $(-2)(-4)(-3)(-1)(-2)(-1) = +48$ Ans   (6 negatives, even number, positive product)
4. $(-2)(-4)(-3)(-1)(-2) = -48$ Ans   (5 negatives, odd number, negative product)
5. $(2)(4)(3)(1)(2) = +48$ Ans   (all positives, positive product)
6. $(2)(-4)(-3)(1)(-2) = -48$ Ans   (3 negatives, odd number, negative product)
7. $(-2)(4)(-3)(-1)(-2) = +48$ Ans   (4 negatives, even number, positive product)

Note: The product of any number or numbers and 0 = 0; for example, $0(9) = 0$; $0(-9) = 0$; $8(-6)(0)(6) = 0$.

DIVISION OF SIGNED NUMBERS

Procedure: To divide two signed numbers

- Divide the absolute values of the numbers.
- Determine the sign of the quotient.

    If both numbers have the same sign (both negative or both positive) the quotient is positive.

    If the two numbers have unlike signs (one positive and one negative) the quotient is negative.

Examples:

1. $\dfrac{-8}{-2} = +4$ Ans

2. $\dfrac{8}{2} = +4$ Ans

3. $15 \div 3 = +5$ Ans

4. $-3 \overline{\smash{)}-15} = +5$ Ans

5. $\dfrac{-30}{5} = -6$ Ans

6. $\dfrac{30}{-5} = -6$ Ans

7. $-21 \div 3 = -7$ Ans

8. $-3 \overline{\smash{)}21} = -7$ Ans

Note: Zero divided by any number = 0; for example, $0 \div (+3) = 0$, $0 \div (-3) = 0$. A number divided by 0 is not allowed. For the purposes of this text, division by 0 has no meaning; for example, $14 \div 0$ and $-14 \div 0$ have no meaning.

⎯POWERS OF SIGNED NUMBERS

Procedure: To raise numbers with positive exponents to a power

- Apply the procedure for multiplying signed numbers to raising signed numbers to powers.

Examples:

1. $3^2 = +9$ Ans

2. $3^3 = +27$ Ans

3. $2^4 = +16$ Ans

4. $2^5 = +32$ Ans

5. $-3^2 = (-3)(-3) = +9$ Ans

6. $-3^3 = (-3)(-3)(-3) = -27$ Ans

7. $-2^4 = (-2)(-2)(-2)(-2) = +16$ Ans

8. $-2^5 = (-2)(-2)(-2)(-2)(-2) = -32$ Ans

Note: • A positive number raised to any power is positive.

- A negative number raised to an even power is positive.

- A negative number raised to an odd power is negative.

Procedure: To raise numbers with negative exponents to a power

- Invert the number.
- Change the negative exponent to a positive exponent.
- Apply procedure for raising numbers with positive exponents to a power.

Examples:

1. $3^{-2} = \dfrac{3^{-2}}{1} = \dfrac{1}{3^2} = \dfrac{1}{9}$ Ans

2. $2^{-3} = \dfrac{2^{-3}}{1} = \dfrac{1}{2^3} = \dfrac{1}{8}$ Ans

3. $-4^{-3} = \dfrac{-4^{-3}}{1} = \dfrac{1}{-4^3} = \dfrac{1}{-64}$ Ans

⎯ROOTS OF SIGNED NUMBERS

When either a positive number or a negative number are squared, a positive number results. For example, $3^2 = 9$ and $(-3)^2 = 9$. Therefore, every positive number has two square roots, one positive root and one negative root. The square roots of 9 are +3 and −3. The expression $\sqrt{9}$ is used to indicate the positive or *principal root*, +3 or 3. The expression $-\sqrt{9}$ is used to indicate the negative root, −3. The expression $\pm\sqrt{9}$ indicates both the positive and negative square roots, ±3. The principal cube root of 8 is 2, $\sqrt[3]{8} = 2$. The principal cube root of −8 is −2, $\sqrt[3]{-8} = -2$. In this book, only principal roots are to be determined or used in problem solving.

Examples:

1. $\sqrt{36} = \sqrt{(6)(6)} = 6$ Ans

2. $\sqrt[4]{16} = \sqrt[4]{(2)(2)(2)(2)} = 2$ Ans

3. $\sqrt[3]{-27} = \sqrt{(-3)(-3)(-3)} = -3$ Ans

4. $\sqrt[5]{32} = \sqrt[5]{(2)(2)(2)(2)(2)} = 2$ Ans

5. $\sqrt[3]{\dfrac{-8}{27}} = \sqrt[3]{\dfrac{(-2)(-2)(-2)}{(3)(3)(3)}} = \dfrac{-2}{3}$ Ans

Note: The square root of a negative number has no solution in the real number system. For example:

- $\sqrt{-4}$ *does not* equal +2 since $\sqrt{(+2)(+2)} = \sqrt{+4}$
- $\sqrt{-4}$ *does not* equal –2 since $\sqrt{(-2)(-2)} = \sqrt{+4}$
- $\sqrt{-4}$ has no solution

EXPRESSING NUMBERS WITH FRACTIONAL EXPONENTS AS RADICALS

Procedure: To simplify numbers with fractional exponents
- Write the numerator of the fractional exponent as the power of the radicand.
- Write the denominator of the fractional exponent as the root index of the radicand.
- Simplify.

Examples:

1. $25^{1/2} = \sqrt[2]{25^1} = \sqrt{25} = \sqrt{(5)(5)} = 5$ Ans
2. $8^{1/3} = \sqrt[3]{8^1} = \sqrt{(2)(2)(2)} = 2$ Ans
3. $8^{2/3} = \sqrt[3]{8^2} = \sqrt[3]{64} = \sqrt[3]{(4)(4)(4)} = 4$ Ans
4. $36^{-1/2} = \dfrac{1}{36^{1/2}} = \dfrac{1}{\sqrt{36}} = \dfrac{1}{\sqrt{(6)(6)}} = \dfrac{1}{6}$ Ans

COMBINED OPERATIONS OF SIGNED NUMBERS

Expressions consisting of two or more operations of signed numbers are solved using the same order of operations as in arithmetic.

Example: Compute the value of $50 + (-2)[6 + (-2)^3(4)]$.

$50 + (-2)[6 + (-2)^3(4)] = 50 + (-2)[6 + (-8)(4)] =$
$50 + (-2)[6 + (-32)] = 50 + (-2)(-26) = 50 + 52 = 102$ Ans

APPLICATION

The Number Scale

1. Refer to the number scale and give the direction (+ or –) and the number of units counted going from the first to the second number.

a. –11 to –2 _____	g. +10 to –10 _____	m. –7.5 to + 10 _____
b. –8 to –1 _____	h. +10 to 0 _____	n. +10 to –7.5 _____
c. –6 to 0 _____	i. +2 to +7 _____	o. –10.8 to –2.3 _____
d. –2 to –8 _____	j. +9 to +1 _____	p. –2.3 to –0.8 _____
e. +2 to –8 _____	k. +11 to 0 _____	q. $+7\frac{1}{2}$ to $2\frac{1}{4}$ _____
f. +3 to +9 _____	l. 0 to –4 _____	r. $+6\frac{7}{8}$ to 0 _____

Comparing Signed Numbers

2. Select the greater of the two signed values and indicate the number of units by which it is greater.

a. +5, –14 _____	d. +8, +13 _____	g. +14.3, +21 _____
b. +7, –3 _____	e. +20, –22 _____	h. –1.8, +1.8 _____
c. –6, –2 _____	f. –18, –4 _____	i. +17.6, –21.9 _____

3. List the following signed numbers in order of increasing value starting with the smallest number.

   a. +17, –1, +2, 0, –18, +4, –22 _____

   b. –5, +5, 0, +13, +27, –21, –2, –19 _____

   c. +10, –10, –7, +7, 0, +25, –25, +14 _____

   d. 0, 15, –3.6, –2.5, –14.9, + 17, + 0.3 _____

   e. –16, +14$\frac{1}{8}$, –13$\frac{7}{8}$, +6, –4$\frac{3}{8}$ _____

Absolute Value

4. Express each of the following pairs of signed numbers as absolute values and subtract the smaller absolute value from the larger absolute value.

   a. +23, –14 _____   c. –6, +6 _____   e. –16, +16 _____

   b. –17, +11 _____   d. +25, +13 _____   f. –32.1, –29.7 _____

Addition of Signed Numbers

5. Add the following signed numbers as indicated.

   a. +15 + (+8) _____   k. –4 + (–31) _____

   b. 7 + (+18) + 2 _____   l. –15.3 + (–3.5) _____

   c. 0 + (+25) _____   m. –15.3 + (+3.5) _____

   d. –8 + (–15) _____   n. –16.4 + (–2.7) _____

   e. –14 + (–4) + (–11) _____   o. –9$\frac{1}{4}$ + $\left(-3\frac{3}{4}\right)$ _____

   f. +12 + (–5) _____   p. 18$\frac{5}{8}$ + $\left(-21\frac{3}{4}\right)$ _____

   g. +18 + (–26) _____   q. –13 + $\left(-\frac{3}{16}\right)$ _____

   h. –20 + (+17) _____   r. –4.25 + (–7) + (–3.22) _____

   i. –23 + 17 _____   s. 18.07 + (–17.64) _____

   j. –25 + 3 _____   t. 8 + 16.7 + (–4.1) + 9.5 _____

Subtraction of Signed Numbers

6. Subtract the following signed numbers as indicated.

   a. –10 – (–8) _____   k. –16.5 – (–14.3) _____

   b. +5 – (–13) _____   l. –50.2 – (+51) _____

   c. –22 – (–14) _____   m. +50.2 – (–51) _____

   d. +17 – (+8) _____   n. 0.03 – (+0.05) _____

   e. +40 – (+40) _____   o. –10$\frac{1}{2}$ – $\left(-7\frac{1}{4}\right)$ _____

   f. –40 – (–40) _____   p. 5$\frac{7}{8}$ – $\left(-4\frac{1}{8}\right)$ _____

   g. –40 – (+40) _____   q. (6 + 10) – (–7 + 8) _____

   h. 0 – (–7) _____   r. (–14 + 5) – (2 – 10) _____

   i. –52 – (–8) _____   s. (7.23 – 6.81) – (–10.73) _____

   j. 16.5 – (+14.3) _____   t. [3 – (–7)] – [14 – (–6)] _____

Multiplication of Signed Numbers

7. Multiply the following signed numbers as indicated.

   a. (–4)(6) _____   f. (–2)(–14) _____

   b. (–4)(–6) _____   g. 0(–16) _____

   c. (+10)(–2) _____   h. (6.5)(–2) _____

   d. (–10)(–2) _____   i. (–3.2)(–0.1) _____

   e. (–5)(7) _____   j. (–0.06)(–0.60) _____

k. $\left(1\frac{1}{2}\right)\left(-\frac{1}{2}\right)$ _____

l. $\frac{1}{4}(0)$ _____

m. $(-2)(-2)(-2)$ _____

n. $(-2)(+2)(+2)$ _____

o. $(-2)(+2)(-2)$ _____

p. $(8)(-2)(3)(0)(-1)$ _____

q. $(-6)(-0.5)(2)(-1)(-0.5)$ _____

r. $(-4)(-0.25)(-2)(-0.5)$ _____

s. $(-0.03)(-100)(-0.10)$ _____

t. $\left(+\frac{1}{4}\right)\left(-8\right)\left(\frac{1}{2}\right)\left(2\frac{3}{8}\right)$ _____

Division of Signed Numbers

8. Divide the following signed numbers as indicated.

a. $-10 \div (-5)$ _____

b. $-10 \div (+5)$ _____

c. $+18 \div (+9)$ _____

d. $-21 \div 3$ _____

e. $-30 \div (-5)$ _____

f. $+48 \div (-6)$ _____

g. $-35 \div 7$ _____

h. $\frac{-16}{-4}$ _____

i. $\frac{0}{-10}$ _____

j. $\frac{-40}{-8}$ _____

k. $\frac{-36}{6}$ _____

l. $\frac{-60}{-0.5}$ _____

m. $\frac{-20}{-2.5}$ _____

n. $\frac{-17.92}{3.2}$ _____

o. $-\frac{1}{2} \div \left(-\frac{1}{2}\right)$ _____

p. $-3 \div \frac{3}{4}$ _____

q. $4\frac{1}{3} \div \left(-2\frac{2}{3}\right)$ _____

r. $-29.96 \div 5.35$ _____

s. $-4.125 \div (-1.5)$ _____

t. $-41.87 \div 7.9$ _____

Powers of Signed Numbers

9. Raise the following signed numbers to the indicated powers.

a. $(-2)^2$ _____
f. $(-2)^5$ _____
k. $(-1.2)^3$ _____
p. $\left(-\frac{3}{4}\right)^3$ _____

b. 2^3 _____
g. $(-5)^2$ _____
l. $(-0.3)^2$ _____
q. $(-2)^{-2}$ _____

c. $(-2)^3$ _____
h. $(-5)^3$ _____
m. $(-0.3)^3$ _____
r. $(+4)^{-2}$ _____

d. $(-3)^3$ _____
i. $(-2)^6$ _____
n. $\left(-\frac{1}{2}\right)^3$ _____
s. $(-5)^{-3}$ _____

e. $(-2)^4$ _____
j. $(-1.5)^2$ _____
o. $\left(+\frac{1}{2}\right)^3$ _____
t. $(-2.1)^{-3}$ _____

Roots of Signed Numbers

10. Determine the indicated root of the following signed numbers.

a. $\sqrt[3]{64}$ _____
f. $\sqrt[5]{-32}$ _____
k. $\sqrt[5]{-1}$ _____
p. $\sqrt[3]{\frac{+64}{-125}}$ _____

b. $\sqrt[3]{-64}$ _____
g. $\sqrt[3]{-125}$ _____
l. $\sqrt[3]{216}$ _____
q. $\frac{\sqrt{16}}{4}$ _____

c. $\sqrt[3]{-27}$ _____
h. $\sqrt[5]{+32}$ _____
m. $\sqrt[3]{\frac{-8}{+27}}$ _____
r. $\frac{16}{\sqrt{4}}$ _____

d. $\sqrt[3]{-1000}$ _____
i. $\sqrt[3]{+1}$ _____
n. $\sqrt[3]{\frac{+8}{-27}}$ _____
s. $\frac{\sqrt[3]{-27}}{8}$ _____

e. $\sqrt[4]{+81}$ _____
j. $\sqrt[3]{-1}$ _____
o. $\sqrt[4]{\frac{+1}{+16}}$ _____
t. $\frac{27}{\sqrt[3]{-8}}$ _____

Expressing Numbers with Fractional Exponents as Radicals

11. Determine the value of the following.

a. $4^{1/2}$ _____
d. $27^{1/3}$ _____
g. $-125^{1/3}$ _____
j. $4^{-1/2}$ _____

b. $81^{1/2}$ _____
e. $-8^{1/3}$ _____
h. $125^{1/3}$ _____
k. $8^{-1/3}$ _____

c. $8^{1/3}$ _____
f. $16^{1/4}$ _____
i. $8^{2/3}$ _____
l. $8^{-2/3}$ _____

Combined Operations of Signed Numbers

Solve each of the following problems using the proper order of operations.

12. $17 - (3)(-2) + (-5)^2$ _____

13. $4 - 5(8 - 10)$ _____

14. $-2(4 + 2) + 3(5 - 7)$ _____

15. $5 - 3(8 - 6) - [1 + (-4)]$ _____

16. $\dfrac{2(-1)(-3) - (6)(5)}{3(7) - 9}$ _____

17. $(-3)^3 + 3^3 - (-6)(3) - \left(\dfrac{-6}{2}\right)$ _____

18. $4^2 + \sqrt[3]{-8} + (-4)(0)(-3)$ _____

19. $[4^2 + (2)(5)(-3)]^2 + 2(-3)^3$ _____

20. $(-2)^3 + \sqrt{16} - (5)(3)(4)$ _____

21. $\dfrac{2(-5)^2}{2(5)} - \dfrac{(-4)^3}{18 + (-2)}$ _____

22. $(-2)^3 + \sqrt{(-3)(-10) - (-6)} - \sqrt[3]{-8}$ _____

23. $10^{-2} + [43 + (9)(-2)]^{1/2}$ _____

24. $\sqrt[3]{-21(2) - (-15)} + (16)^{1/2}$ _____

25. $[(15 - 40)(5)]^{-1/3} + 10^{-2}$ _____

Substitute the given numbers for letters in the following expressions and solve.

26. Find $6xy + 15 - xy$ when $x = -2$ and $y = 5$. _____

27. Find $\dfrac{-3ab - 2bc}{abc - 35}$ when $a = -3$, $b = 10$, and $c = -4$. _____

28. Find $(x - y)(3x - 2y)$ when $x = -5$ and $y = -7$. _____

29. Find $\dfrac{d^3 + 4f - fh}{h^2 - (2 + d)}$ when $d = -2$, $f = -3$, and $h = 4$. _____

30. Find $\dfrac{x^2}{n} - \dfrac{21 + y^3}{xy}$ when $n = 5$, $x = -5$, and $y = -1$. _____

31. Find $\sqrt{6(ab - 6)} - (b)^3$ when $a = -6$ and $b = -2$. _____

UNIT 28 *ALGEBRAIC OPERATIONS OF ADDITION, SUBTRACTION, AND MULTIPLICATION*

OBJECTIVES

After studying this unit you should be able to

- Perform the basic algebraic operations of addition, subtraction, and multiplication.

A knowledge of basic algebraic operations is essential in order to solve equations. For certain applications, formulas given in machine trade handbooks cannot be used directly as given, but must be rearranged. Formulas are rearranged by using the principles of algebraic operations.

DEFINITIONS

It is important to understand the following definitions in order to apply procedures which are required for solving problems involving basic operations.

A *term* of an algebraic expression is that part of the expression which is separated from the rest by a plus or a minus sign. For example, $4x + \dfrac{ab}{2x} - 12 + 3ab^2x - 8a\sqrt{b}$ is an expression that consists of five terms: $4x$, $\dfrac{ab}{2x}$, 12, $3ab^2x$, and $8a\sqrt{b}$.

A *factor* is one of two or more literal and/or numerical values of a term that are multiplied. For example, 4 and x are each factors of $4x$; 3, a, b^2 and x are each factors of $3ab^2x$; 8, a, and $\sqrt{b}$ are each factors of $8a\sqrt{b}$.

A *numerical coefficient* is the number factor of a term. The letter factors of a term are the *literal factors*. For example, in the term $5x$, 5 is the numerical coefficient; x is the literal factor.

In the term $\dfrac{1}{3}ab^2c^3$, $\dfrac{1}{3}$ is the numerical coefficient; a, b^2, and c^3 are the literal factors.

Like terms are terms that have identical literal factors including exponents. The numerical coefficients do not have to be the same. For example, $6x$ and $13x$ are like terms; $15ab^2c^3$, $3.2ab^2c^3$, and $\frac{1}{8}ab^2c^3$ are like terms.

Unlike terms are terms which have different literal factors or exponents. For example, $12x$ and $12y$ are unlike terms. The terms $15xy$, $3x^2y$, and $4x^2y^2$ are unlike terms. Although the literal factors are x and y in each of the terms, these literal factors are raised to different powers.

ADDITION

Terms must be like terms to be added. The addition of unlike terms can only be indicated. As in arithmetic, like things can be added, but unlike things cannot be added. For example, 4 inches + 5 inches = 9 inches. Both values are inches; therefore, they can be added. But 4 inches + 5 pounds cannot be added because they are unlike things.

Procedure: To add like terms

- Add the numerical coefficients applying the procedure for addition of signed numbers. If a term does not have a numerical coefficient, the coefficient 1 is understood: $x = 1x$, $abc = 1abc$, $n^2rs^3 = 1n^2rs^3$.
- Leave the literal factors unchanged.

Examples: Add the following like terms.

1.	2.	3.	4.	5.
$3x$	x	$-5xy^2$	$6x^2y^3$	$2(a+b)$
$\underline{12x}$	$\underline{-14x}$	$\underline{+5xy^2}$	$\underline{-13x^2y^3}$	$-3(a+b)$
$15x$ Ans	$-13x$ Ans	0 Ans	$-7x^2y^3$ Ans	$\underline{7(a+b)}$
				$6(a+b)$ Ans

Procedure: To add unlike terms

- The addition of unlike terms can only be indicated.

Examples: Add the following unlike terms.

1.	2.	3.	4.
15	$7x$	$3x$	$8a$
$\underline{x}$	$\underline{8y}$	$\underline{-7x^2}$	$-6b$
$15+x$ Ans	$7x+8y$ Ans	$3x+(-7x^2)$ Ans	$\underline{2c}$
			$8a+(-6b)+2c$ Ans

Procedure: To add two or more expressions that consist of two or more terms

- Group the like terms in the same column.
- Add like terms and indicate the addition of unlike terms.

Examples: Add the following expressions.

1. $12x - 2xy + 6x^2y^3$ and $-4x - 7xy + 5x^2y^3$

    Group like terms in the same column.

    Add like terms.

    $$\begin{array}{l} 12x - 2xy +\ \ 6x^2y^3 \\ \underline{-4x - 7xy +\ \ 5x^2y^3} \\ \ \ 8x - 9xy + 11x^2y^3 \quad \text{Ans} \end{array}$$

2. $6a - 7b$ and $18b - 3ab + a$ and $-14a + ab^2 - 5ab$

    Group like terms.

    Add like terms and indicate the addition of unlike terms.

    $$\begin{array}{l} 6a -\ \ 7b \\ \ \ a + 18b - 3ab \\ \underline{-14a \qquad\quad - 5ab + ab^2} \\ -7a + 11b - 8ab + ab^2 \quad \text{Ans} \end{array}$$

SUBTRACTION

As in addition, terms must be like terms to be subtracted. The subtraction of unlike terms can only be indicated. The same principles apply in arithmetic. For example, 8 feet − 3 feet = 5 feet, but 8 feet − 3 ounces cannot be subtracted because they are unlike things.

Procedure: To subtract like terms
- Subtract the numerical coefficients applying the procedure for subtraction of signed numbers.
- Leave the literal factors unchanged.

Examples: Subtract the following like terms as indicated.
1. $18ab - 7ab = 11ab$ Ans
2. $bx^2 y^3 - 13bx^2 y^3 = -12bx^2 y^3$ Ans
3. $-5x^2 y - 8x^2 y = -13x^2 y$ Ans
4. $-24dmr - (-24dmr) = 0$ Ans

Procedure: To subtract unlike terms
- The subtraction of unlike terms can only be indicated.

Examples: Subtract the following unlike terms as indicated.
1. $3x^2 - (+2x) = 3x^2 - 2x$ Ans
2. $-13abc - (+8abc^2) = -13abc - 8abc^2$ Ans
3. $-2xy - (-7y) = -2xy + 7y$ Ans

Procedure: To subtract expressions that consist of two or more terms
- Group like terms in the same column.
- Subtract like terms and indicate the subtractions of the unlike terms.

Note: Each term of the subtrahend is subtracted following the procedure for subtraction of signed numbers.

Examples: Subtract the following expressions as indicated.
1. Subtract $7a + 3b - 3d$ from $8a - 7b + 5d$

    Group like terms in the same column.
    Change the sign of each term in the subtrahend and follow the procedure for addition of signed numbers.

$$\begin{array}{rcl} 8a - 7b + 5d & = & 8a - 7b + 5d \\ - (7a + 3b - 3d) & = & + (-7a - 3b + 3d) \\ \hline & & a - 10b + 8d \quad \text{Ans} \end{array}$$

2. Subtract as indicated: $(3x^2 + 5x - 12xy) - (7x^2 - x - 3x^3 + 6y)$

$$\begin{array}{rcl} 3x^2 + 5x - 12xy & = & 3x^2 + 5x - 12xy \\ -(7x^2 - x \quad - 3x^3 + 6y) & = & +(-7x^2 + x \quad + 3x^3 - 6y) \\ \hline & & -4x^2 + 6x - 12xy + 3x^3 - 6y \quad \text{Ans} \end{array}$$

MULTIPLICATION

Procedure: To multiply two or more terms
- Multiply the numerical coefficients following the procedure for multiplication of signed numbers.
- Add the exponents of the same literal factors.
- Show the product as a combination of all numerical and literal factors.

Examples: Multiply as indicated.
1. Multiply $(-3x^2)(6x^4)$

    Multiply numerical coefficients $(-3)(6) = -18$

    Add exponents of like literal factors $(x^2)(x^4) = x^{2+4} = x^6$

    Show product as combination of all numerical and literal factors. $(-3x^2)(6x^4) = -18x^6$ Ans

2. $(3a^2 b^3)(7ab^3) = (3)(7)(a^{2+1})(b^{3+3}) = 21a^3 b^6$ Ans

3. $(-4a)(-7b^2 c^2)(-2ac^3 d^3) = (-4)(-7)(-2)(a^{1+1})(b^2)(c^{2+3})d^3 = -56a^2 b^2 c^5 d^3$ Ans

Procedure: To multiply expressions that consist of more than one term within an expression

- Multiply each term of one expression by each term or the other expression.
- Combine like terms.

This multiplication procedure is consistent with arithmetic.

Examples in Arithmetic:

1. Multiply. $3(4 + 2)$

    From arithmetic: $3(4 + 2) = 3(6) = 18$ Ans

    From algebra:
    Multiply each term of one expression by $3(4 + 2) = 3(4) + 3(2) = 12 + 6 = 18$ Ans
    each term of the other expression.

    Combine like terms.

2. Multiply. $(5 + 3)(2 + 4)$

    From arithmetic: $(5 + 3)(2 + 4) = (8)(6) = 48$ Ans

    From algebra: Multiply each term of one
    expression by each term of the other expression.

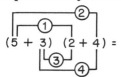

	Step 1	Step 2	Step 3	Step 4
	↓	↓	↓	↓
	5(2) +	5(4) +	3(2) +	3(4) =
	↓	↓	↓	↓
	10 +	20 +	6 +	12 = 48 Ans

    $(5 + 3)\ (2 + 4) =$

    Combine like terms.

Examples in Algebra:

1. $3a(6 + 2a^2) = (3a)(6) + 3a(2a^2) = 18a + 6a^3$ Ans
2. $-5x^2 y(3xy - 4x^3 y^2 + 5y) = -5x^2 y(3xy) - 5x^2 y(-4x^3 y^2) - 5x^2 y(5y) =$
    $-15x^3 y^2 + 20x^5 y^3 - 25x^2 y^2$
3. $(3c + 5d^2)(4d^2 - 2c)$

    Multiply each term of one expression by each term of the other expression.

    $(3c + 5d^2)\ (4d^2 - 2c) =$

Step 1	Step 2	Step 3	Step 4
↓	↓	↓	↓
$3c(4d^2) +$	$3c(-2c) +$	$5d^2(4d^2) +$	$5d^2(-2c) =$
↓	↓	↓	↓
$12cd^2 +$	$(-6c^2) +$	$20d^4 +$	$(-10cd^2)$

    Combine like terms.

    ┌──────Combine──────┐
    $12cd^2 + (-6c^2) + 20d^4 + (-10cd^2) = 2cd^2 + (-6c^2) + 20d^4$ or $2cd^2 - 6c^2 + 20d^4$ Ans

APPLICATION

Addition of Single Terms

Add the terms in the following expressions.

1. $20y + y$ _____
2. $15xy + 7xy$ _____
3. $-15xy + (-7xy)$ _____
4. $25m^2 + (-m^2)$ _____
5. $-5x^2 y + 5x^2 y$ _____

6. $4c^3 + 0$ _____
7. $-7pt + (-pt)$ _____
8. $0.4x + (-0.8x)$ _____
9. $8.3a^2 b + 6.9a^2 b$ _____
10. $-0.02y + 0.07y$ _____

11. $\frac{1}{2}xy + \frac{3}{4}xy$ _____

12. $2\frac{7}{8}c^2 d + \left(-3\frac{1}{8}c^2 d\right)$ _____

13. $-2.06gh^3 + (-0.85gh^3)$ _____

14. $-50.6abc + 50.5abc$ _____

15. $4P + (-6P) + P + 7P$ _____

16. $-0.3dt^2 + (-1.7dt^2) + (-dt^2)$ _____

17. $\frac{1}{4}xy + \frac{7}{8}xy + xy + (-4xy)$ _____

18. $20.06D + (-19.97D) + (-0.9D)$ _____

19. $6M + 0.6M + 0.06M + 0.006M$ _____

20. $-3xy^2 + 8xy^2 + 7.8xy^2$ _____

21. $3.2a^2 b^2 c + 6.7a^2 b^2 c + (-4.4a^2 b^2 c)$ _____

Addition of Expressions with Two or More Terms

Add the following expressions.

22. $\begin{aligned} -5x + \ 7xy - \ 8y \\ \underline{-9x - 12xy + 13y} \end{aligned}$

23. $\begin{aligned} 3a - 11d - 8m \\ \underline{- \ a + 11d - 3m} \end{aligned}$

24. $\begin{aligned} -6ab - \ 7a^2 b^2 - \ 3a^3 b \\ -5ab + 14a^2 b^2 - 12a^3 b \\ -9ab - \ 7a^2 b^2 + \ \ a^3 b \\ \underline{\ \ ab \ \ \ \ \ \ \ \ \ \ \ \ - \ 2a^3 b} \end{aligned}$

25. $(3xy^2 + x^2 y - x^2 y^2), (2x^2 y + x^2 y^2)$ _____

26. $(10a - 5b), (-12a - 7b), (17a + b)$ _____

27. $(x^3 + 5), (3x - 7x^2 + 7), (x - 3x^3)$ _____

28. $(b^4 + 4b^3 c - 2b^2 c), (4b^3 c - 7bc)$ _____

29. $(x^2 - 4xy), (4xy - y^2), (-x^2 + y^2)$ _____

30. $(1.3M - 3N), (-8M + 0.5N), (20M + 0.7N)$ _____

31. $(c + 3.6cd - 5.7d), (-1.4c + 8.6d)$ _____

Subtraction of Single Terms

Subtract the following terms as indicated.

32. $7xy^2 - (-13xy^2)$ _____

33. $3xy - xy$ _____

34. $-3xy - xy$ _____

35. $-3xy - (-xy)$ _____

36. $9ab - (-9ab)$ _____

37. $-5a^2 - (5a^2)$ _____

38. $0.7a^2 b^2 - 1.5a^2 b^2$ _____

39. $0 - (-8mn^3)$ _____

40. $-8mn^3 - 0$ _____

41. $\frac{7}{8}x^2 - \left(-\frac{3}{8}x^2\right)$ _____

42. $13a - 7a^2$ _____

43. $-13a - (-7a^2)$ _____

44. $0.6xy - 0.9xy^2$ _____

45. $-ax^2 - ax^2$ _____

46. $\frac{1}{2}dt - \left(-\frac{3}{8}dt\right)$ _____

47. $\frac{1}{2}d^2 t^2 - \left(-\frac{1}{2}d^2 t^2\right)$ _____

48. $18 - 3x$ _____

49. $3x - 18$ _____

50. $-3.2d - 6.4d$ _____

51. $-1.4xy - (-1.4xy)$ _____

Subtraction of Expressions with Two or More Terms

Subtract the following expressions as indicated.

52. $(2a^2 - 3a) - (7a^2 - 8a)$ _____

53. $(4x^2 + 8xy) - (3x^2 + 5xy)$ _____

54. $(9b^2 + 1) - (9b^2 - 1)$ _____

55. $(9b^2 - 1) - (9b^2 - 1)$ _____

56. $(xy^2 - x^2 y^2 + x^3 y^2) - 0$ _____

57. $(2a^3 - 0.5a^2) - (-a^3 + a^2 - a)$ _____

58. $(5x + 3xy - 7y) - (3y^2 - x^2 y)$ _____

59. $(-d^2 - dt + dt^2) - (-4 + dt)$ _____

60. $(15L - 12H) - (-12L + 6H - 4)$ _____

61. $(11.09e + 14.76f) - (e - f - 10.03)$ _____

Multiplication of Single Terms

Multiply the following terms as indicated.

62. $(-5b^2c)(3b^3)$ _____

63. $(x)(x^2)$ _____

64. $(-3a^2)(-4a^3)$ _____

65. $(8ab^2c)(7a^3bc^2)$ _____

66. $(-x^3y^3)(3x^2y^4)$ _____

67. $(-3xy)(0)$ _____

68. $(7ab^4)(3a^4b)$ _____

69. $(-3d^5r^4)(-d^3)$ _____

70. $(-3d^5r^4)(-d^3)(-1)$ _____

71. $(0.3x^2y^4)(0.4x^5)$ _____

72. $\left(\frac{1}{4}a^3\right)\left(\frac{3}{8}a^2\right)$ _____

73. $(-5x)(0)(-5x)$ _____

74. $(m^2t)(st)$ _____

75. $(-1.6bc)(2.1)$ _____

76. $(abc^3)(c^3d)$ _____

77. $(2x^6y^6)(-2x^2)$ _____

78. $\left(-\frac{2}{3}mt\right)(t^4)$ _____

79. $(7ab^3)(-7a^3b)$ _____

80. $(-0.3a^3b^2)(-5b^4)$ _____

81. $(-x^2y)(-xy)(-x)$ _____

82. $(d^4m^2)(-1)(-m^3)$ _____

Multiplication of Expressions with Two or More Terms

Multiply the following expressions as indicated and combine like terms where possible.

83. $-7x^2y^3(2xy^2 - 3x^4)$ _____

84. $3a^2(-a^2 + a^3b)$ _____

85. $-2a^3b^2(4ab^3 - b^2 - 4)$ _____

86. $xy^2(x^2 + y^3 + xy)$ _____

87. $-4(dt + t^2 - 1)$ _____

88. $(m^2t^3s^4)(-m^4s^2 + m - s^5)$ _____

89. $(3x + 7)(x^2 + 8)$ _____

90. $(7x^2 - y^3)(-2x^3 + y^2)$ _____

91. $(5ax^3 + bx)(2a^2x^3 + b^2x)$ _____

92. $(-4a^2b^3 + 5xy^2)(4a^2b^3 - 5xy)$ _____

¯UNIT 29 *ALGEBRAIC OPERATIONS OF DIVISION, POWERS, AND ROOTS* _____

¯OBJECTIVES_____

After studying this unit you should be able to

- Perform the basic algebraic operations of division, powers, and roots.
- Remove parentheses which are preceded by a plus or minus sign.
- Simplify algebraic expressions which involve combined operations.

¯DIVISION_____

As with multiplication, the exponents of the literal factors do not have to be the same to divide the values.

Procedure: To divide two terms

- Divide the numerical coefficients following the procedure for division of signed numbers.
- Subtract the exponents of the literal factors of the divisor from the exponents of the same letter factors of the dividend.
- Combine numerical and literal factors.

This division procedure is consistent with arithmetic.

Example in Arithmetic: Divide. $\frac{2^5}{2^2}$

From arithmetic:

$$\frac{2^5}{2^2} = \frac{(2)(2)(2)(2)(2)}{(2)(2)} = (2)(2)(2) = 8 \quad \text{Ans}$$

From algebra:

$$\frac{2^5}{2^2} = 2^{5-2} = 2^3 = 8 \quad \text{Ans}$$

Examples in Algebra:

1. Divide $-16x^3$ by $8x$.

    Divide the numerical coefficients following the procedure for signed numbers.

    $$-16 \div 8 = -2$$

    Subtract the exponents of the literal factors in the divisor from the exponents of the same letter factors in the dividend.

    $$x^3 \div x = x^{3-1} = x^2$$

    Combine the numerical and literal factors.

    $$\frac{-16x^3}{8x} = -2x^2 \quad \text{Ans}$$

2. $\frac{-30a^3b^5c^2}{-5a^2b^3} = \left(\frac{-30}{-5}\right)(a^{3-2})(b^{5-3})(c^2) = 6ab^2c^2 \quad \text{Ans}$

In arithmetic, any number except 0 divided by itself equals 1. For example, $4 \div 4 = 1$. Applying the division procedure $4 \div 4 = 4^{1-1} = 4^0$. Therefore, $4^0 = 1$. Any number except 0 raised to the zero power equals 1.

Example 1: $\frac{5^3}{5^3} = 5^{3-3} = 5^0 = 1 \quad \text{Ans}$

Example 2: $\frac{a^3b^2c}{a^3b^2c} = (a^{3-3})(b^{2-2})(c^{1-1}) = a^0b^0c^0 = (1)(1)(1) = 1 \quad \text{Ans}$

Procedure: To divide when the divisor consists of one term and the dividend consists of more than one term

- Divide each term of the dividend by the divisor following the procedure for division of signed numbers.
- Combine terms.

This division procedure is consistent with arithmetic.

Example in Arithmetic: Divide. $\frac{6+8}{2}$

From arithmetic:

$$\frac{6+8}{2} = \frac{14}{2} = 7 \quad \text{Ans}$$

From algebra:

$$\frac{6+8}{2} = \frac{6}{2} + \frac{8}{2} = 3 + 4 = 7 \quad \text{Ans}$$

Example in Algebra: Divide. $\frac{-20xy^2 + 15x^2y^3 + 35x^3y}{-5xy}$

$$\frac{-20xy^2 + 15x^2y^3 + 35x^3y}{-5xy} = \frac{-20xy^2}{-5xy} + \frac{15x^2y^3}{-5xy} + \frac{35x^3y}{-5xy} = 4y - 3xy^2 - 7x^2 \quad \text{Ans}$$

⌐POWERS

Procedure: To raise a single term to a power

- Raise the numerical coefficients to the indicated power following the procedure for powers of signed numbers.
- Multiply each of the literal factor exponents by the exponent of the power to which it is raised.
- Combine numerical and literal factors.

This power procedure is consistent with arithmetic.

Example in Arithmetic: Raise to the indicated power. $(2^2)^3$

From arithmetic: $(2^2)^3 = (4)^3 = (4)(4)(4) = 64 \quad \text{Ans}$

From algebra: $(2^2)^3 = 2^{2(3)} = 2^6 = (2)(2)(2)(2)(2)(2) = 64 \quad \text{Ans}$

Examples in Algebra:

1. Raise to the indicated power. $(5x^3)^2$

    Raise the numerical coefficient to the indicated power following the procedure for powers of signed numbers.

    $5^2 = 25$

    Multiply each literal factor exponent by the exponent of the power to which it is to be raised.

    $(x^3)^2 = x^{3(2)} = x^6$

    Combine numerical and literal factors.

    $(5x^3)^2 = 25x^6$ Ans

Note: $(x^3)^2$ is not the same as $x^3 x^2$.

$$(x^3)^2 = (x^3)(x^3) = (x)(x)(x)(x)(x)(x) = x^6$$
$$x^3 x^2 = (x)(x)(x)(x)(x) = x^5$$

2. $(-3a^2 b^4 c)^3 = (-3)^3 a^{2(3)} b^{4(3)} c^{1(3)} = -27a^6 b^{12} c^3$ Ans

3. $\left[-\frac{1}{2} x^3 (yd^2)^3 r^4\right]^2 = \left[-\frac{1}{2} x^3 y^3 d^6 r^4\right]^2 = \frac{1}{4} x^6 y^6 d^{12} r^8$ Ans

Procedure: To raise two or more terms to a power

*   Apply the procedure for multiplying expressions that consist of more than one term.

Example: Solve. $(2x + 4y^3)^2$

$$(2x + 4y^3)^2 = (2x + 4y^3)(2x + 4y^3)$$

$$= 2x(2x) + 2x(4y^3) + 4y^3(2x) + 4y^3(4y^3) =$$

$4x^2 + 8xy^3 + 8xy^3 + 16y^6 = 4x^2 + 16xy^3 + 16y^6$ Ans

∠—Combine⌐

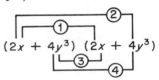

─ROOTS

Procedure: To extract the root of a term

*   Determine the root of the numerical coefficient following the procedure for roots of signed numbers.

*   The roots of the literal factors are determined by dividing the exponent of each literal factor by the index of the root.

*   Combine the numerical and literal factors.

This procedure for extracting roots is consistent with arithmetic.

Example in Arithmetic: Find the indicated root. $\sqrt{2^6}$

From arithmetic: $\sqrt{2^6} = \sqrt{(2)(2)(2)(2)(2)(2)} = \sqrt{64} = 8$ Ans

From algebra: $\sqrt{2^6} = 2^{6 \div 2} = 2^3 = (2)(2)(2) = 8$ Ans

Examples in Algebra:

1. $\sqrt{25a^6 b^4 c^8} = \sqrt{25}(a^{6 \div 2})(b^{4 \div 2})(c^{8 \div 2}) = 5a^3 b^2 c^4$ Ans

2. $\sqrt[3]{-27d^3 x^9 y^2} = \sqrt[3]{-27}(d^{3 \div 3})(x^{9 \div 3})\sqrt[3]{y^2} = -3dx^3 \sqrt[3]{y^2}$ Ans

3. $\sqrt[4]{\frac{16}{81} d^8 t^{12} y^2} = \sqrt[4]{\frac{16}{81}} (d^{8 \div 4})(t^{12 \div 4})(y^{2 \div 4}) = \frac{2}{3} d^2 t^3 y^{1/2} = \frac{2}{3} d^2 t^3 \sqrt{y}$ Ans

Note: Roots of expressions that consist of two or more terms *cannot* be extracted by this procedure. This fact is consistent with arithmetic.

$\sqrt{3^2 + 4^2} = \sqrt{9 + 16} = \sqrt{25} = 5$ but $\sqrt{3^2} + \sqrt{4^2} = 3 + 4 = 7$.

5 does *not* equal 7 therefore $\sqrt{3^2 + 4^2} \neq \sqrt{3^2} + \sqrt{4^2}$.

─REMOVAL OF PARENTHESES

In certain expressions terms are enclosed within parentheses which are preceded by a plus or minus sign. In order to combine like terms, it is necessary to first remove parentheses.

Procedure: To remove parentheses preceded by a plus sign

- Remove the parentheses without changing the signs of any terms within the parentheses.
- Combine like terms.

Example: $5a + (4b + 7a - 3d) = 5a + 4b + 7a - 3d = 12a + 4b - 3d$ Ans

Procedure: To remove parentheses preceded by a minus sign

- Remove the parentheses and change the sign of each term within the parentheses.
- Combine like terms.

Example: $-(7a^2 + b - 3) + 12 - (-b + 5) = -7a^2 - b + 3 + 12 + b - 5 = -7a^2 + 10$ Ans

COMBINED OPERATIONS

Procedure: To solve expressions consisting of two or more different operations

- Apply the proper order of operations.

Examples:

1. $10x - 3x(2 + x - 4x^2) = 10x - 6x - 3x^2 + 12x^3 = 4x - 3x^2 + 12x^3$ Ans

2. $15a^6 b^3 + (2a^2 b)^3 - \dfrac{a^7 (b^3)^2}{ab^3} = 15a^6 b^3 + 8a^6 b^3 - \dfrac{a^7 b^6}{ab^3} = 15a^6 b^3 + 8a^6 b^3 - a^6 b^3 = 22a^6 b^3$ Ans

3. $-4a[15 - 3(2a + ab) + a] - 2a^2 b = -4a(15 - 6a - 3ab + a) - 2a^2 b = -60a + 24a^2 + 12a^2 b - 4a^2 - 2a^2 b = -60a + 20a^2 + 10a^2 b$ Ans

APPLICATION

Division of Single Terms

Divide the following terms as indicated.

1. $\dfrac{4x}{2x}$ _____

2. $\dfrac{-16a^4 b^5}{4ab^3}$ _____

3. $\dfrac{FS^2}{-FS^2}$ _____

4. $\dfrac{-FS^2}{-FS^2}$ _____

5. $0 \div 14mn$ _____

6. $(-42a^5 d^2) \div (-7a^2 d^2)$ _____

7. $(-3.6H^2 P) \div (0.6HP)$ _____

8. $DM^2 \div (-1)$ _____

9. $2.6ab \div ab$ _____

10. $0.8PV^2 \div (-0.2V)$ _____

11. $1\frac{1}{4} c^2 d^3 \div \frac{1}{4} cd^2$ _____

12. $\left(-\frac{1}{2} x^3 y^3\right) \div \frac{1}{8} x^3$ _____

13. $-6g^3 h^2 \div \left(-\frac{3}{4} gh\right)$ _____

14. $-24x^2 y^5 \div (-0.5x^2 y^4)$ _____

15. $x^2 y^3 z^4 \div xy^3 z$ _____

16. $18a^2 bc^2 y \div (-a^2)$ _____

17. $0.25P^2 V \div 0.0625$ _____

18. $-0.08xy \div 0.4y$ _____

19. $-\frac{3}{4} FS^3 \div (-3S)$ _____

20. $-9.6x^2 yz \div (-1.2x)$ _____

Division of Expressions with Two or More Terms in the Dividend

Divide the following expressions as indicated.

21. $(8x^3 + 12x^2) \div 2x$ _____

22. $(12x^3 y^3 - 8x^2 y^2) \div 4xy$ _____

23. $(9x^6 y^3 - 6x^2 y^5) \div (-3xy^2)$ _____

24. $(2x - 4y) \div 2$ _____

25. $(15a^2 + 25a^5) \div (-a)$ _____

26. $(-18a^2 b^7 - 12a^5 b^5) \div (-6a^2 b^5)$ _____

27. $(7cd^2 - 35c^2 d - 7) \div (-7)$ _____

28. $(0.8x^5 y^6 + 0.2x^4 y^7) \div (2x^2 y^4)$ _____

29. $(-0.9a^2 x - 0.3ax^2 + 0.6) \div (-0.3)$ _____

30. $(5y^2 - 25xy^2 - 10y^5) \div 5y^2$ _____

31. $\left(\frac{1}{2} a^2 c - \frac{3}{4} a^3 c^2 - ac^3\right) \div \frac{1}{8} ac$ _____

32. $(-2.5e^2 f - 0.5ef^2 + e^2 f^2) \div 0.5f$ _____

Powers of Single Terms

Raise the following terms to indicated powers.

33. $(3ab)^2$ _____

34. $(-4xy)^2$ _____

35. $(2x^2y)^3$ _____

36. $(4a^4b^3)^2$ _____

37. $(-3c^3d^2e^4)^3$ _____

38. $(2MS^2)^4$ _____

39. $(-7x^4y^5)^2$ _____

40. $(-3N^2P^2T^3)^4$ _____

41. $(a^2bc^3)^3$ _____

42. $(-2a^2bc^3)^3$ _____

43. $(-x^4y^5z)^3$ _____

44. $(9C^4F^2H)^2$ _____

45. $(0.4x^3y)^3$ _____

46. $(-0.5c^2d^3e)^3$ _____

47. $(3.2M^3NP^2)^2$ _____

48. $\left(\frac{3}{4}abc^3\right)^3$ _____

49. $[-8(a^2b^3)^2c]^2$ _____

50. $[-3x^2(y^2)^2z^3]^3$ _____

51. $[0.6d^3(ef^2)^3]^2$ _____

52. $[(-2x^2y)^2(xy^2)^2]^3$ _____

Powers of Expressions of Two or More Terms

Raise the following terms to the indicated powers and combine like terms where possible.

53. $(3x^2-5y^3)^2$ _____

54. $(a^3+b^4)^2$ _____

55. $(5t^2-6x)^2$ _____

56. $(a^2b^3+ab^3)^2$ _____

57. $(0.6d^3t^2-0.2t)^2$ _____

58. $(-0.4x^2y-y^4)^2$ _____

59. $\left(\frac{2}{3}c^2d+\frac{3}{4}cd^2\right)^2$ _____

60. $[(x^2)^3-(y^3)^2]^2$ _____

61. $[(-a^4b)^2+(x^2y)^3]^2$ _____

Roots

Determine the roots of the following terms.

62. $\sqrt{16c^2d^6}$ _____

63. $\sqrt{m^4n^2s^6}$ _____

64. $\sqrt[3]{64x^3y^9}$ _____

65. $\sqrt{81x^8y^6}$ _____

66. $\sqrt[3]{8p^6t^3w^9}$ _____

67. $\sqrt[3]{-27x^6y^{12}}$ _____

68. $\sqrt{0.25h^4y^2}$ _____

69. $\sqrt{0.64a^6c^8f^2}$ _____

70. $\sqrt{\frac{4}{9}a^2b^4c^6}$ _____

71. $\sqrt{\frac{1}{16}x^2y^2}$ _____

72. $\sqrt[3]{\frac{8}{27}m^3n^6}$ _____

73. $\sqrt[3]{-64d^6t^9}$ _____

74. $\sqrt[4]{16x^4y^8}$ _____

75. $\sqrt[5]{32h^{10}}$ _____

76. $\sqrt{25a^2b}$ _____

77. $\sqrt[3]{64a^3c}$ _____

78. $\sqrt[3]{\frac{1}{-64}x^3y^6z^2}$ _____

79. $\sqrt{\frac{4}{9}a^4b^2c}$ _____

80. $\sqrt[3]{27d^3e^6f^2}$ _____

81. $\sqrt[5]{-32a^5b^3}$ _____

Removal of Parentheses

Remove parentheses and combine like terms where possible.

82. $4a+(3a-2a^2+a^3)$ _____

83. $9b-(15b^2-c+d)$ _____

84. $15+(x^2-10)$ _____

85. $-(ab+a^2b-6a)$ _____

86. $-10c^3-(-8c^3-d+12)$ _____

87. $-(16+xy-x)+(-x)$ _____

88. $-15a^2b-(-2a^2b-a+b^2)$ _____

89. $15-(r^2+r)+(r^2-14)$ _____

90. $-(a^2+b^2)+(a^2+b^2)$ _____

91. $-(3x+xy-6)+12+(x+xy)$ _____

92. $20+(cd-c^2d+d)+14-(cd+d)$ _____

93. $20-(cd-c^2d+d)-14+(cd+d)$ _____

Combined Operations

Simplify the following expressions.

94. $15 - 2(3xy)^2 + x^2y^2 - 3$ _____

95. $5(a^2 - b) + a^2 - b$ _____

96. $(2 - c^2)(2 + c^2) + c$ _____

97. $\frac{ab}{a} - \left(\frac{-a^2b}{a^2} - \frac{a^3b}{a^3} \right)$ _____

98. $\frac{4 - 8x + 16x^2}{2} + \frac{3x^4}{x^2}$ _____

99. $\frac{16xy^8}{2xy^2} - (y^2)^3 + 15$ _____

100. $\frac{\sqrt{25x^2}}{-5}(3xy^3) - (-4)$ _____

101. $\sqrt{\frac{64d^6}{9}} \div d^2$ _____

102. $\frac{12x^6 + 16x^4y}{(2x)^2} - (16x^4y^2)^{1/2}$ _____

103. $-4a(-8 + (ab^2)^3 - 12)$ _____

104. $5a[-6 + (ab^2)^3 - 10]$ _____

105. $(10f^6 + 12f^4h) \div \sqrt{4f^4}$ _____

‾UNIT 30 *INTRODUCTION TO EQUATIONS*‗

‾OBJECTIVES‗

After studying this unit you should be able to

- Express word problems as equations.
- Express problems given in graphic form as equations.
- Solve simple equations using logical reasoning.

It is essential that the skilled machine technician understand equations and their applications. The solution of equations is required to compute problems using trade handbook formulas. Often machine shop problems are solved using a combination of equations, with elements of geometry and trigonometry.

‾EXPRESSION OF EQUALITY‗

An *equation* is a mathematical statement of equality between two or more quantities and always contains the equal sign (=). The value of all quantities on the left side of the equal sign equals the value of all quantities on the right side of the equal sign. A *formula* is a particular type of an equation which states a mathematical rule.

The following are examples of simple equations:

$$7 + 2 = 5 + 4 \qquad\qquad \frac{12}{3} + 2 \times 5 = 18 - 4$$

$$3 \times 5\frac{1}{2}'' = 16\frac{1}{2}'' \qquad\qquad 360° = 5 \times 80° - 40°$$

$$a + b = c - d \qquad\qquad \frac{xy}{2} = x + y$$

Because it expresses the equality of the quantities on the left and on the right of the equal sign, an equation is a balanced mathematical statement. An equation may be considered similar to a balanced scale as illustrated in A. The total weight on the left side of the scale equals the total weight on the right side; therefore, the scale balances.

$$3 \text{ pounds} + 5 \text{ pounds} + 2 \text{ pounds} = 4 \text{ pounds} + 6 \text{ pounds}$$
$$10 \text{ pounds} = 10 \text{ pounds}$$

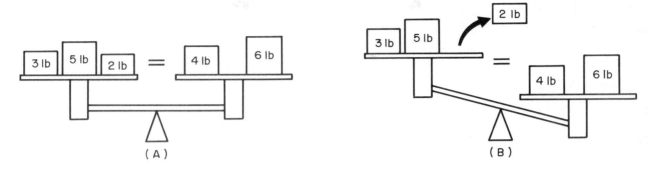

(A)   (B)

When the 2-pound weight is removed from the scale, the scale is no longer in balance as illustrated in B.

$$3 \text{ pounds} + 5 \text{ pounds} \neq 4 \text{ pounds} + 6 \text{ pounds}$$
$$8 \text{ pounds} \neq 10 \text{ pounds}$$

THE UNKNOWN QUANTITY

In general, an equation is used to determine the numerical value of an unknown quantity. Although any letter or symbol can be used to represent the unknown quantity, the letter x is commonly used.

The first letter of the unknown quantity is often used to represent a quantity. Some common letter designations are

L to represent length	P to represent pressure
A to represent area	F to represent feet of cutter
t to represent time	W to represent weight
D to represent diameter	h to represent height

WRITING EQUATIONS FROM WORD STATEMENTS

An equation asks a question. It asks for the value of the unknown which makes the left side of the equation equal to the right side. The question asked may not be in equation form; instead it may be expressed in words.

It is important to develop the ability to express word statements as mathematical symbols, or equations. A problem must be fully understood before it can be written as an equation.

Whether the word problem is simple or complex, a definite logical procedure should be followed to analyze the problem. A few or all of the following steps may be required, depending on the complexity of the particular problem.

- Carefully read the entire problem, several times if necessary.
- Break the problem down into simpler parts.
- It is sometimes helpful to draw a simple picture as an aid in visualizing the various parts of the problem.
- Identify and list the unknowns. Give each unknown a letter name, such as x.
- Decide where the equal sign should be, and group the parts of the problem on the proper side of the equal sign.
- Check. Are the statements on the left equal to the statements on the right of the equal sign?
- After writing the equation, check it against the original problem, step-by-step. Does the equation state mathematically what the problem states in words?

The following examples illustrate the method of writing equations from given word statements. After each equation is written the value of the unknown quantity is obtained. No specific procedures are given at this time in solving for the unknowns. The unknown quantity values are determined by logical reasoning.

Example 1: What weight must be added to 15 pounds so that it will be in balance with a 22-pound weight?

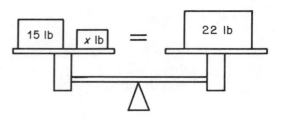

Ask the question: 15 pounds + what weight = 22 pounds?

To help visualize the problem, a picture is shown.

Let x represent the unknown weight.

Write an equation. 15 lb + x = 22 lb

Ask the question: What number added to 15 pounds equals 22 pounds? Since 7 pounds added to 15 pounds equals 22 pounds, x = 7 lb   Ans

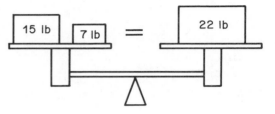

Check the answer by substituting 7 pounds for x in the original equation. 15 lb + 7 lb = 22 lb
$$22 \text{ lb} = 22 \text{ lb}   \text{Ck}$$

The equation is balanced.

Example 2: A $9\frac{1}{2}$-inch piece is cut from a 12-inch length of bar stock. Find the length of the unused piece. Make no allowance for thickness of the cut.

Ask the question: What number subtracted from 12 inches = $9\frac{1}{2}$ inches?

A picture of the problem is shown. All dimensions are in inches. Let x represent the number of inches cut off.

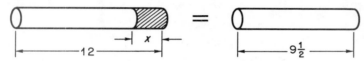

Express the problem as an equation. $12'' - x = 9\frac{1}{2}''$

Since $2\frac{1}{2}$ inches subtracted from 12 inches is equal to $9\frac{1}{2}$ inches, $x = 2\frac{1}{2}''$ Ans

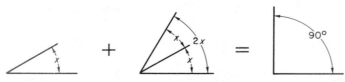

Check the answer by substituting $2\frac{1}{2}$ inches for x in the original equation.
$$12'' - 2\frac{1}{2}'' = 9\frac{1}{2}''$$
$$9\frac{1}{2}'' = 9\frac{1}{2}''   \text{Ck}$$
The equation is balanced.

Example 3: The sum of two angles equals 90°. One angle is twice as large as the other. What is the size of the smaller angle?

An angle + an angle twice as large = 90°.

A picture of the problem is shown.

Let x represent the smaller angle. Let $2x$ represent the larger angle.

Express the problem as an equation. $x + 2x = 90°$ or $3x = 90°$

Ask the question: What number multiplied by 3 = 90°? Since 3 multiplied by 30° = 90°, $x = 30°$

The smaller angle $x = 30°$ and the larger angle = $2x$ or $60°$ as shown.   Ans

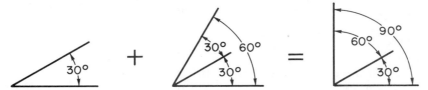

Check the answer by substituting $30°$ for x in the original equation.

$$30° + 2(30°) = 90°$$
$$90° = 90° \quad \text{Ck}$$

The equation is balanced.

Example 4: Three gage blocks are used to tilt a sine plate. The total height of the three blocks is 2.75 inches. The bottom block is 4 times as thick as the middle block. The middle block is twice as thick as the top block. How thick is each block?

Convert the problem from word form to equation form.

Let x represent the thickness of the thinnest block, the top block.
The middle block is twice as thick as the top block, or $2x$.
The bottom block is four times as thick as the middle block, or $(4)(2x) = 8x$.
The sum of the three blocks = 2.75".
Therefore, $x + 2x + 8x = 2.75"$, or $11x = 2.75"$.

A picture of the problem is shown. All dimensions are in inches.

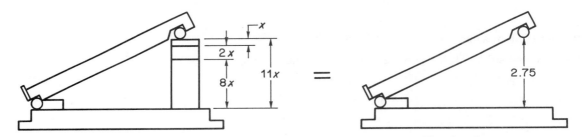

Ask the question: What number multiplied by 11 = 2.75"? Since 11 × 0.25" = 2.75", $x = 0.25"$.

The top block is x or 0.25".   Ans
The middle block is $2x$ or $2(0.25") = 0.50"$.   Ans
The bottom block is $8x$ or $8(0.25") = 2.00"$.   Ans

The thickness of each block is shown. All dimensions are in inches.

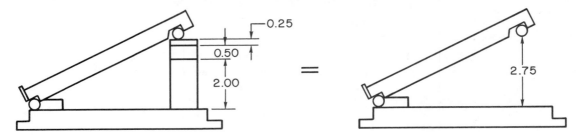

Check the answer by substituting 0.25" for x in the original equation:

$$x + 2x + 4(2x) = 2.75"$$
$$0.25" + 2(0.25") + 4(2)(0.25") = 2.75"$$
$$0.25 + 0.50" + 2.00" = 2.75"$$
$$2.75" = 2.75" \quad \text{Ck}$$

The equation is balanced.

In many cases the problems to be solved in actual machine shop applications will be more difficult than the preceding examples. It is essential, therefore, to be able to use the procedure shown to analyze the problem, determine the unknowns, and set up the equation.

⌐CHECKING THE EQUATION

In the final step in each of the preceding examples, the value found for the unknown was substituted in the original equation to prove that it was the correct value. If an equation is properly written and if both sides of the equation are equal, the equation is balanced and the solution is correct.

All work in a machine shop should be checked and rechecked to prevent errors. It is important that you check your computations. When working with equations on the job, checking your work is essential. Errors in computation can often be costly in terms of time, labor, and materials.

⌐APPLICATION

Express each of the following word problems as equations. Let the unknown number equal x and find the value of the unknown. Check the equation by comparing it to the word problem. Does the equation state mathematically what the problem states in words? Check whether the equation is balanced by substituting the value of the unknown in the equation.

1. A number plus 18 equals 32. Find the number. _____

2. A number less 7 equals 15. Find the number. _____

3. Five times a number equals 55. Find the number. _____

4. A number divided by 3 equals 9. Find the number. _____

5. Thirty-two divided by a number equals 8. Find the number. _____

6. A number plus twice the number equals 36. Find the number. _____

7. Six times a number minus the number equals 45. Find the number. _____

8. Seven times a number plus eight times the number equals 60. Find the number. _____

9. Sixty divided by 3 times a number equals 4. Find the number. _____

10. A piece of bar stock 16 inches long is cut into two unequal lengths. One piece is 3 times as long as the other. How long is each piece? _____

11. Three blocks are used to tilt a sine plate. The total height of the three blocks is 4.5 inches. The first block is 3 times as thick as the second block. The second block is twice as thick as the third block. How thick is each block? _____

12. Five holes are drilled in a steel plate on a bolt circle. There are 300° between hole 1 and hole 5. The number of degrees between any two consecutive holes doubles in going from hole 1 to hole 5. Find the number of degrees between the indicated holes.

    a. 1 and 2   _____

    b. 2 and 3   _____

    c. 3 and 4   _____

    d. 4 and 5   _____

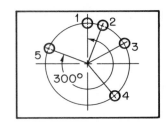

13. The total amount of stock milled off an aluminum casting in two cuts is 7.62 millimeters. The roughing cut is 6.35 millimeters greater than the finish cut. What is the depth of the finish cut? _____

In each of the following problems, refer to the corresponding figure. Write an equation, solve for x, and check.

14. All dimensions are in inches.

_____ x = _____

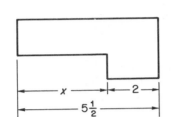

15. All dimensions are in millimeters.

_____ x = _____

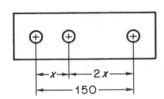

16. All dimensions are in inches.

_____ x = _____

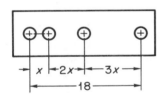

17. _____ x = _____

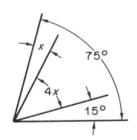

18. All dimensions are in inches.

_____ x = _____

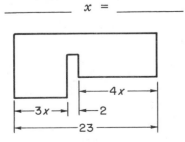

19. _____ x = _____

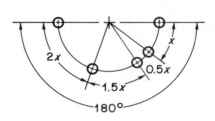

20. All dimensions are in millimeters.

_____ x = _____

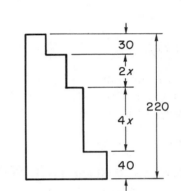

21. All dimensions are in inches.

_____ x = _____

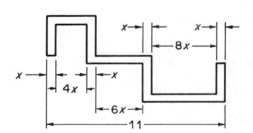

For each of the following problems, refer to the given figure, solve for the unknowns, and check.

22. Find the distances between the indicated holes.   All dimensions are in millimeters.

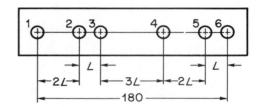

    a. Hole 1 to Hole 2   _____

    b. Hole 2 to Hole 3   _____

    c. Hole 3 to Hole 4   _____

    d. Hole 4 to Hole 5   _____

    e. Hole 5 to Hole 6   _____

    f. Hole 2 to Hole 4   _____

    g. Hole 3 to Hole 6   _____

23. Find the distances between the indicated points. All dimensions are in inches.

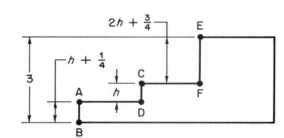

    a. A and B   _____

    b. C and D   _____

    c. E and F   _____

24. Find the value of each of the four angles.

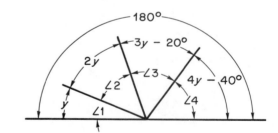

    a. $\angle$ 1   _____

    b. $\angle$ 2   _____

    c. $\angle$ 3   _____

    d. $\angle$ 4   _____

Solve for the unknown values in the following equations.

25. $x + 4x = 30$ _____

26. $x + 3 = 12$ _____

27. $2y + 5y + 3y = 70$ _____

28. $32 = 21 + y$ _____

29. $18 - a = 12$ _____

30. $b - 13 = 80$ _____

31. $3b + 5b - 2b = 48$ _____

32. $3(5a) = \frac{6(30)}{2}$ _____

33. $\frac{1}{2}x = 42$ _____

34. $\frac{x}{4} = 12$ _____

35. $\frac{27}{x} = 9$ _____

36. $\frac{d}{6} + 4 = 9$ _____

37. $0.75x - 0.5x = \frac{18 + 30}{4}$ _____

38. $6(2.5x) + 5x = 60$ _____

39. $\frac{2y + 4y + 6y}{3} = 80$ _____

40. $27 - (3)(6) = b + 7$ _____

UNIT 31 *SOLUTION OF EQUATIONS BY THE SUBTRACTION, ADDITION, AND DIVISION PRINCIPLES OF EQUALITY* _____

OBJECTIVES _____

After studying this unit you should be able to

- Solve equations using the subtraction principle of equality.
- Solve equations using the addition principle of equality.
- Solve equations using the division principle of equality.
- Solve equations using transposition.

There are specific procedures for solving equations using the fundamental principles of equality. Equations are solved more directly and efficiently by application of these principles.

SOLUTION OF EQUATIONS BY THE SUBTRACTION PRINCIPLE OF EQUALITY _____

The subtraction principle of equality states that if the same number is subtracted from both sides of an equation, the sides remain equal, and the equation remains balanced. The subtraction principle is used to solve an equation in which a number is added to the unknown, such as $x + 15 = 20$.

The values on each side of an equation are equal and an equation is balanced. If the same value is subtracted from both sides, the equation remains balanced. The equation 8 pounds + 4 pounds = 12 pounds is pictured. If 4 pounds are removed from the left side only, the scale is not in balance. If 4 pounds are removed from both the left and right sides, the scale remains in balance.

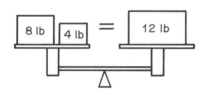

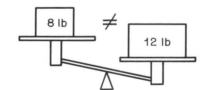

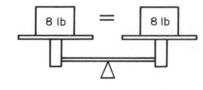

8 lb + 4 lb = 12 lb	8 lb + 4 lb − 4 lb ≠ 12 lb	8 lb + 4 lb − 4 lb = 12 lb − 4 lb
12 lb = 12 lb	8 lb ≠ 12 lb	8 lb = 8 lb

Procedure: To solve an equation in which a number is added to the unknown

- Subtract the number which is added to the unknown from both sides of the equation.
- Check.

Examples:

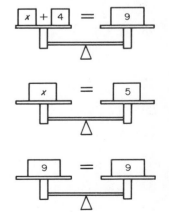

1. $x + 4 = 9$. Solve for x.

$$\begin{array}{rcl} x + 4 &=& 9 \\ -4 &=& -4 \\ \hline x &=& 5 \quad \text{Ans} \end{array}$$

Subtract 4 from both sides of the equation.

Check.

$$\begin{array}{rcl} x + 4 &=& 9 \\ 5 + 4 &=& 9 \\ \hline 9 &=& 9 \quad \text{Ck} \end{array}$$

2. In the part shown, determine dimension y.
    All dimensions are in inches.

    Write an equation $5.5'' + y \; = \; 17''$

    Subtract 5.5″ from both sides. $\dfrac{- \; 5.5'' \qquad\;\; = \; -5.5''}{y \; = \; 11.5'' \quad \text{Ans}}$

    Check. $5.5'' + y \; = \; 17''$
    $5.5'' + 11.5'' \; = \; 17''$
    $17'' \; = \; 17'' \qquad \text{Ck}$

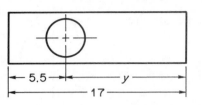

3. $-39 = P + 18$. Solve for P.

    $-39 \; = \; P + 18$ Check. $-39 \; = \; P + 18$
    $\dfrac{-18 \; = \; \quad - 18}{-57 \; = \; P \quad \text{Ans}}$ $-39 \; = \; -57 + 18$
    $-39 \; = \; -39 \quad \text{Ck}$

4. $W + 4\frac{3}{4} = 12$. Solve for W.   Check. $W + 4\frac{3}{4} \; = \; 12$

    $\dfrac{-\;4\frac{3}{4} \; = \; -4\frac{3}{4}}{W \; = \; \quad 7\frac{1}{4} \quad \text{Ans}}$ $7\frac{1}{4} + 4\frac{3}{4} \; = \; 12$
    $12 \; = \; 12 \quad \text{Ck}$

TRANSPOSITION

With your instructor's permission an alternate method of solving certain equations may be used. The alternate method is called transposition. *Transposition* or transposing a term means that a term is moved from one side of an equation to the opposite side with the sign changed.

Transposition is not a mathematical process although it is based on the addition and subtraction principles of equality. Transposition should only be used after the principles of equality are fully understood and applied.

Transposition is a quick and convenient means of solving equations in which a term is added to or subtracted from the unknown. The purpose of using transposition is the same as that of using the addition and subtraction principles of equality. Both methods involve getting the unknown term to stand alone on one side of the equation in order to determine the value of the unknown.

The following example is solved by applying the subtraction principle of equality and transposition. Notice that when applying the subtraction principle of equality, a term is eliminated on one side of the equation and appears on the other side with the sign changed.

Example 1: Solve for x.

$x + 15 \; = \; 25$

METHOD 1. The Subtraction Principle of Equality

$x + 15 \; = \; \quad 25$
$\dfrac{- \; 15 \; = \; -15}{x \; = \; \quad 25 - 15} \longleftarrow$ Observe that +15 is eliminated from the left side of the
$x \; = \; \quad 10 \quad \text{Ans}$ equation and appears as −15 on the right side.

METHOD 2. Transposition

$x + 15 \; = \; 25$
$x \; \boxed{+ \; 15} \; = \; 25 - 15$
$x \; = \; 25 - 15 \longleftarrow$ Observe that this expression is identical to the expression
$x \; = \; 10 \quad \text{Ans}$ obtained when applying the subtraction principle of
equality.

The following examples are solved by transposition.

Examples:

1. $y + 10.7 = 18$. Solve for y.

    Move +10.7 from the left side of the equation to the right side and change to –10.7.

$$y + 10.7 = 18$$
$$y = 18 - 10.7$$
$$y = 7.3 \quad \text{Ans}$$

2. $T + 6\frac{1}{8} = -19$. Solve for T.

    Move $+6\frac{1}{8}$ from the left side of the equation to the right side and change to $-6\frac{1}{8}$.

$$T + 6\frac{1}{8} = -19$$
$$T = -19 - 6\frac{1}{8}$$
$$T = -25\frac{1}{8} \quad \text{Ans}$$

SOLUTION OF EQUATIONS BY THE ADDITION PRINCIPLE OF EQUALITY

The addition principle of equality states that if the same number is added to both sides of an equation, the sides remain equal and the equation remains balanced. The addition principle is used to solve an equation in which a number is subtracted from the unknown, such as $x - 17 = 30$.

Procedure: To solve an equation in which a number is subtracted from the unknown

- Add the number, which is subtracted from the unknown, to both sides of the equation.
- Check.

Examples:

1. $x - 6 = 15$. Solve for x.

    Add 6 to both sides of the equation.

$$
\begin{aligned}
x - 6 &= 15 \\
+ 6 &= +6 \\
\hline
x &= 21 \quad \text{Ans}
\end{aligned}
$$

Check.

$$
\begin{aligned}
x - 6 &= 15 \\
21 - 6 &= 15 \\
15 &= 15 \quad \text{Ck}
\end{aligned}
$$

2. A 7-inch piece is cut from the height of a block as shown. The remaining block is 10 inches high. What is the height of the original block? All dimensions are in inches. Make no allowance for thickness of cut.

    Let y = the height of the original block.

    Write an equation.

    Add 7″ to both sides of the equation.

$$
\begin{aligned}
y - 7'' &= 10'' \\
+ 7'' &= +7'' \\
\hline
y &= 17'' \quad \text{Ans}
\end{aligned}
$$

    Check. $y - 7'' = 10''$
    $$17'' - 7'' = 10''$$
    $$10'' = 10'' \quad \text{Ck}$$

3. $-35 = P - 20.4$. Solve for P.

$$
\begin{aligned}
-35 &= P - 20.4 \\
+20.4 &= + 20.4 \\
\hline
-14.6 &= P \quad \text{Ans}
\end{aligned}
$$

    Check. $-35 = P - 20.4$
    $$-35 = -14.6 - 20.4$$
    $$-35 = -35 \quad \text{Ck}$$

The following examples are solved by transposition.

Examples:

1. $x - 4 = 19$.   Solve for x.

    Move –4 from the left side of the equation to the right and change to +4.

$$x - 4 = 19$$
$$x = 19 + 4$$
$$x = 23 \quad \text{Ans}$$

2. $y - 16.9 = 30$ Solve for y.

Move −16.9 from the left side of the equation to the
right and change to +16.9.

$$y - 16.9 = 30$$
$$y = 30 + 16.9$$
$$y = 46.9 \quad \text{Ans}$$

SOLUTION OF EQUATIONS BY THE DIVISION PRINCIPLE OF EQUALITY

The division principle of equality states that if both sides of an equation are divided by
the same number, the sides remain equal and the equation remains balanced. The division
principle is used to solve an equation in which a number is multiplied by the unknown such
as $3x = 18$.

Procedure: To solve an equation in which a number is multiplied by the unknown

• Divide both sides of the equation by the number which multiplies the unknown.

• Check.

Examples:

1. $6x = 24$. Solve for x.

Divide both sides of the equation by 6.

$$6x = 24$$
$$\frac{6x}{6} = \frac{24}{6}$$
$$x = 4 \quad \text{Ans}$$

Check.

$$6x = 24$$
$$6(4) = 24$$
$$24 = 24 \quad \text{Ck}$$

2. A part is shown. Solve for y. All dimensions are in millimeters.

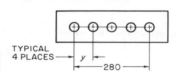

TYPICAL
4 PLACES → y ←
← 280 →

Write an equation.

Divide both sides of the equation by 4.

$$4y = 280 \text{ mm}$$
$$\frac{4y}{4} = \frac{280}{4} \text{ mm}$$
$$y = 70 \text{ mm} \quad \text{Ans}$$

Check. $4y = 280$
$4(70 \text{ mm}) = 280 \text{ mm}$
$280 \text{ mm} = 280 \text{ mm} \quad \text{Ck}$

3. $-14.4 = 3.2F$. Solve for F.

$$-14.4 = 3.2F$$
$$\frac{-14.4}{3.2} = \frac{3.2F}{3.2}$$
$$-4.5 = F \quad \text{Ans}$$

Check. $-14.4 = 3.2F$
$-14.4 = 3.2(-4.5)$
$-14.4 = -14.4 \quad \text{Ck}$

4. $7\frac{1}{4}A = 21\frac{3}{4}$ Solve for A.

$$7\frac{1}{4}A = 21\frac{3}{4}$$

$$\frac{7\frac{1}{4}A}{7\frac{1}{4}} = \frac{21\frac{3}{4}}{7\frac{1}{4}}$$

$$A = 3 \quad \text{Ans}$$

Check. $7\frac{1}{4}A = 21\frac{3}{4}$

$$7\frac{1}{4}(3) = 21\frac{3}{4}$$

$$21\frac{3}{4} = 21\frac{3}{4} \quad \text{Ck}$$

APPLICATION

Solution by the Subtraction Principle of Equality

Solve each of the following equations using the subtraction principle of equality.
Check each answer.

1. $P + 15 = 25$ _____

2. $x + 18 = 27$ _____

3. $M + 24 = 43$ _____

4. $y + 50 = 82$ _____

5. $13 = T + 9$ _____

6. $37 = D + 2$ _____

7. $55 = a + 19$ _____

8. $y + 16 = 15$ _____

9. $C + 34 = 12$ _____

10. $x + 6 = -11$ _____

11. $y + 30 = -23$ _____

12. $x + 63 = 17$ _____

13. $10 + R = 44$ _____

14. $51 = 48 + E$ _____

15. $-36 = 14 + x$ _____

16. $H + 7.6 = 15.2$ _____

17. $22.5 = L + 3.7$ _____

18. $-36.2 = y + 6.2$ _____

19. $86.04 = x + 61.95$ _____

20. $F + 0.007 = 1.006$ _____

21. $T + 9.07 = 9.07$ _____

22. $H + 3\frac{1}{4} = 7\frac{1}{2}$ _____

23. $-\frac{7}{8} = x + \frac{3}{4}$ _____

24. $20\frac{3}{16} = A + 17\frac{1}{8}$ _____

25. $39\frac{5}{8} = y + 42\frac{7}{8}$ _____

26. $1\frac{7}{16} = W + \frac{9}{16}$ _____

27. $x + 13\frac{1}{8} = -10$ _____

28. $0.015 = 1.009 + H$ _____

29. $-14.067 = 3.034 + x$ _____

30. $20.863 = D + 25.942$ _____

Write an equation for each of the following problems, solve for the unknown, and check.

31. All dimensions are in inches.
Find x.

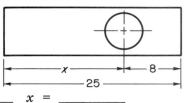

_____ $x = $ _____

32. All dimensions are in millimeters.
Find y.

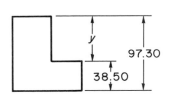

_____ $y = $ _____

33. All dimensions are in inches.
Find r.

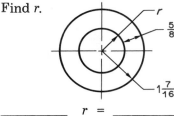

_____ $r = $ _____

34. All dimensions are in inches.
Find T.

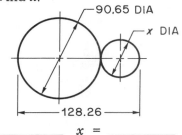

_____ $T = $ _____

35. All dimensions are in millimeters.
Find x.

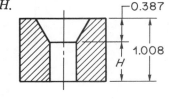

_____ $x = $ _____

36. All dimensions are in inches.
Find H.

_____ $H = $ _____

37. The height of 2 gage blocks is 0.8504 inch. One block is 0.750 inch thick. What is the thickness of the other block?   _____

38. Three holes are drilled on a horizontal line in a housing. The center distance between the first hole and the second hole is 193.75 millimeters and the center distance between the first hole and the third hole is 278.12 millimeters. What is the distance between the second hole and the third hole?   _____

39. A milling cut of 9/32 inch is required to provide a reference surface on a rough casting which is 7 5/8 inches high. What is the height of the casting after the milling cut?   _____

40. A shaft rotates in a bearing which is 0.3873 inch in diameter. The total clearance between the shaft and bearing is 0.0008 inch. What is the diameter of the shaft?   _____

For each of the following problems, substitute the given values in the formula and solve for the unknown. Check.

41. One of the formulas used in computing spur gear dimensions is $D_o = D + 2a$. Determine D when $a = 0.1429$ inch and $D_o = 4.7144$ inches.   _____

42. A formula used to compute the dimensions of a ring is $D = d + 2T$. Determine d when $D = 52$ millimeters and $T = 8.60$ millimeters.   _____

43. A formula used in relation to the depth of a gear tooth is $WD = a + d$. Determine d when $WD = 0.3082$ inch and $a = 0.1429$ inch.   _____

Solution by the Addition Principle of Equality

Solve each of the following equations using the addition principle of equality. Check each answer.

44. $T - 12 = 28$ _____

45. $x - 9 = -19$ _____

46. $B - 4 = 9$ _____

47. $P - 50 = 87$ _____

48. $y - 23 = -20$ _____

49. $16 = M - 12$ _____

50. $-35 = E - 21$ _____

51. $47 = R - 36$ _____

52. $h - 8 = 12$ _____

53. $T - 19 = -6$ _____

54. $-22 = x - 31$ _____

55. $39 = F - 39$ _____

56. $W - 16 = 33$ _____

57. $N - 2.4 = 6.9$ _____

58. $A - 0.8 = 0.5$ _____

59. $x - 10.09 = -13.78$ _____

60. $4.93 = r - 3.07$ _____

61. $-30.003 = x - 29.998$ _____

62. $91.96 = L - 13.74$ _____

63. $P - 0.02 = 0.07$ _____

64. $G - 59.875 = 49.986$ _____

65. $x - 8.12 = -13.01$ _____

66. $D - \frac{1}{2} = \frac{1}{2}$ _____

67. $y - \frac{7}{8} = -\frac{3}{4}$ _____

68. $15\frac{5}{8} = H - 2\frac{7}{8}$ _____

69. $-46\frac{3}{32} = x - 29\frac{15}{16}$ _____

70. $C - 5\frac{7}{16} = -5\frac{7}{16}$ _____

71. $W - 10.0039 = 9.0583$ _____

72. $-14\frac{15}{32} = y - 14\frac{7}{16}$ _____

73. $E - 29.8936 = 18.3059$ _____

Write an equation for each of the following problems, solve for the unknown, and check.

74. The bushing shown has a body diameter of 44.45 millimeters which is 14.29 millimeters less than the head diameter. What is the size of the head diameter? All dimensions are in millimeters.

_____

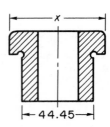

75. The flute length of the reamer shown is 1 1/8 inches, which is 3 3/8 inches less than the shank length. How long is the shank? All dimensions are in inches.

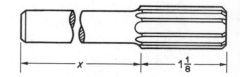

_____

76. A hole is countersunk as shown to a depth of 0.275 inch. The depth of the countersink is 1.650 inches less than the depth of the 0.625-inch hole. Find the depth of the countersink. All dimensions are in inches.

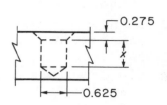

_____

For each of the following problems, substitute the given values in the formula and solve for the unknown. Check each answer.

77. The total taper of a shaft equals the diameter of the large end minus the diameter of the small end, $T = D - d$. Determine D when $T = 22.5$ millimeters and $d = 30.8$ millimeters. _____

78. Using the spur gear formula, $D_R = D - 2d$, compute the pitch diameter (D) when the root diameter $(D_R) = 3.0118$ inches and the dedendum $(d) = 0.1608$ inch. _____

79. Using a sheet metal formula, $W = L.S. - 4S$, determine the length size $(L.S.)$ when $W = 382$ millimeters and $S = 106$ millimeters. _____

Solution by the Division Principle of Equality

Solve each of the following equations using the division principle of equality. Check each answer.

80. $4D = 32$ _____

81. $7x = -21$ _____

82. $15M = 60$ _____

83. $54 = 9P$ _____

84. $-27 = 3y$ _____

85. $30 = 6x$ _____

86. $10y = 0.80$ _____

87. $18T = 41.4$ _____

88. $-12x = 42$ _____

89. $-x = 19$ _____

90. $0 = 7H$ _____

91. $-5C = 0$ _____

92. $7.1E = 42.6$ _____

93. $0.6L = 12$ _____

94. $-2.7x = 23.76$ _____

95. $0.1y = -0.09$ _____

96. $13.2W = 0$ _____

97. $-x = -19.75$ _____

98. $0.125P = 0.875$ _____

99. $9.37R = 103.07$ _____

100. $-0.66x = 4.752$ _____

101. $\frac{1}{4}D = 8$ _____

102. $24 = \frac{3}{8}B$ _____

103. $-\frac{1}{2}y = 36$ _____

104. $1\frac{5}{8}L = 8\frac{1}{8}$ _____

105. $-48\frac{3}{8} = 10\frac{3}{4}x$ _____

106. $-\frac{7}{16} = -\frac{7}{16}y$ _____

107. $50.98W = 10.196$ _____

108. $-0.002x = 4.938$ _____

109. $-\frac{3}{16} = -1\frac{1}{16}y$ _____

Write an equation for each of the following problems, solve for the unknown.

110. All dimensions are in millimeters.
Find x.

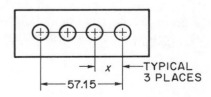

_____ $x =$ _____

111. Find x.

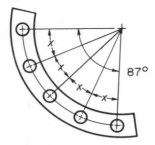

_____ x = _____

112. The feed of a drill is the depth of material that the drill penetrates in one revolution. The total depth of penetration equals the product of the number of revolutions and the feed. Compute the feed of a drill which cuts to a depth of 3.300 inches while turning 500 revolutions.

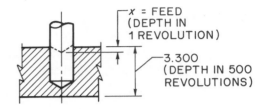

_____

For each of the following problems, substitute the given values in the formula and solve for the unknown. Check each answer.

113. The circumference of a circle (C) equals π (approximately 3.14) times the diameter (d) of the circle, $C = \pi d$. Determine d when $C = 376.80$ millimeters. _____

114. The depth (d) of a sharp V-thread is equal to 0.866 times the pitch (p). $d = 0.866p$. Determine p when $d = 0.125$ inch. _____

115. The length of cut (L) in inches of a workpiece in a lathe is equal to the product of the cutting time (T) in minutes, the tool feed (F) in inches per revolution, and the number of revolutions per minute (N) of the workpiece, $L = TFN$. Determine N when $L = 18$ inches, $T = 3$ minutes, and $F = 0.050$ inch per revolution. _____

UNIT 32 *SOLUTION OF EQUATIONS BY THE MULTIPLICATION, ROOT, AND POWER PRINCIPLES OF EQUALITY* _____

OBJECTIVES

After studying this unit you should be able to

• Solve equations using the multiplication principle of equality.

• Solve equations using the root principle of equality.

• Solve equations using the power principle of equality.

SOLUTION OF EQUATIONS BY THE MULTIPLICATION PRINCIPLE OF EQUALITY

The multiplication principle of equality states that if both sides of an equation are multiplied by the same number, the sides remain equal, and the equation remains balanced.

The multiplication principle is used to solve an equation in which the unknown is divided by a number, such as $\frac{x}{4} = 10$.

Procedure: To solve an equation in which the unknown is divided by a number

• Multiply both sides of the equation by the number which divides the unknown.

• Check.

Examples:

1. $\frac{x}{3} = 7$. Solve for x. $\frac{x}{3} = 7$

    Multiply both sides of the equation by 3. $3\left(\frac{x}{3}\right) = 3(7)$
    $$x = 21 \quad \text{Ans}$$

    Check. $\frac{x}{3} = 7$

    $$\frac{21}{3} = 7$$
    $$7 = 7 \quad \text{Ck}$$

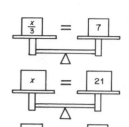

2. The length of bar stock shown is cut into 5 equal pieces. Each
    piece is 4.5 inches long. Find y, the length of the bar before
    it was cut. All dimensions are in inches. Make no allowance
    for thickness of cuts.

    Write the equation. $\frac{y}{5} = 4.5''$

    Multiply both sides of the equation by 5. $5\left(\frac{y}{5}\right) = 5(4.5'')$
    $$y = 22.5'' \quad \text{Ans}$$

    Check. $\frac{y}{5} = 4.5''$

    $$\frac{22.5''}{5} = 4.5''$$
    $$4.5'' = 4.5'' \quad \text{Ck}$$

3. $6\frac{1}{8} = \frac{F}{-5}$. Solve for F.   Check. $6\frac{1}{8} = \frac{F}{-5}$

    $$6\frac{1}{8} = \frac{F}{-5}$$ $6\frac{1}{8} = \frac{-30\frac{5}{8}}{-5}$

    $$-5\left(6\frac{1}{8}\right) = -5\left(\frac{F}{-5}\right)$$ $6\frac{1}{8} = 6\frac{1}{8} \quad \text{Ck}$

    $$-30\frac{5}{8} = F \quad \text{Ans}$$

SOLUTION OF EQUATIONS BY THE ROOT PRINCIPLE OF EQUALITY

The root principle of equality states that if the same root of both sides of an equation is
taken, the sides remain equal, and the equation remains balanced.

The root principle is used to solve an equation that contains an unknown which is raised
to a power, such as $x^2 = 36$.

Procedure: To solve an equation in which an unknown is raised to a power

- Extract the root of both sides of the equation which leaves the unknown with an
  exponent of one.
- Check.

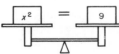

Examples:

1. $x^2 = 9$. Solve for x. $x^2 = 9$

    Extract the square root of both sides $\sqrt{x^2} = \sqrt{9}$
    of the equation. $x = 3 \quad \text{Ans}$

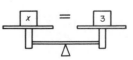

    Check. $x^2 = 9$
    $$3^2 = 9$$
    $$9 = 9 \quad \text{Ck}$$

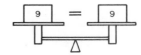

2. The area of a square piece of sheet steel shown equals 16 square feet. What is the length of each side(s)?

Write an equation.

$$s^2 = 16 \text{ sq ft}$$

Extract the square root of both sides of the equation.

$$\sqrt{s^2} = \sqrt{16} \text{ sq ft}$$
$$s = 4 \text{ ft} \quad \text{Ans}$$

Check.

$$s^2 = 16 \text{ sq ft}$$
$$(4 \text{ ft})^2 = 16 \text{ sq ft}$$
$$16 \text{ sq ft} = 16 \text{ sq ft} \quad \text{Ck}$$

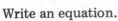

3. Solve for T.

$$T^3 = -64$$
$$\sqrt[3]{T^3} = \sqrt[3]{-64}$$
$$T = -4 \quad \text{Ans}$$

Check.
$$T^3 = -64$$
$$(-4)^3 = -64$$
$$-64 = -64 \quad \text{Ck}$$

4. Solve for V.

$$V^2 = \frac{9}{64}$$
$$\sqrt{V^2} = \sqrt{\frac{9}{64}}$$
$$V = \frac{3}{8} \quad \text{Ans}$$

Check.
$$V^2 = \frac{9}{64}$$
$$\left(\frac{3}{8}\right)^2 = \frac{9}{64}$$
$$\frac{9}{64} = \frac{9}{64} \quad \text{Ck}$$

SOLUTION OF EQUATIONS BY THE POWER PRINCIPLE OF EQUALITY

The power principle of equality states that if both sides of an equation are raised to the same power, the sides remain equal and the equation remains balanced. The power principle is used to solve an equation that contains a root of the unknown, such as $\sqrt{x} = 8$.

Procedure: To solve an equation which contains a root of the unknown

- Raise both sides of the equation to the power which leaves the unknown with an exponent of one.

- Check.

Examples:

1. $\sqrt{x} = 8$. Solve for x.

Square both sides of the equation.

$$\sqrt{x} = 8$$
$$(\sqrt{x})^2 = 8^2$$
$$x = 64 \quad \text{Ans}$$

Check.

$$\sqrt{x} = 8$$
$$\sqrt{64} = 8$$
$$8 = 8 \quad \text{Ck}$$

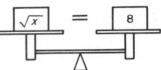

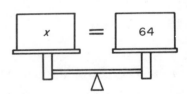

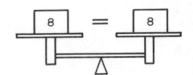

2. The length of a side of the cube shown equals 2 inches. The cube root of the volume equals the length of a side. Find the volume of the cube.

Let V = the volume of the cube.

Write the equation.

$$\sqrt[3]{V} = 2 \text{ in}$$

Cube both sides of the equation.

$$(\sqrt[3]{V})^3 = (2 \text{ in})^3$$
$$V = 8 \text{ cu in} \quad \text{Ans}$$

Check.

$$\sqrt[3]{V} = 2 \text{ in}$$
$$\sqrt[3]{8} \text{ cu in} = 2 \text{ in}$$
$$2 \text{ in} = 2 \text{ in} \quad \text{Ck}$$

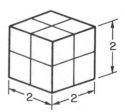

─APPLICATION──────────────────────────────

Solution by the Multiplication Principle of Equality

Solve each of the following equations using the multiplication principle of equality. Check each answer.

1. $\frac{P}{5} = 4$ _____

2. $\frac{M}{12} = 5$ _____

3. $D \div 9 = 7$ _____

4. $3 = L \div 6$ _____

5. $3 = W \div 9$ _____

6. $\frac{N}{12} = -2$ _____

7. $\frac{C}{14} = 0$ _____

8. $\frac{x}{-10} = 7$ _____

9. $\frac{E}{-2} = -18$ _____

10. $13 = y \div (-4)$ _____

11. $\frac{F}{3.6} = 5$ _____

12. $\frac{A}{-0.5} = 24$ _____

13. $S \div (7.8) = 3$ _____

14. $x \div (-0.4) = 16$ _____

15. $-20 = \frac{y}{0.3}$ _____

16. $\frac{T}{-1.8} = 2.4$ _____

17. $0 = H \div (-3.8)$ _____

18. $M \div 9.5 = -10$ _____

19. $\frac{y}{-0.1} = -0.01$ _____

20. $\frac{R}{12.6} = 0.002$ _____

21. $1.04 = \frac{H}{0.08}$ _____

22. $\frac{B}{\frac{1}{2}} = 7$ _____

23. $V \div 1\frac{1}{4} = 3$ _____

24. $\frac{x}{\frac{3}{8}} = -\frac{1}{2}$ _____

25. $D \div \left(-\frac{1}{16}\right) = -32$ _____

26. $4 = y \div \left(-\frac{7}{8}\right)$ _____

27. $\frac{1}{4} = \frac{T}{1\frac{1}{2}}$ _____

28. $H \div (-2) = 7\frac{9}{16}$ _____

29. $\frac{M}{0.009} = 100$ _____

30. $x \div (6.004) = -0.17$ _____

Write an equation for each of the following problems, solve for the unknown and check.

31. All dimensions are in millimeters.
    Find x.

_____ $x =$ _____

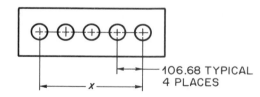

106.68 TYPICAL
4 PLACES

32. Find x.

_____ $x =$ _____

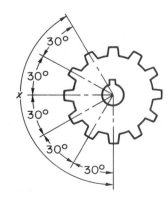

33. A 10-inch sine plate is tilted at an angle of 45° as shown. The gage block height divided by 10 equals 0.70711 inch. Compute the height of the gage blocks. All dimensions are in inches.

_____

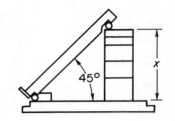

34. The width of a rectangular sheet of metal shown is equal to the area of the sheet divided by its length. Compute the area of a sheet which is 3 1/4 feet wide and 5 1/2 feet long.

_____

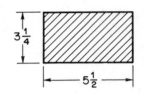

35. The depth of an American Standard thread shown divided by 0.6495 is equal to the pitch. Compute the depth of a thread with a 0.050-inch pitch. All dimensions are in inches.

_____

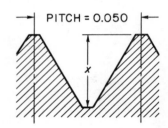

For each of the following problems, substitute the given values in the formula and solve for the unknown. Check each answer.

36. In mechanical energy applications, force (F) in pounds equals work (W) in foot-pounds divided by distance (D) in feet, $F = \frac{W}{D}$. Determine W when F = 150 pounds and D = 7.5 feet.

_____

37. The diameter (D) of a circle equals the circle circumference (C) divided by 3.1416, $D = \frac{C}{3.1416}$. Determine C when D = 100 millimeters.

_____

38. The pitch (P) of a spur gear equals the number of gear teeth (N) divided by the pitch diameter (D), $P = \frac{N}{D}$. Determine N when P = 5 teeth per inch and D = 5.6000 inches.

_____

Solution by the Root Principle of Equality

Solve each of the following equations using the root principle of equality.

39. $S^2 = 16$ _____

40. $P^2 = 36$ _____

41. $81 = M^2$ _____

42. $49 = B^2$ _____

43. $D^3 = 27$ _____

44. $x^3 = -27$ _____

45. $144 = F^2$ _____

46. $-64 = y^3$ _____

47. $L^3 = 125$ _____

48. $T^3 = 0$ _____

49. $10000 = L^2$ _____

50. $-125 = x^3$ _____

51. $\frac{4}{9} = W^2$ _____

52. $C^2 = \frac{1}{16}$ _____

53. $P^2 = \frac{9}{25}$ _____

54. $M^3 = \frac{1}{8}$ _____

55. $-\frac{1}{8} = y^3$ _____

56. $D^3 = \frac{64}{27}$ _____

57. $G^3 = \frac{64}{125}$ _____

58. $x^3 = \frac{-64}{125}$ _____

59. $E^2 = 0.09$ _____

60. $0.64 = H^2$ _____

61. $W^2 = 2.25$ _____

62. $0.0001 = R^2$ _____

63. $N^3 = 0.064$ _____

64. $-0.125 = x^3$ _____

65. $7.84 = F^2$ _____

66. $T^2 = 88.36$ _____

67. $y^3 = 0.027$ _____

68. $-0.027 = y^3$ _____

Write an equation for each of the following problems, solve for the unknown, and check.

69. The area of a square equals the length of a side squared, $A = s^2$. For each area of a square given, compute the length of a side.

   a. 36 square inches _____ $s =$ _____

   b. $\frac{25}{81}$ square foot _____ $s =$ _____

   c. 1.44 square meters _____ $s =$ _____

   d. 64 square meters _____ $s =$ _____

   e. 0.0049 square foot _____ $s =$ _____

70. The volume of a cube equals the length of a side cubed, $V = s^3$. For each volume of a cube given, compute the length of a side.

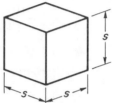

   a. 125 cubic inches _____ $s =$ _____

   b. $\frac{64}{125}$ cubic foot _____ $s =$ _____

   c. 0.064 cubic meter _____ $s =$ _____

   d. 1000 cubic millimeters _____ $s =$ _____

   e. 0.027 cubic foot _____ $s =$ _____

Solution by the Power Principle of Equality

Solve each of the following equations using the power principle of equality. Check all answers.

71. $\sqrt{C} = 8$ _____

72. $\sqrt{T} = 12$ _____

73. $\sqrt{P} = 1.2$ _____

74. $0.7 = \sqrt{M}$ _____

75. $0.82 = \sqrt{F}$ _____

76. $\sqrt[3]{V} = 3$ _____

77. $\sqrt[3]{H} = 2.3$ _____

78. $\sqrt[3]{x} = -4$ _____

79. $-0.1 = \sqrt[3]{y}$ _____

80. $\sqrt[4]{M} = 2$ _____

81. $\sqrt{A} = 0$ _____

82. $\sqrt[5]{N} = 1$ _____

83. $-2 = \sqrt[5]{y}$ _____

84. $0.2 = \sqrt[4]{D}$ _____

85. $\sqrt[3]{x} = -0.6$ _____

86. $\sqrt[4]{P} = 0.1$ _____

87. $0.1 = \sqrt[3]{B}$ _____

88. $\frac{1}{2} = \sqrt{A}$ _____

89. $\sqrt{R} = \frac{3}{8}$ _____

90. $\sqrt[3]{V} = \frac{2}{3}$ _____

91. $\sqrt[4]{F} = \frac{1}{2}$ _____

92. $-\frac{3}{5} = \sqrt[3]{y}$ _____

93. $\frac{5}{8} = \sqrt{H}$ _____

94. $\sqrt{P} = 1\frac{1}{4}$ _____

95. $\sqrt[3]{B} = 2\frac{1}{2}$ _____

96. $\sqrt[5]{x} = -\frac{1}{2}$ _____

97. $1\frac{3}{4} = \sqrt[3]{y}$ _____

98. $-1\frac{3}{4} = \sqrt[3]{y}$ _____

99. $\sqrt[3]{x} = -\frac{3}{10}$ _____

100. $\sqrt[5]{x} = -0.2$ _____

Write an equation for each of the following problems, solve for the unknown, and check.

101. The length of a side of a square equals the square root of the area, $s = \sqrt{A}$.
For each side of a square given, compute the area.

a. 2.8″   _____ $A =$ _____
b. 0.75′   _____ $A =$ _____
c. 0.6 m   _____ $A =$ _____
d. 220 mm   _____ $A =$ _____
e. 12.9″   _____ $A =$ _____

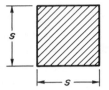

102. The length of a side of a cube equals the cube root of the volume, $s = \sqrt[3]{V}$.
For each side of a cube given, compute the volume.

a. 3.3″   _____ $V =$ _____
b. 0.9′   _____ $V =$ _____
c. 0.75 m   _____ $V =$ _____
d. 40 mm   _____ $V =$ _____
e. 12.8″   _____ $V =$ _____

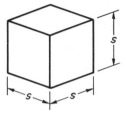

⌐ UNIT 33 *SOLUTION OF EQUATIONS CONSISTING OF COMBINED OPERATIONS AND REARRANGEMENT OF FORMULAS* ____

⌐OBJECTIVES____

After studying this unit you should be able to

- Solve equations involving several operations.
- Rearrange formulas in terms of any letter value.
- Substitute values in formulas and solve for unknowns.

Often in actual occupational applications, the formulas used result in complex equations. These equations require the use of two or more principles of equality for their solutions. For example,

$$0.13x - 4.73(x + 6.35) = 5.06x - 2.87$$

requires a definite procedure in determining the value of x. Use of proper procedure results in the unknown standing alone on one side of the equation with its value on the other.

⌐PROCEDURE FOR SOLVING EQUATIONS CONSISTING OF COMBINED OPERATIONS____

It is essential that the steps used in solving an equation be taken in the following order. Some or all of these steps may be used depending upon the particular equation.

- Remove parentheses.
- Combine like terms on each side of the equation.
- Apply the addition and subtraction principles of equality to get all unknown terms on one side of the equation and all known terms on the other side.
- Combine like terms.
- Apply the multiplication and division principles of equality.
- Apply the power and root principles of equality.

Note: Always solve for a positive unknown. The solution may be $x = -7$ but *not* $-x = 7$. The unknown may equal a negative value but a negative unknown is not the solution.

Examples:

1. $5x + 7 = 22$. Solve for x.

    The operations involved are multiplication and addition. Follow the procedure for solving equations consisting of combined operations.

    Apply the subtraction principle. Subtract 7 from both sides of the equation.

    Apply the division principle. Divide both sides of the equation by 5.

$$5x + 7 = 22$$
$$\underline{ - 7 = -7}$$
$$5x = 15$$
$$\frac{5x}{5} = \frac{15}{5}$$
$$x = 3 \quad \text{Ans}$$

    Check.

$$5x + 7 = 22$$
$$5(3) + 7 = 22$$
$$15 + 7 = 22$$
$$22 = 22 \quad \text{Ck}$$

2. $6x + 4x = 3x - 5x + 19 + 5$. Solve for x.

    Combine like terms on each side of the equation.

    Apply the addition principle. Add $2x$ to both sides of the equation.

    Apply the division principle. Divide both sides of the equation by 12.

$$6x + 4x = 3x - 5x + 19 + 5$$
$$10x = -2x + 24$$
$$\underline{+2x = +2x}$$
$$12x = 24$$
$$\frac{12x}{12} = \frac{24}{12}$$
$$x = 2 \quad \text{Ans}$$

    Check.

$$6x + 4x = 3x - 5x + 19 + 5$$
$$6(2) + 4(2) = 3(2) - 5(2) + 19 + 5$$
$$12 + 8 = 6 - 10 + 19 + 5$$
$$20 = 20 \quad \text{Ck}$$

3. $9x + 7(x + 3) = 25$. Solve for x.

    Remove parentheses.

    Combine like terms.

    Apply the subtraction principle. Subtract 21 from both sides of the equation.

    Apply the division principle. Divide both sides of the equation by 16.

$$9x + 7(x + 3) = 25$$
$$9x + 7x + 21 = 25$$
$$16x + 21 = 25$$
$$\underline{- 21 = -21}$$
$$16x = 4$$
$$\frac{16x}{16} = \frac{4}{16}$$
$$x = \frac{1}{4} \quad \text{Ans}$$

    Check.

$$9x + 7(x + 3) = 25$$
$$9\left(\tfrac{1}{4}\right) + 7\left(\tfrac{1}{4} + 3\right) = 25$$
$$2\tfrac{1}{4} + 22\tfrac{3}{4} = 25$$
$$25 = 25 \quad \text{Ck}$$

4. $-x = 14$. Solve for x.

    Apply the multiplication principle. Multiply both sides of the equation by -1.

$$-x = 14$$
$$(-1)(-x) = (-1)(14)$$
$$x = -14 \quad \text{Ans}$$

    Check.

$$-x = 14$$
$$-(-14) = 14$$
$$14 = 14 \quad \text{Ck}$$

5. $\frac{x^2}{4} - 32 = -23$. Solve for x.

Apply the addition principle. Add 32 to both sides of the equation.

Apply the multiplication principle. Multiply both sides of the equation by 4.

Apply the root principle. Extract the square root of both sides of the equation.

Check.

$$\frac{x^2}{4} - 32 = -23$$

$$\underline{\quad + 32 = +32\quad}$$

$$\frac{x^2}{4} = 9$$

$$4\left(\frac{x^2}{4}\right) = 4(9)$$

$$x^2 = 36$$

$$\sqrt{x^2} = \sqrt{36}$$

$$x = 6 \quad \text{Ans}$$

$$\frac{x^2}{4} - 32 = -23$$

$$\frac{6^2}{4} - 32 = -23$$

$$\frac{36}{4} - 32 = -23$$

$$9 - 32 = -23$$

$$-23 = -23 \quad \text{Ck}$$

6. $6\sqrt[3]{x} = 4(\sqrt[3]{x} + 1.5)$. Solve for x.

Remove parentheses.

Apply the subtraction principle. Subtract $4\sqrt[3]{x}$ from both sides of the equation.

Apply the division principle. Divide both sides of the equation by 2.

Apply the power principle. Raise both sides of the equation to the third power.

Check.

$$6\sqrt[3]{x} = 4(\sqrt[3]{x} + 1.5)$$

$$6\sqrt[3]{x} = 4\sqrt[3]{x} + 6$$

$$\underline{-4\sqrt[3]{x} = -4\sqrt[3]{x}\quad}$$

$$2\sqrt[3]{x} = 6$$

$$\frac{2\sqrt[3]{x}}{2} = \frac{6}{2}$$

$$\sqrt[3]{x} = 3$$

$$(\sqrt[3]{x})^3 = 3^3$$

$$x = 27 \quad \text{Ans}$$

$$6\sqrt[3]{27} = 4(\sqrt[3]{27} + 1.5)$$

$$6(3) = 4(3 + 1.5)$$

$$18 = 4(4.5)$$

$$18 = 18 \quad \text{Ck}$$

⎯SUBSTITUTING VALUES AND SOLVING FORMULAS⎯⎯⎯⎯⎯⎯⎯⎯⎯⎯

   Manufacturing applications often require solving formulas in which all but one numerical value for letter values is known. The unknown letter value can appear anywhere within the formula. To determine the numerical value of the unknown, write the original formula, substitute the known number values for their respective letter values, and simplify. Then follow the procedure given for solving equations consisting of combined operations.

Example: An open belt pulley system is shown. The larger pulley diameter is 6 inches and the smaller pulley diameter is 4 inches. The belt length is 56 inches. Find the distance between pulley centers using this formula found in a trade handbook.

$L = 3.14(0.5D + 0.5d) + 2x$ 　　　 where　L = belt length

D = the diameter of the larger pulley

d = the diameter of the smaller pulley

x = the distance between pulley centers

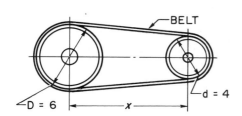

Write the formula.	$L = 3.14(0.5D + 0.5d) + 2x$
Substitute the known numerical values for their respective letter values and simplify.	$56'' = 3.14[0.5(6'') + 0.5(4'')] + 2x$
	$56'' = 3.14(5'') + 2x$
	$56'' = 15.70'' + 2x$
	$56.00'' = 15.70'' + 2x$
Apply the subtraction principle. Subtract 15.70″ from both sides of the equation.	$\dfrac{-15.70'' = -15.70''}{40.30'' = 2x}$
Apply the division principle. Divide both sides of the equation by 2.	$\dfrac{40.30''}{2} = \dfrac{2x}{2}$
	$20.15'' = x$　Ans
Check.	$L = 3.14(0.5D + 0.5d) + 2x$
	$56'' = 3.14[0.5(6'') + 0.5(4'')] + 2(20.15'')$
	$56'' = 15.70'' + 40.30''$
	$56'' = 56''$

REARRANGING FORMULAS

A formula that is given in terms of a particular value is sometimes rearranged to solve for another value before substituting the given numerical values. The formula $A = \frac{1}{2} bh$ which is given in terms of A may be rearranged to solve for h if the height is to be calculated for more than one problem.

Consider the letter to be solved for as the unknown term and the other letters in the formula as the known values. The formula must be rearranged so that the unknown term is on one side of the equation and all other values are on the other side. The formula is rearranged using the same procedure that is used for solving equations consisting of combined operations.

Examples:　Given the following formulas, rearrange and solve for the designated letter.

1. $A = bh$. Solve for h. 　　　　　　　　　　　　$A = bh$

   Apply the division principle. Divide both sides of the equation by b. 　　　$\dfrac{A}{b} = \dfrac{bh}{b}$

   　　　　　　　　　　　　　　　　　　　　$\dfrac{A}{b} = h$　Ans

2. In the figure shown $L = a + b$. Solve for a. 　　　$L = a + b$

   Apply subtraction principle. Subtract b from both sides of the equation. 　　$\dfrac{-b = \quad -b}{L - b = a}$　Ans

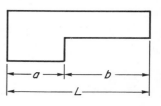

3. The area of the ring shown is computed using this formula.

    $A = \pi(R^2 - r^2)$ where A = area
    π = pi
    R = radius of larger circle
    r = radius of smaller circle

Rearrange the formula and solve for R.

Remove parentheses.

Apply addition principle.
Add πr^2 to both sides of the equation.

Apply division principle.
Divide both sides of the equation by π.

Apply root principle. Extract the square root
of both sides of the equation.

$$A = \pi(R^2 - r^2)$$

$$A = \pi R^2 - \pi r^2$$

$$\begin{array}{r} +\pi r^2 = + \pi r^2 \\ \hline A + \pi r^2 = \pi R^2 \end{array}$$

$$\frac{A + \pi r^2}{\pi} = \frac{\pi R^2}{\pi}$$

$$\frac{A + \pi r^2}{\pi} = R^2$$

$$\sqrt{\frac{A + \pi r^2}{\pi}} = \sqrt{R^2}$$

$$\sqrt{\frac{A + \pi r^2}{\pi}} = R \quad \text{Ans}$$

4. A screw thread is checked using a micrometer and 3 wires as shown. The measurement is checked using the following formula. Solve the formula for W.

    $M = D - 1.5155P + 3W$ where M = measurement over the wires
    D = major diameter
    P = pitch
    W = wire size

THREAD CHECKING

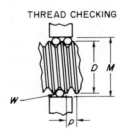

Apply subtraction principle. Subtract D from both sides of the equation.

Apply addition principle. Add $1.5155P$ to both sides of the equation.

Apply division principle. Divide both sides of the equation by 3.

$$\begin{array}{r} M = D - 1.5155P + 3W \\ -D = -D \\ \hline M - D = -1.5155P + 3W \end{array}$$

$$\begin{array}{r} +1.5155P = +1.5155P \\ \hline M - D + 1.5155P = 3W \end{array}$$

$$\frac{M - D + 1.5155P}{3} = \frac{3W}{3}$$

$$\frac{M - D + 1.5155P}{3} = W \quad \text{Ans}$$

5. A flat is to be ground on the circular cross section part shown. The formula used to determine the width of the flat is given. Solve for W.

    $R - D = \sqrt{R^2 - 0.25W^2}$ where R = radius of the part
    D = depth
    W = width of the flat

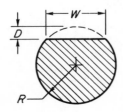

Note: Observe that both terms, R^2 and $0.25W^2$ are enclosed within the radical sign. Neither term can be removed until the radical sign is eliminated.

$$R - D = \sqrt{R^2 - 0.25W^2}$$

Apply the power principle. Square both sides of the equation to eliminate the radical sign.

$$(R - D)^2 = (\sqrt{R^2 - 0.25W^2})^2$$
$$(R - D)^2 = R^2 - 0.25W^2$$

Apply the subtraction principle. Subtract R^2 from both sides of the equation.

$$\frac{- R^2 = -R^2}{(R - D)^2 - R^2 = -0.25W^2}$$

Apply the division principle. Divide both sides of the equation by -0.25.

$$\frac{(R - D)^2 - R^2}{-0.25} = \frac{-0.25\,W^2}{-0.25}$$
$$\frac{(R - D)^2 - R^2}{-0.25} = W^2$$

Apply the root principle. Extract the square root of both sides of the equation.

$$\sqrt{\frac{(R - D)^2 - R^2}{-0.25}} = \sqrt{W^2}$$
$$\sqrt{\frac{(R - D)^2 - R^2}{-0.25}} = W \quad \text{Ans}$$

APPLICATION

Equations Consisting of Combined Operations

Solve for the unknown and check each of the following combined operations equations.

1. $5x - 33 = 7$ _____

2. $10M + 5 + 4M = 89$ _____

3. $8E - 14 = 2E + 28$ _____

4. $4B - 7 = B + 14$ _____

5. $7T - 14 = 0$ _____

6. $6N + 4 = 84 + N$ _____

7. $2.5A + 8 = 15 - 6.5$ _____

8. $12 - (-x + 8) = 18$ _____

9. $3H + (2 - H) = 20$ _____

10. $6 = -(2 + C) - (4 + 2C)$ _____

11. $-5(R + 6) = 10(R - 2)$ _____

12. $0.29E = 9.39 - 0.01E$ _____

13. $7.2F + 5(F - 8.1) = 0.6F + 15.18$ _____

14. $\frac{P}{7} + 8 = 5.9$ _____

15. $\frac{1}{4} W + (W - 8) = \frac{3}{4}$ _____

16. $\frac{1}{8} D - 3(D - 7) = 5\frac{1}{8} D - 3$ _____

17. $0.58y = 18.78 - 0.02y$ _____

18. $2H^2 - 20 = (H + 4)(H - 4)$ _____

19. $4A^2 + 3A + 36 = 8A^2 + 3A$ _____

20. $x(3 + x) + 20 = x^2 - (x - 5)$ _____

21. $\left(\frac{b}{2}\right)^3 + 34 = 42$ _____

22. $3F^3 + F(F + 8) = 8F + F^2 + 81$ _____

23. $9 + y^2 = (y - 4)(y - 1)$ _____

24. $\frac{1}{4}(2B - 12) + B^2 = \frac{1}{2} B + 22$ _____

25. $-4(y - 3) = 2\sqrt{y} - 4y$ _____

26. $14\sqrt{x} = 6(\sqrt{x} + 8) + 16$ _____

27. $8P^2 + 6P + 72 = 16P^2 + 6P$ _____

28. $7\sqrt{x} = 3(\sqrt{x} + 8) - 4$ _____

29. $\sqrt{B^2} - 2B = -2(B - 3)$ _____

30. $(2y)^3 - 2.8(5 + 3y) = -22 - 8.4$ _____

Substituting Values and Solving Formulas

The following formulas are used in the machine trades. Substitute the given values in each formula and solve for the unknown. Round answers to 3 decimal places when necessary.

31. $F = 2.380P + 0.250$
     Given: $F = 1.750$.
     Solve for P. _____

32. $a = 3H \div 8$
     Given: $a = 0.1760$.
     Solve for H. _____

33. $H.P. = 0.000016MN$
     Given $H.P. = 22$, $N = 50.8$.
     Solve for M. _____

34. $N = 0.707DP_n$
     Given: $N = 24$, $P_n = 8$.
     Solve for D. _____

35. $S = T - \frac{1.732}{N}$

    Given: $S = 0.4134, N = 20$.
    Solve for T.   _____

36. $a = \frac{D_2 - D_1}{2}$

    Given: $a = 0.250, D_1 = 0.875$.
    Solve for D_2.   _____

37. $S = \frac{0.290W}{t^2}$

    Given: $S = 1200, t = 0.750$.
    Solve for W.   _____

38. $W = St(0.55d^2 - 0.25d)$

    Given: $W = 1150, d = 0.750$.
    Solve for St.   _____

39. $M = E - 0.866P + 3W$

    Given: $M = 3.3700, E = 3.2200, P = 0.125$.
    Solve for W.   _____

40. $S = \frac{L_1}{L_2}\left[\frac{1}{2}(D_1 - D_2)\right]$

    Given: $S = \frac{1}{4}, L_1 = 16, L_2 = 4, D_2 = 2\frac{1}{2}$.
    Solve for D_1.   _____

41. $C = \frac{\pi D N}{12}$

    Given: $C = 225, D = 6, \pi = 3.1416$.
    Solve for N.   _____

42. $D_o = \frac{P_c(N + 2)}{\pi}$

    Given: $D_o = 4.3750, \pi = 3.1416, P_c = 0.3927$.
    Solve for N.   _____

43. $S = \sqrt{\frac{d^2}{4} + h^2}$

    Given: $S = 12.700, d = 6$.
    Solve for h.   _____

44. $C = 2\sqrt{h(2r - h)}$

    Given: $C = 7.600, h = 3.750$.
    Solve for r.   _____

Rearranging Formulas

The following formulas are used in machine trade calculations. Rearrange the formulas in terms of the designated values.

45. The dimensions shown can be found
    using these two formulas.

    (1) $A = ab$

    (2) $d = \sqrt{a^2 + b^2}$

    a. Solve formula (1) for a.   _____

    b. Solve formula (1) for b.   _____

    c. Solve formula (2) for a.   _____

    d. Solve formula (2) for b.   _____

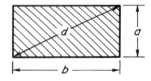

46. The radii shown in this figure can be
    found using these two formulas.

    (1) $R = 1.155r$

    (2) $A = 2.598R^2$

    a. Solve formula (1) for r.   _____

    b. Solve formula (2) for R.   _____

47. The dimensions shown can be found
    using these two formulas.

    (1) $FW = \sqrt{D_o^2 - D^2}$

    (2) $D_o = 2C - d + 2a$

    a. Solve formula (1) for D_o.   _____

    b. Solve formula (1) for D.   _____

    c. Solve formula (2) for d.   _____

    d. Solve formula (2) for C.   _____

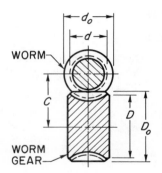

48. $A = \pi(R^2 - r^2)$

    a. Solve for R.

    ———

    b. Solve for r.

    ———

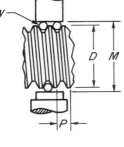

51. $L = 3.14(0.5D + 0.5d) + 2x$

    a. Solve for D.

    ———

    b. Solve for d.

    ———

    c. Solve for x.

    ———

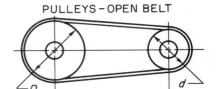

PULLEYS – OPEN BELT

49. $M = D - 1.5155P + 3W$

    a. Solve for D.

    ———

    b. Solve for P.

    ———

    c. Solve for W.

    ———

52. $By(F - 1) = Cx$

    a. Solve for x.

    ———

    b. Solve for B.

    ———

    c. Solve for F.

    ———

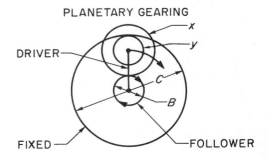

PLANETARY GEARING

50. $\angle A + \angle B + \angle C = 180°$

    a. Solve for $\angle A$.

    ———

    b. Solve for $\angle B$.

    ———

    c. Solve for $\angle C$.

    ———

53. $Ca = S(C - F)$

    a. Solve for S.

    ———

    b. Solve for F.

    ———

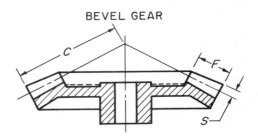

BEVEL GEAR

UNIT 34 *RATIO AND PROPORTION*

OBJECTIVES

After studying this unit you should be able to

- Write comparisons as ratios.
- Express ratios in lowest terms.
- Solve for the unknown term of a proportion.
- Substitute given numerical values for symbols in a proportion and solve for the unknown term.

The ability to solve practical machine shop problems using ratio and proportion is a requirement for the skilled machinist. Ratio and proportion are used for calculating gear and pulley speeds and sizes, for computing thread cutting values on a lathe, for computing taper dimensions, and for determining machine cutting times.

DESCRIPTION OF RATIO

Ratio is the comparison of two like quantities by division.

Examples:

1. Two pulleys are shown. What is the ratio of the diameter of the small pulley to the diameter of the larger pulley? All dimensions are in inches.

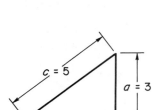

   The ratio is 3 to 5.   Ans

2. A triangle with given lengths of 3 meters, 4 meters, and 5 meters for sides *a*, *b*, and *c* is shown.

   a. What is the ratio of side *a* to side *b*?

   The ratio is 3 to 4.   Ans

   b. What is the ratio of side *a* to side *c*?

   The ratio is 3 to 5.   Ans

   c. What is the ratio of side *b* to side *c*?

   The ratio is 4 to 5.   Ans

The terms of a ratio are the two numbers that are compared. *Both terms of a ratio must be expressed in the same units of measure.*

Example: Two pieces of bar stock are shown. What is the ratio of the short piece to the long piece?

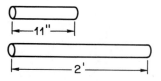

   The terms cannot be compared as a ratio until the 2-foot length is expressed as 24 inches.

   The ratio is 11 to 24.   Ans

It is impossible to express two quantities as ratios if the terms have unlike units that cannot be expressed as like units. Inches and pounds as shown cannot be compared as ratios.

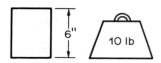

Expressing Ratios: Ratios are expressed in the following ways.

• With a colon between the two terms, such as 4:7. The ratio 4:7 is read as 4 to 7.

• With a division sign separating the two numbers, such as 4 ÷ 7 or as a fraction, $\frac{4}{7}$.

ORDER OF TERMS

The terms of a ratio must be compared in the order in which they are given. The first term is the numerator of a fraction and the second is the denominator.

Examples:

1. 1 to 3 = 1 ÷ 3 = $\frac{1}{3}$ Ans

2. 3 to 1 = 3 ÷ 1 = $\frac{3}{1}$ Ans

3. $x{:}y = x \div y = \frac{x}{y}$ Ans

4. $y{:}x = y \div x = \frac{y}{x}$ Ans

EXPRESSING RATIOS IN LOWEST TERMS

Generally, a ratio should be expressed in lowest fractional terms.

Examples:

1. $3{:}9 = \frac{3}{9} = \frac{1}{3}$ Ans

2. $40{:}15 = \frac{40}{15} = \frac{8}{3}$ Ans

3. $\frac{3}{8}{:}\frac{9}{16} = \frac{3}{8} \div \frac{9}{16} = \frac{3}{8} \times \frac{16}{9} = \frac{2}{3}$ Ans

4. $10{:}\frac{5}{6} = 10 \div \frac{5}{6} = \frac{10}{1} \times \frac{6}{5} = \frac{12}{1}$ Ans

5. $x^3{:}x^2 = \frac{x^3}{x^2} = \frac{x}{1}$ Ans

6. $4ab{:}6a = \frac{4ab}{6a} = \frac{2b}{3}$ Ans

DESCRIPTION OF PROPORTIONS

A *proportion* is an expression that states the equality of two ratios.

Expressing Proportions: Proportions are expressed in the following two ways.

- 3:4::6:8, which is read as 3 is to 4 as 6 is to 8.
- $\frac{3}{4} = \frac{6}{8}$. This equation form is generally the way that proportions are used.

A proportion consists of four terms. The first and the fourth term are called *extremes* and the second and third terms are called *means.*

Examples:

1. 2:3::4:6   2 and 6 are the extremes; 3 and 4 are the means.   Ans

2. $\frac{5}{6} = \frac{10}{12}$ 5 and 12 are the extremes; 6 and 10 are the means.   Ans

In a proportion the product of the means equals the product of the extremes. If the terms are cross multiplied, their products are equal.

Examples:

1. $\frac{3}{4} = \frac{6}{8}$

   Cross multiply, $\frac{3}{4} \diagup\!\!\!\!\diagdown \frac{6}{8}$

   $$3 \times 8 = 4 \times 6$$
   $$24 = 24$$

2. $\frac{a}{b} = \frac{c}{d}$

   Cross multiply, $\frac{a}{b} \diagup\!\!\!\!\diagdown \frac{c}{d}$

   $$a \times d = b \times c$$
   $$ad = bc$$

The method of cross multiplying is used in solving proportions which have an unknown term. Since a proportion is an equation, the principles used for solving equations are applied in determining the value of the unknown after the terms have been cross multiplied.

Examples: Solve for the value of x.

1. $\frac{3}{4} = \frac{x}{16}$

   Cross multiply.

   Apply the division principle of equality. Divide both sides of the equation by 4.

   Check.

   $\frac{3}{4} = \frac{x}{16}$

   $$4x = 3(16)$$
   $$4x = 48$$
   $$\frac{4x}{4} = \frac{48}{4}$$
   $$x = 12 \quad \text{Ans}$$
   $$\frac{3}{4} = \frac{x}{16}$$
   $$\frac{3}{4} = \frac{12}{16}$$
   $$\frac{3}{4} = \frac{3}{4} \quad \text{Ck}$$

2. $\frac{7}{x} = \frac{8}{15}$

   $$8x = 7(15)$$
   $$8x = 105$$
   $$\frac{8x}{8} = \frac{105}{8}$$
   $$x = 13\frac{1}{8} \quad \text{Ans}$$
   Check.
   $$\frac{7}{x} = \frac{8}{15}$$
   $$\frac{7}{13\frac{1}{8}} = \frac{8}{15}$$
   $$\frac{8}{15} = \frac{8}{15} \quad \text{Ck}$$

3. $\frac{x}{7.5} = \frac{23.4}{20}$

   $$20x = 7.5(23.4)$$
   $$20x = 175.5$$
   $$\frac{20x}{20} = \frac{175.5}{20}$$
   $$x = 8.775 \quad \text{Ans}$$
   Check.
   $$\frac{x}{7.5} = \frac{23.4}{20}$$
   $$\frac{8.775}{7.5} = \frac{23.4}{20}$$
   $$1.17 = 1.17 \quad \text{Ck}$$

4. $\frac{a}{b} = \frac{c}{x}$

   $$ax = bc$$
   $$\frac{ax}{a} = \frac{bc}{a}$$
   $$x = \frac{bc}{a} \quad \text{Ans}$$
   Check.
   $$\frac{a}{b} = \frac{c}{x}$$
   $$\frac{a}{b} = \frac{c}{\frac{bc}{a}}$$
   $$\frac{a}{b} = \frac{a}{b} \quad \text{Ck}$$

APPLICATION

Ratios

Express the following ratios in lowest fractional form.

1. 6:15 _____

2. 15:6 _____

3. 2:11 _____

4. 7:21 _____

5. 12″:46″ _____

6. 3 lb:18 lb _____

7. 17 mi:9 mi _____

8. 156 mm:200 mm _____

9. $3a^2 b:6ab$ _____

10. $xy:x^2 y$ _____

11. $\frac{2}{3}:\frac{1}{2}$ _____

12. $\frac{1}{2}:\frac{2}{3}$ _____

Related Ratio Problems

13. Length A in this figure is 3 inches and length B is 1.5 feet. Determine the ratio of length A to length B in lowest fractional form.

_____

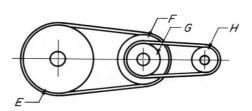

14. The diameters of pulleys E, F, G, and H are given in the table. Determine the ratios in lowest fractional form.

_____

	Diameters (inches)				Ratios							
	E	F	G	H	$\frac{E}{F}$	$\frac{E}{G}$	$\frac{E}{H}$	$\frac{F}{G}$	$\frac{F}{H}$	$\frac{G}{H}$	$\frac{G}{E}$	$\frac{H}{F}$
a.	8	6	4	3								
b.	10	6	5	4								
c.	12	9	6	3								
d.	15	12	10	6								

15. Refer to the hole locations given for the plate. Determine the ratios in lowest fractional form. All dimensions are in millimeters.

a. Dimension A to dimension B. _____

b. Dimension A to dimension C. _____

c. Dimension C to dimension D. _____

d. Dimension C to dimension E. _____

e. Dimension D to dimension F. _____

f. Dimension F to dimension B. _____

g. Dimension F to dimension C. _____

h. Dimension E to dimension A. _____

i. Dimension D to dimension B. _____

j. Dimension C to dimension F. _____

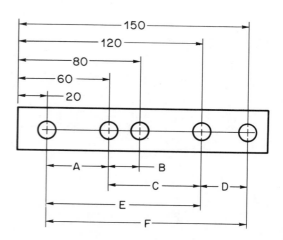

16. Gear A is turning at 120 revolutions per minute and gear B is turning at 18 revolutions per second. Determine the ratio of the speed of gear A to the speed of gear B.

GEAR A
120 r/min

GEAR B
18 r/s

_____

Proportions

Solve for the unknown value in each of the following proportions. Check each answer.

17. $\frac{x}{4} = \frac{6}{24}$ _____

18. $\frac{3}{A} = \frac{15}{30}$ _____

19. $\frac{7}{9} = \frac{E}{45}$ _____

20. $\frac{6}{13} = \frac{24}{y}$ _____

21. $\frac{15}{c} = \frac{5}{4}$ _____

22. $\frac{P}{18} = \frac{1}{3}$ _____

23. $\frac{6}{7} = \frac{15}{F}$ _____

24. $\frac{12}{H} = \frac{4}{25}$ _____

25. $\frac{T}{6.6} = \frac{7.5}{22.5}$ _____

26. $\frac{2.4}{3} = \frac{M}{0.8}$ _____

27. $\frac{4}{4.1} = \frac{8}{L}$ _____

28. $\frac{3.4}{y} = \frac{1}{-7}$ _____

29. $\frac{A}{5} = \frac{3.2}{A}$ _____

30. $\frac{\frac{3}{8}}{N} = \frac{\frac{1}{2}}{4}$ _____

31. $\frac{3}{1\frac{1}{2}} = \frac{5}{F}$ _____

32. $\frac{G}{\frac{1}{4}} = \frac{\frac{7}{8}}{\frac{3}{8}}$ _____

33. $\frac{7}{\frac{-1}{8}} = \frac{x}{\frac{9}{16}}$ _____

34. $\frac{4}{R} = \frac{2R}{12.5}$ _____

35. $\frac{11}{8} = \frac{E + 3}{6}$ _____

36. $\frac{M - 5}{12} = \frac{15}{9}$ _____

37. $\frac{8.5}{2} = \frac{16.15}{7.6F}$ _____

38. $\frac{5}{x + 9.2} = \frac{10}{13}$ _____

39. $\frac{D + 8}{-4} = \frac{7}{D - 8}$ _____

40. $\frac{6}{3H^2 - 12} = \frac{5}{12.5}$ _____

Related Proportion Problems

41. The proportion $\frac{A}{B} = \frac{C}{D}$ compares the sides of the two illustrated similar triangles. Determine the missing values in the table.

	A	B	C	D
a.	18"	4.5"		4"
b.	6 1/2"	1 5/8"	4 1/2"	
c.	87.5 mm		75 mm	62.5 mm
d.		25.8 mm	20.6 mm	16.4 mm

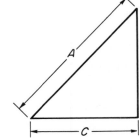

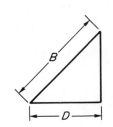

42. Where machine parts are doweled in position, it is good practice to extend the pin 1 to 1 1/2 times its diameter into the mating part. Use the following proportion to determine the value of each unknown in the table.

$$\frac{N}{1} = \frac{L}{D}$$

where

N = the number of times the pin extension is greater than the pin diameter
L = the length of the pin extension
D = the pin diameter

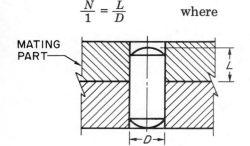

	N	D	L
a.	1.250	7.94 mm	
b.	1 1/8	1/2″	
c.	1 1/2		3/4″
d.	1.375		8.73 mm
e.	1.250	15.88 mm	

	N	D	L
f.	1.375		1.032″
g.	1.250	0.875″	
h.	1.500	3.12 mm	
i.	1.125		0.281″
j.	1.000	7.50 mm	

43. It is sometimes impractical to make engineering drawings full size. If the part to be drawn is very large or small, a scale drawing is generally made. The scale which is shown on the drawing compares the lengths of the lines on the drawing to the dimensions on the part. A scale on a drawing which states 1/4″ = 1″ means the drawing is one-quarter the size of the part. It is expressed as a ratio of 1:4 or 1/4. A scale drawing which states 2″ = 1″ means that the drawing is double the size of the part. It is expressed as a ratio of 2:1 or 2/1. The actual dimensions of a steel support are given in the figure. All dimensions are in inches. Using the formula given, compute the lengths on a drawing for each unknown in the table.

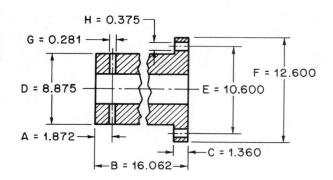

$$\frac{\text{numerator of scale ratio}}{\text{denominator of scale ratio}} = \frac{\text{drawing length}}{\text{part dimension}}$$

	Scale	Drawing Length
a.	1/4″ = 1″	B =
b.	4″ = 1″	G =
c.	1/2″ = 1″	B =
d.	2″ = 1″	C =
e.	1 1/2″ = 1″	A =
f.	3/4″ = 1″	E =
g.	3″ = 1″	H =
h.	1/8″ = 1″	F =

	Scale	Drawing Length
i.	1/2″ = 1″	E =
j.	6″ = 1″	G =
k.	3/4″ = 1″	F =
l.	1 1/2″ = 1″	C =
m.	1/4″ = 1″	F =
n.	3″ = 1″	G =
o.	1/2″ = 1″	B =
p.	2″ = 1″	A =

44. This figure shows the relationship of gears in a lathe using a simple gear train. The proportion given is used for lathe thread cutting computations using simple gearing. The fixed stud gear and the spindle gear have the same number of teeth. Determine the missing values for each of the following problems.

$$\frac{N_L}{N_C} = \frac{T_S}{T_L}$$ where N_L = number of threads per inch on the lead screw
N_C = number of threads per inch to be cut
T_S = number of teeth on stud gear
T_L = number of teeth on lead screw gear

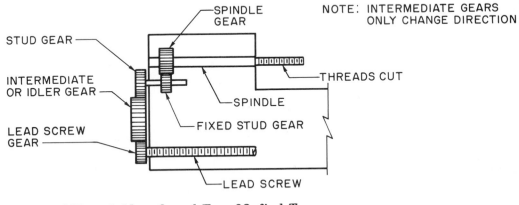

a. If $N_L = 4$, $N_C = 8$, and $T_S = 32$, find T_L.   _____
b. If $N_L = 7$, $T_S = 35$, and $N_C = 15$, find T_L.   _____
c. If $N_C = 10$, $N_L = 6$, and $T_L = 40$, find T_S.   _____
d. If $N_L = 8$, $T_L = 42$, and $T_S = 28$, find N_C.   _____

UNIT 35 *DIRECT AND INVERSE PROPORTIONS*

OBJECTIVES

After studying this unit you should be able to

- Analyze problems to determine whether quantities are directly or inversely proportional.

- Set up and solve direct and inverse proportions.

Many shop problems are solved by the use of proportions. A machinist may be required to express word statements or other given data as proportions. Generally, three of the four terms of a proportion must be known in order to solve the proportion. When setting up a proportion it is important that the terms be placed in their proper positions.

A problem which is to be set up and solved as a proportion must first be analyzed in order to determine where the terms are to be placed. Depending upon the positions of the terms, proportions are either direct or inverse.

DIRECT PROPORTIONS

Two quantities are *directly proportional* if a change in one produces a change in the other in the same direction. If an increase in one produces an increase in the other, or if a decrease in one produces a decrease in the other, the two quantities are directly proportional. The proportions discussed will be those that change at the same rate. An increase or decrease in one quantity produces the same rate of increase or decrease in the other quantity.

When setting up a direct proportion in fractional form, the numerator of the first ratio must correspond to the numerator of the second ratio. The denominator of the first ratio must correspond to the denominator of the second ratio.

Example 1: If 120 parts are produced in 2 hours, how many parts are produced in 3 hours?

Analyze the problem. An increase in time (from 2 hours to 3 hours) will produce an increase in the number of pieces produced. Production increases as time increases. The proportion is direct.

Set up the direct proportion. Let x represent the number of parts that are produced in 3 hours. The numerator of the first ratio must correspond to the numerator of the second ratio; 2 hours corresponds to 120 parts. The denominator of the first ratio must correspond to the denominator of the second ratio; 3 hours corresponds to x.

Solve for x.

$$\frac{2\ \text{hours}}{3\ \text{hours}} = \frac{120\ \text{parts}}{x}$$
$$2x = 3(120\ \text{parts})$$
$$2x = 360\ \text{parts}$$
$$x = 180\ \text{parts}\quad\text{Ans}$$

Check.

$$\frac{2\ \text{hours}}{3\ \text{hours}} = \frac{120\ \text{parts}}{x}$$
$$\frac{2\ \text{hours}}{3\ \text{hours}} = \frac{120\ \text{parts}}{180\ \text{parts}}$$
$$\frac{2}{3} = \frac{2}{3}\quad\text{Ck}$$

Example 2: A tapered shaft is one that varies uniformly in diameter along its length. The shaft shown is 15 inches long with a 1.200-inch diameter on the large end. A 9-inch piece is cut from the shaft. Determine the diameter at the large end of the 9-inch piece.

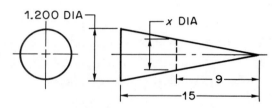

Analyze the problem. As the length decreases from 15 inches to 9 inches, the diameter also decreases at the same rate. The proportion is direct.

Set up the proportion. Let x represent the diameter at the large end of the 9-inch piece. The numerator of the first ratio must correspond to the numerator of the second ratio; the 15-inch piece has a 1.200-inch diameter at the large end. The denominator of the first ratio must correspond to the denominator of the second ratio; the 9-inch piece has a diameter of x at the large end.

Solve for x.

$$\frac{\overset{5}{\cancel{15\ \text{inches}}}}{\underset{3}{\cancel{9\ \text{inches}}}} = \frac{1.200\text{-inch DIA}}{x\ \text{DIA}}$$
$$5x = 3(1.200\ \text{inches})$$
$$5x = 3.600\ \text{inches}$$
$$x = 0.720\ \text{inch}\quad\text{Ans}$$

Check.

$$\frac{15\ \text{inches}}{9\ \text{inches}} = \frac{1.200\text{-inch DIA}}{x\ \text{DIA}}$$
$$\frac{15\ \text{inches}}{9\ \text{inches}} = \frac{1.200\text{-inch DIA}}{0.720\text{-inch DIA}}$$
$$1.67 = 1.67\quad\text{Ck}$$

INVERSE PROPORTIONS

Two quantities are *inversely or indirectly proportional* if a change in one produces a change in the other in the opposite direction. If an increase in one produces a decrease in the other, or if a decrease in one produces an increase in the other, the two quantities are inversely proportional. For example, if one quantity increases by 4 times its original value, the other quantity decreases by 4 times its value or is 1/4 of its original value. Notice 4 or 4/1 inverted is 1/4.

When setting up an inverse proportion in fractional form, the numerator of the first ratio must correspond to the denominator of the second ratio. The denominator of the first ratio must correspond to the numerator of the second ratio.

Example 1: Two gears in mesh are shown. The driver gear has 40 teeth and revolves at 360 revolutions per minute. Determine the number of revolutions per minute of a driven gear with 16 teeth.

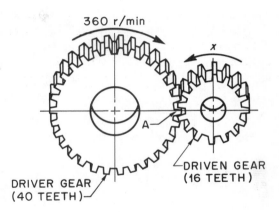

Analyze the problem. When the driver turns one revolution, 40 teeth pass point A. The same number of teeth on the driven gear must pass point A. Therefore, the driven gear turns more than one revolution for each revolution of the driver gear. The gear with 16 teeth (driven gear) revolves at greater revolutions per minute than the gear with 40 teeth (driver gear). A decrease in the number of teeth produces an increase in revolutions per minute. The proportion is inverse.

Set up the proportion. Let *x* represent the revolutions per minute of the gear with 16 teeth. The numerator of the first ratio must correspond to the denominator of the second ratio; the gear with 40 teeth revolves at 360 r/min. The denominator of the first ratio must correspond to the numerator of the second ratio; the gear with 16 teeth revolves at *x*.

Solve for *x*.

$$\frac{\overset{5}{\cancel{40 \text{ teeth}}}}{\underset{2}{\cancel{16 \text{ teeth}}}} = \frac{x}{360 \text{ r/min}}$$

$$2x = 1800 \text{ r/min}$$

$$x = 900 \text{ r/min} \quad \text{Ans}$$

Check.

$$\frac{40 \text{ teeth}}{16 \text{ teeth}} = \frac{x}{360 \text{ r/min}}$$

$$\frac{5}{2} = \frac{900 \text{ r/min}}{360 \text{ r/min}}$$

$$\frac{5}{2} = \frac{5}{2} \quad \text{Ck}$$

Example 2: A balanced lever is shown. A 40-pound weight is placed 6 feet from the fulcrum. Determine the weight required 15 feet from the fulcrum in order to balance the lever.

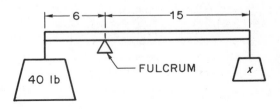

Analyze the problem. The 40-pound weight is closer to the fulcrum than the unknown weight. An increase in the distance from the fulcrum produces a decrease in weight required to balance the lever. The proportion is inverse.

Set up the proportion. Let *x* represent the weight 15 feet from the fulcrum. The numerator of the first ratio must correspond to the denominator of the second ratio; the 40-pound weight is 6 feet from the fulcrum. The denominator of the first ratio must correspond to the numerator of the second ratio; the unknown weight is 15 feet from the fulcrum.

Solve for *x*.

$$\frac{\overset{2}{\cancel{6 \text{ feet}}}}{\underset{5}{\cancel{15 \text{ feet}}}} = \frac{x}{40 \text{ pounds}}$$

$$5x = 80 \text{ pounds}$$

$$x = 16 \text{ pounds} \quad \text{Ans}$$

Check.

$$\frac{6 \text{ feet}}{15 \text{ feet}} = \frac{x}{40 \text{ pounds}}$$

$$\frac{2}{5} = \frac{16 \text{ pounds}}{40 \text{ pounds}}$$

$$\frac{2}{5} = \frac{2}{5} \quad \text{Ck}$$

──APPLICATION────────────────────────────────

Tapers

Taper is the difference between the diameters at each end of a part. Tapers are expressed as the difference in diameters for a particular length along the centerline of a part.

Note: All dimensions are in millimeters.   Note: All dimensions are in inches.

 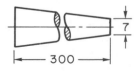   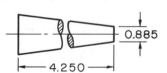

20 mm – 7 mm = 13 mm taper per 300 mm   1.187″ – 0.885″ = 0.302″ taper per 4.250″

1. A plug gage tapers 3.10 mm along a 38 mm length. Set up a proportion and determine the amount of taper in the workpiece for each of the following problems. Express the answers to 2 decimal places.

	Workpiece Thickness	Proportion	Taper in Workpiece
a.	20.30 mm		
b.	31.75 mm		
c.	14.28 mm		
d.	28.58 mm		
e.	23.80 mm		

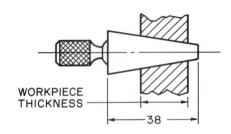

2. A reamer tapers 0.130″ along a 4.250″ length. Set up a proportion and determine length A for each of the following problems. Express the answers to 3 decimal places.

	Taper in Length A	Proportion	Length A
a.	0.030″		
b.	0.108″		
c.	0.075″		
d.	0.008″		
e.	0.093″		

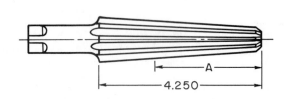

3. A micrometer reading is made at dimension D on a tapered shaft. For each of the problems use the dimensions given in the table, compute the taper, set up a proportion, and determine diameter C to 3 decimal places.

	Length of Shaft	Diameter A	Diameter B	Dimension D	Diameter C
a.	10.000″	1.500″	0.700″	6.500″	
b.	8.750″	1.250″	0.375″	4.875″	
c.	550.000 mm	106.250 mm	62.500 mm	337.500 mm	
d.	147.500 mm	22.500 mm	10.000 mm	112.500 mm	
e.	9.200″	1.325″	0.410″	8.620″	

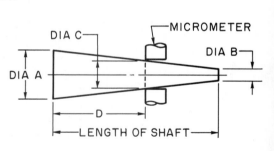

Proportions

Analyze each of the following problems to determine whether the problem is a direct or inverse proportion. Set up the proportion and solve.

4. A sheet of steel 8 1/4 feet long weighs 350 pounds. A piece 2 1/2 feet long is sheared from the sheet. Determine the weight of the 2 1/2-foot piece to 2 decimal places.   _____

5. If 1350 parts are produced in 6.75 hours, find the number of parts produced in 8.25 hours.   _____

6. The production rate for each of 3 machines is the same. Using these 3 machines, 720 parts are produced in 1.6 hours. How many hours will it take 2 of these machines to produce 720 parts?   _____

7. Two forgings are made of the same stainless steel alloy. A forging which weighs 76 kilograms contains 0.38 kilogram of chromium. How many kilograms of chromium does the second forging contain if it weighs 98 kilograms? Express the answer to 2 decimal places.   _____

Gears and Pulleys

8. A belt connects a 10-inch diameter pulley which rotates at 160 rpm with a 6.5-inch diameter pulley. An 8-inch diameter pulley is fixed to the same shaft as the 6.5-inch pulley. A belt connects the 8-inch pulley with a 3.5-inch diameter pulley. Determine the revolutions per minute of the 3.5-inch diameter pulley. Express the answer to 1 decimal place.   _____

9. Of two gears that mesh, the one which has the greater number of teeth is called the gear, and the one which has the fewer teeth is called the pinion. For each of the problems, set up a proportion, and determine the unknown value, x.

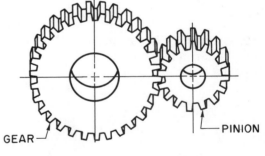

	Number of Teeth on Gear	Number of Teeth on Pinion	Speed of Gear (rpm)	Speed of Pinion (rpm)
a.	48	20	120	$x =$
b.	32	24	$x =$	210
c.	35	$x =$	160	200
d.	$x =$	15	150	250
e.	54	28	80	$x =$

10. The figure shows a compound gear train. Gears B and C are keyed to the same shaft; therefore, they turn at the same speed. Gear A and gear C are driving gears. Gear B and gear D are driven gears. Set up a proportion for each problem and determine the unknown values, x, y, and z.

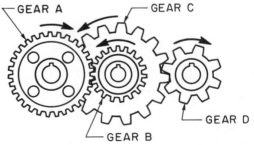

	Number of Teeth				Speed (rpm)			
	Gear A	Gear B	Gear C	Gear D	Gear A	Gear B	Gear C	Gear D
a.	80	30	50	20	120	$x =$	$y =$	$z =$
b.	60	$x =$	45	$y =$	100	300	$z =$	450
c.	$x =$	24	60	36	144	$y =$	$z =$	280
d.	55	25	$x =$	15	$y =$	$z =$	175	350

UNIT 36 APPLICATIONS OF FORMULAS TO CUTTING SPEED, REVOLUTIONS PER MINUTE, AND CUTTING TIME

OBJECTIVES

After studying this unit you should be able to

- Solve cutting speed, revolutions per minute, and cutting time problems by substitution in given formulas.
- Solve production time and cutting feed problems by rearranging and combining formulas.

In order to perform cutting operations efficiently, a machine must be run at the proper cutting speed. Proper cutting speed is largely determined by the type of material that is being cut, the feed and depth of cut, the cutting tool, and the machine characteristics. The machinist must be able to determine proper cutting speeds by using trade handbook data and formulas.

CUTTING SPEED USING ENGLISH UNITS OF MEASURE

Cutting speeds or surface speeds for lathes, drills, milling cutters, and grinding wheels are computed using the same formula. On the lathe, the workpiece revolves. On drill presses, milling machines, and grinders the tool revolves. Speeds are computed in reference to the tool rather than the workpiece. The speed of a revolving object equals the product of the circumference times the number of revolutions per minute made by the object. Generally, diameters are expressed in inches. In order to express inches per minute as feet per minute, it is necessary to divide by 12.

$$C = \frac{3.1416DN}{12} \qquad \text{where} \qquad \begin{aligned} C &= \text{cutting speed in feet per minute (fpm)} \\ D &= \text{diameter in inches} \\ N &= \text{revolutions per minute (rpm)} \end{aligned}$$

Lathe

The *cutting speed* of a lathe is the number of feet that the revolving workpiece travels past the cutting edge of the tool in one minute.

Example: A steel shaft 2.5 inches in diameter is turned in a lathe at 184 rpm. Determine the cutting speed to 1 decimal place.

$$C = \frac{3.1416DN}{12} = \frac{3.1416(2.5)(184)}{12} = 120.4 \text{ fpm} \quad \text{Ans}$$

Milling Machine, Drill Press, and Grinder

The *cutting speed* or surface speed of a drill press, milling machine, and grinder is the number of feet that a point on the circumference of the tool travels in 1 minute.

Example 1: A 10-inch diameter grinding wheel runs at 1910 rpm. Determine the surface speed to the nearer whole number.

$$C = \frac{3.1416DN}{12} = \frac{3.1416(10)(1910)}{12} = 5000 \text{ fpm} \quad \text{Ans}$$

Example 2: Determine the cutting speed to the nearer whole number of a 3 1/2-inch diameter milling cutter revolving at 120 rpm.

$$C = \frac{3.1416DN}{12} = \frac{3.1416(3.5)(120)}{12} = 110 \text{ fpm} \quad \text{Ans}$$

172

REVOLUTIONS PER MINUTE USING ENGLISH UNITS OF MEASURE

The cutting speed formula is rearranged in terms of N in order to determine the revolutions per minute of a workpiece or tool.

$$C = \frac{3.1416DN}{12}$$
$$12C = 3.1416DN$$
$$\frac{12C}{3.1416D} = N \quad \text{or} \quad N = \frac{12C}{3.1416D}$$

Lathe

Example: An aluminum cylinder with a 6″ outside diameter is turned in a lathe at a cutting speed of 225 feet per minute. Determine the revolutions per minute to 1 decimal place.

$$N = \frac{12C}{3.1416D} = \frac{12(225)}{3.1416(6)} = 143.2 \text{ rpm} \quad \text{Ans}$$

Milling Machine, Drill Press, and Grinder

Example 1: A 1/2-inch diameter twist drill has a cutting speed of 60 feet per minute. Determine the revolutions per minute to 1 decimal place.

$$N = \frac{12C}{3.1416D} = \frac{12(60)}{3.1416(0.5)} = 458.4 \text{ rpm} \quad \text{Ans}$$

Example 2: A 6-inch diameter grinding wheel operates at a cutting speed of 6000 feet per minute. Determine the revolutions per minute to 1 decimal place.

$$N = \frac{12C}{3.1416D} = \frac{12(6000)}{3.1416(6)} = 3819.7 \text{ rpm} \quad \text{Ans}$$

CUTTING TIME USING ENGLISH UNITS OF MEASURE

The same formula is used to compute cutting times for machines which have a revolving workpiece, such as the lathe, as is used for machines which have a revolving tool, such as the milling machine, drill press and grinder. Cutting time is determined by the length or depth to be cut in inches, the revolutions per minute of the revolving workpiece or revolving tool, and the tool feed in inches for each revolution of the workpiece or tool.

$$T = \frac{L}{FN} \quad \text{where} \quad \begin{aligned} T &= \text{cutting time per cut in minutes} \\ L &= \text{length of cut in inches} \\ F &= \text{tool feed in inches per revolution} \\ N &= \text{speed of revolving workpiece or tool in revolutions} \\ &\quad \text{per minute} \end{aligned}$$

Lathe

Example 1: How many minutes are required to take one cut 22 inches in length on a steel shaft when the lathe feed is 0.050 inch per revolution and the shaft turns 152 rpm?

$$T = \frac{L}{FN} = \frac{22}{0.050(152)} = 2.9 \text{ min} \quad \text{Ans}$$

Example 2: A 3.250-inch diameter cast iron sleeve which is 20 inches long is turned in a lathe to a 2.450-inch diameter. Roughing cuts are each made to a 0.125-inch depth of cut. One finish cut using a 0.025-inch depth of cut is made. The feed is 0.100 inch per revolution for roughing and 0.030 inch for finishing. Roughing cuts are made at 150 rpm and the finish cut at 200 rpm. What is the total cutting time required?

Compute the total depth of cut.	$\dfrac{3.250'' - 2.450''}{2} = \dfrac{0.800''}{2} = 0.400''$
Compute the number of roughing cuts required.	$\dfrac{0.400'' - 0.025''}{0.125''} = 3$
Compute the time required for one roughing cut.	$T = \dfrac{20}{(0.100)(150)} = 1.33$ min
Compute the total time for roughing.	3×1.33 min = 4.0 min
Compute the time required for finishing.	$T = \dfrac{20}{(0.030)(200)} = 3.3$ min
Compute the total cutting time.	4.0 min + 3.3 min = 7.3 min Ans

Milling Machine, Drill Press, and Grinder

Example 1: Determine the cutting time required to drill through a workpiece which is 3.600 inches thick with a drill revolving 300 rpm and a feed of 0.025 inch per revolution.

$$T = \frac{L}{FN} = \frac{3.600}{0.025(300)} = 0.48 \text{ min}\quad \text{Ans}$$

Example 2: A milling machine cutter makes 460 rpm with a table feed of 0.020 inch per revolution. Four cuts are required to mill a slot in an aluminum plate 28 inches long. Compute the total cutting time.

Compute cutting time per cut.	$T = \dfrac{L}{FN} = \dfrac{28}{0.020(460)} = 3.04$ min
Compute total cutting time.	4(3.04 min) = 12.16 min Ans

CUTTING SPEED USING METRIC UNITS OF MEASURE

Diameters are expressed in millimeters. Cutting speeds are expressed in meters per minute. The symbol for meters per minute is m/min. In order to express speed in millimeters per minute as meters per minute, it is necessary to divide by 1000 or to move the decimal point 3 places to the left.

$$C = \frac{3.1416DN}{1000} \quad \text{where} \quad \begin{aligned} C &= \text{cutting speed in meters per minute} \\ D &= \text{diameter in millimeters} \\ N &= \text{revolutions per minute} \end{aligned}$$

Example: A medium-steel shaft is cut in a lathe using a high-speed tool. The shaft has a diameter of 55 millimeters and is turning at 260 revolutions per minute. Determine the cutting speed to the nearer whole number.

$$C = \frac{3.1416DN}{1000} = \frac{3.1416(55)(260)}{1000} = 45 \text{ m/min}\quad \text{Ans}$$

REVOLUTIONS PER MINUTE USING METRIC UNITS OF MEASURE

In the metric system, the symbol for revolutions per minute is r/min. The cutting speed formula is rearranged in terms of N in order to determine the revolutions per minute of a workpiece or tool.

$$C = \frac{3.1416DN}{1000}$$

$$1000C = 3.1416DN$$

$$\frac{1000C}{3.1416D} = N \text{ or } N = \frac{1000C}{3.1416D}$$

Example: A high-speed steel milling cutter with a 45 millimeter diameter and a cutting speed of 12 meters per minute is used for a roughing operation on an annealed chromium-nickel steel workpiece. Determine the revolutions per minute to the nearer whole number.

$$N = \frac{1000C}{3.1416D} = \frac{1000(12)}{3.1416(45)} = 85 \text{ r/min} \quad \text{Ans}$$

CUTTING TIME USING METRIC UNITS OF MEASURE

Cutting time is determined by the length or depth to be cut in millimeters, the revolutions per minute of the revolving workpiece or revolving tool, and the tool feed in millimeters for each revolution of the workpiece or tool.

$$T = \frac{L}{FN} \quad \text{where} \quad \begin{aligned} T &= \text{cutting time per cut in minutes} \\ L &= \text{length of cut in millimeters} \\ F &= \text{tool feed in millimeters per revolution} \\ N &= \text{r/min of revolving workpiece or tool} \end{aligned}$$

Example: An 88-millimeter diameter cast iron cylinder is turned in a lathe at 260 revolutions per minute. Each length of cut is 700 millimeters and 5 cuts are required. A carbide tool is fed into the workpiece at 0.40 millimeter per revolution. What is the total cutting time?

Calculate the time required for one cut. $\quad T = \frac{L}{FN} = \frac{700}{0.40(260)} = 6.73 \text{ min}$

Calculate the total cutting time. $\quad\quad\quad 5(6.73 \text{ min}) = 33.65 \text{ min} \quad \text{Ans}$

USING DATA FROM A CUTTING SPEED TABLE

Tables of cutting speeds have been developed which are used in determining machine spindle speed (revolutions per minute) settings. The tables take into consideration the material to be cut and the tool material.

In addition to the material being cut and the type of tool used, other factors must be taken into consideration. Variables are considered, such as the depth and width of cut, the design of the cutting tool, the rate of feed, the coolant used, and the finish required.

Because cutting speed depends upon many factors, data given in cutting speed tables should be considered as recommended values. Generally it is not possible to set machines to an exact calculated spindle speed. Therefore, a simplified spindle speed formula is used in computing revolutions per minute. In the simplified formula, 3.1416 is rounded to 3.

$$N = \frac{12C}{3D} = \frac{4C}{D}$$

Comprehensive detailed cutting speed tables are available which list cutting speeds for specific materials based on material alloy composition, hardness, and condition. Some tables list cutting speeds separately for rough and finish cuts.

Selected materials are listed in the following table with their respective cutting speeds using high-speed steel and carbide tools.

| | Cutting Speeds: Feet Per Minute (fpm) | | | | | | |
| | Turning | | Milling | | Drilling | Reaming | |
Material	High-Speed Steel Tool	Carbide Tool	High-Speed Steel Tool	Carbide Tool	High-Speed Steel Tool	High-Speed Steel Tool	Carbide Tool
Carbon Steel (1020), BHN 175–225	100	350	70–130	200–400	70	40	175
Alloy Steel (4320), BHN 220–275	70	300	50–100	225–450	60	40	150
Malleable Cast Iron (32510), BHN 110–160	200	600	130–225	400–800	130	90	240
Stainless Steel (305), BHN 225–275	60	200	50–80	175–275	40	25	100
Aluminum (5052)	600	1200	500–800	1000–1800	250	250	700
Brass, annealed	300	650	250–450	500–900	160	160	320
Manganese Bronze, cold drawn	250	550	200–350	450–650	140	120	275
Beryllium Copper, annealed	100	200	80–140	180–275	60	50	180

Revolutions per minute are generally computed using table cutting speeds with the simplified revolutions per minute formula. Where a range of cutting speed table values is listed, use the average of the low and high speeds given. For example, the cutting speed for milling the alloy steel shown in the table with a high-speed steel cutter is listed as 50–100 feet per minute. Use the average cutting speed of 75 feet per minute $\left(\frac{50 + 100}{2} = 75\right)$.

After revolutions per minute are calculated, generally, the machine spindle speed is set to the closest spindle speed below the calculated revolutions per minute. The spindle speed may then be increased or decreased depending on the performance of the operation.

Example 1: Calculate the revolutions per minute required to turn a 3.5-inch diameter piece of stainless steel using a carbide toolbit. Express the answer to the nearer revolution per minute.

Refer to the table of cutting speeds. The recommended cutting speed is 200 feet per minute. $N = \frac{4C}{D} = \frac{4(200)}{3.5} = 229$ rpm   Ans

Example 2: A carbon steel plate is milled using a 2.75-inch diameter high-speed steel cutter. Compute, to the nearer whole number, the revolutions per minute.

Refer to the table of cutting speeds. The recommended cutting speed is 100 feet per minute $\left(\frac{70 + 130}{2} = 100\right)$. $N = \frac{4C}{D} = \frac{4(100)}{2.75} = 145$ rpm   Ans

APPLICATION

Cutting Speeds

Given the workpiece or tool diameters and the revolutions per minute, determine the cutting speeds in the following tables to the nearer whole number. Use $C = \frac{3.1416DN}{12}$ for   English units and $C = \frac{3.1416DN}{1000}$ for metric units.

	Workpiece or Tool Diameter	Revolutions per Minute	Cutting Speed (fpm)		Workpiece or Tool Diameter	Revolutions per Minute	Cutting Speeds (m/min)
1.	0.500″	460		6.	196.00 mm	59	
2.	2.750″	50		7.	53.98 mm	764	
3.	4.000″	86		8.	3.20 mm	1525	
4.	0.875″	175		9.	133.35 mm	254	
5.	1.750″	218		10.	6.35 mm	4584	

Revolutions Per Minute

Given the cutting speed and the tool or workpiece diameter, determine the revolutions per minute in the following tables to the nearer whole number. Use $N = \frac{12C}{3.1416D}$ for English units and $N = \frac{1000C}{3.1416D}$ for metric units.

	Cutting Speed (fpm)	Workpiece or Tool Diameter	Revolutions per Minute			Cutting Speed (m/min)	Workpiece or Tool Diameter	Revolutions per Minute
11.	70	2.375"			16.	130	25.50 mm	
12.	120	0.750"			17.	100	66.70 mm	
13.	90	8.000"			18.	25	6.35 mm	
14.	180	8.000"			19.	180	15.80 mm	
15.	225	0.375"			20.	150	114.30 mm	

Cutting Time

Given the number of cuts, the length of cut, the revolutions per minute of the workpiece or tool, and the tool feed, determine the total cutting time in the table to 1 decimal place. Use $T = \frac{L}{FN}$.

	Number of Cuts	Feed (per revolution)	Length of Cut	Revolutions per Minute	Total Cutting Time (Minutes)
21.	1	0.002"	20"	2100	
22.	1	0.12 mm	925 mm	610	
23.	4	0.008"	8"	335	

Cutting Speed and Surface Speed Problems

Compute the following problems. Express the answers to the nearer whole number. Use $C = \frac{3.1416DN}{12}$ for English units and $C = \frac{3.1416DN}{1000}$ for metric units.

24. A 3 1/2-inch diameter high-speed steel cutter, running at 54.5 rpm, is used to rough mill a steel casting. What is the cutting speed? _____

25. A 50-millimeter diameter carbon steel drill running at 286 r/min is used to drill an aluminum plate. Find the cutting speed. _____

26. What is the surface speed of a 16-inch diameter surface grinder wheel running at 1194 rpm? _____

27. A medium-steel shaft is cut in a lathe using a high-speed steel tool. Determine the cutting speed if the shaft is 2.125 inches in diameter, and is turning at 262 rpm. _____

28. A finishing cut is taken on a brass workpiece using a 100-millimeter diameter carbon steel milling cutter. What is the cutting speed when the cutter is run at 86 r/min? _____

Revolutions Per Minute Problems

Compute the following problems. Express the answers to the nearer whole number. Use $N = \frac{12C}{3.1416D}$ for English units and $N = \frac{1000C}{3.1416D}$ for metric units.

29. Grooves are cut in a stainless steel plate using a 3.750-inch diameter carbide milling cutter with a cutting speed of 180 feet per minute. Determine the revolutions per minute. _____

30. An annealed cast iron housing is drilled with a cutting speed of 20 meters per minute using a 20-millimeter diameter carbon steel drill. Find the revolutions per minute. _____

31. A grinding operation is performed using a 150-millimeter diameter wheel with a cutting speed of 1800 meters per minute. Determine the revolutions per minute.   _____

32. Determine the revolutions per minute of an aluminum alloy rod 1 inch in diameter with a cutting speed of 550 feet per minute.   _____

33. A high-speed steel milling cutter with a 1.750-inch diameter and a cutting speed of 40 feet per minute is used for a roughing operation on an annealed chromium-nickel steel workpiece. Find the revolutions per minute.   _____

Cutting Time Problems

Compute the following problems. Express the answers to 1 decimal place. Use $T = \frac{L}{FN}$.

34. Cast iron, 3 1/2 inches in diameter, is turned in a lathe at 270 rpm. Each length of cut is 27 inches and five cuts are required. A carbide tool is fed into the work at 0.015 inch per revolution. What is the total cutting time?   _____

35. A slot 812.00 millimeters long is cut into a carbon steel baseplate with a feed of 0.80 millimeter per revolution. Find the cutting time using a 75-millimeter diameter carbide milling cutter running at 640 r/min.   _____

36. Fifteen 3.20 millimeter diameter holes each 57.15 millimeters deep are drilled in an aluminum workpiece. The high-speed steel drill runs at 9200 r/min with a feed of 0.05 millimeter per revolution. Determine the total cutting time.   _____

37. Thirty 2-inch diameter stainless steel shafts are turned in a lathe at 240 rpm. Two cuts each 14.5 inches long are required using a feed of 0.020 inch per revolution. Setup and handling time averages 3 minutes per piece. Calculate the total production time.   _____

38. Seven brass plates 9 inches wide and 21 inches long are machined with a milling cutter along the length of the plates. The entire top face of each plate is milled. The width of each cut allowing for overlap is 2 1/4 inches. Using a feed of 0.020 inch per revolution and 525 rpm, determine the total cutting time.   _____

Complex Problems

The solution of the following problems requires more than one formula and the rearrangement of formulas.

Use $C = \frac{3.1416DN}{12}$ for English units and $C = \frac{3.1416DN}{1000}$ for metric units.

$N = \frac{12C}{3.1416D}$ for English units and $N = \frac{1000C}{3.1416D}$ for metric units.

$T = \frac{L}{FN}$.

39. A 3-inch diameter cylinder is turned for an 11.5-inch length of cut. The cutting speed is 300 feet per minute and the cutting time is 1.02 minutes. Calculate the tool feed in inches per revolution. Round the answer to 3 decimal places.   _____

40. A combination drilling and countersinking operation on bronze round stock is performed on an automatic screw machine. The length of cut per piece is 1 3/4 inches. The total cutting time for 2300 pieces is 6 1/2 hours running at 1600 rpm. What is the tool feed in inches per revolution? Round the answer to 3 decimal places.   _____

41. Steel shafts, 1 1/4 inches in diameter, are turned on an automatic machine. One finishing operation is required for a 16.5-inch length of cut. The tool

feed is 0.015 inch per revolution using a cutting speed of 200 feet per minute. Determine the number of hours of cutting time required for 1500 shafts. Round the answer to 2 decimal places. _____

42. A carbide milling cutter is used for machining a 560-millimeter length of stainless steel. The cutting time is 11.95 minutes, the cutting speed is 60 meters per minute, and the feed is 0.25 millimeter per revolution. What is the diameter of the carbide milling cutter? Round the answer to 2 decimal places. _____

43. Aluminum baseplates are produced that are 1 1/2 inches thick. Six 1/4-inch diameter holes are drilled in each plate using a feed of 0.004 inch per revolution and a cutting speed of 300 feet per minute. Setup and handling time is estimated at 0.5 minute per piece. What is the total number of hours required to produce 850 aluminum baseplates? Round the answer to 2 decimal places. _____

Cutting Speed Table

Refer to cutting speed table on page 176. Use the table values and the simplified revolutions per minute formula, $N = \frac{4C}{D}$. Compute the revolutions per minute to the nearer revolution for each problem in the following table.

	Material Machined	Cutting Operation	Tool Material	Tool or Workpiece Diameter (inches)	Speed (rpm)
44.	Aluminum (5052)	Milling	High-Speed Steel	3.500	
45.	Stainless Steel (305), BHN 225–275	Turning	Carbide	5.200	
46.	Alloy Steel (4320), BHN 220–275	Reaming	High-Speed Steel	0.500	
47.	Manganese Bronze, cold drawn	Drilling	High-Speed Steel	0.375	
48.	Brass, annealed	Milling	High-Speed Steel	4.000	
49.	Carbon Steel (1020), BHN 175–225	Turning	Carbide	6.100	
50.	Beryllium Copper, annealed	Drilling	High-Speed Steel	1.000	
51.	Malleable Cast Iron (32510), BHN 110–160	Milling	Carbide	3.000	
52.	Alloy Steel (4320), BHN 220–275	Milling	Carbide	2.500	
53.	Aluminum (5052)	Turning	High-Speed Steel	5.800	
54.	Carbon Steel (1020), BHN 175–225	Milling	Carbide	4.500	
55.	Brass, annealed	Turning	High-Speed Steel	2.750	
56.	Stainless Steel (305), BHN 225–275	Reaming	Carbide	0.625	
57.	Malleable Cast Iron (32510), BHN 110–160	Turning	Carbide	7.000	
58.	Carbon Steel (1020), BHN 175–225	Drilling	High-Speed Steel	0.375	

UNIT 37 APPLICATIONS OF FORMULAS TO SPUR GEARS

OBJECTIVES

After studying this unit you should be able to

- Identify the proper gear formula to use depending on the unknown and the given data.
- Compute gear part dimensions by substituting known values directly into formulas.
- Compute gear part dimensions by rearranging given formulas in terms of the unknowns.
- Compute gear part dimensions by the application of two or more formulas in order to determine an unknown.

Gears have wide application in machine technology. They are basic to the design and operation of machinery. Most machine shops are equipped to cut gears, and some shops specialize in gear design and manufacture. It is essential that the machinist and drafter have an understanding of gear parts and the ability to determine gear dimensions by the use of trade handbook formulas.

DESCRIPTION OF GEARS

Gears are used for transmitting power by rotary motion between shafts. Gears are designed to prevent slippage and to insure positive motion while maintaining a high degree of accuracy of the speed ratios between driving and driven gears. The shape of the gear tooth is of primary importance in providing a smooth transmission of motion. The shape of most gear teeth is an *involute curve*. This curve is formed by the path of a point on a straight line as it rolls along a circle. *Spur gears* are gears that are in mesh between parallel shafts. Of two gears in mesh, the smaller gear is called the *pinion* and the larger gear is called the *gear*.

SPUR GEAR DEFINITIONS

Spur gears and the terms that are applied to these gears are shown. It is essential to study the figures and gear terms before computing gear problems by the use of formulas.

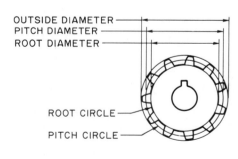

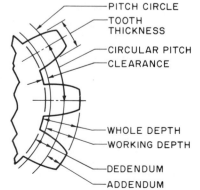

Pitch Circles are the imaginary circles of two meshing gears that make contact with each other. The circles are the basis of gear design and gear calculations.

Pitch Diameter is the diameter of the pitch circle.

Root Circle is a circle which coincides with the bottoms of the tooth spaces.

Root Diameter is the diameter of the root circle.

Outside Diameter is the diameter measured to the tops of the gear teeth.

Addendum is the height of the tooth above the pitch circle.

Dedendum is the depth of the tooth space below the pitch circle.

Whole Depth is the total depth of the tooth space. It is equal to the addendum plus the dedendum.

Working Depth is the total depth of mating teeth when two gears are in mesh. It is equal to twice the addendum.

Clearance is the distance between the top of a tooth and the bottom of the mating tooth space of two gears in mesh. It is equal to the whole depth minus the working depth.

Tooth Thickness (Circular) is the length of the arc, on the pitch circle, between the two sides of a tooth.

Circular Pitch is the length of the arc measured on the pitch circle between the centers of two adjacent teeth. It is equal to the circumference of the pitch circle divided by the number of teeth on the gear.

Diametral Pitch (Pitch) is the ratio of the number of gear teeth to the number of inches of pitch diameter. It is equal to the number of gear teeth for each inch of pitch diameter.

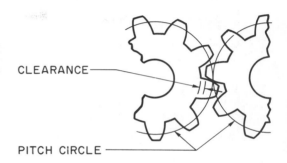

When the pitch of a gear is mentioned, the reference is to diametral pitch, rather than circular pitch. For example, if a gear has 28 teeth and a pitch diameter of 4 inches, it has a pitch (diametral pitch) of 28/4 or 7. It has 7 teeth per inch of pitch diameter, and it is called a 7-pitch gear. It will only mesh with other 7-pitch gears. Gears must have the same pitch in order to mesh.

GEARING — DIAMETRAL PITCH SYSTEM

The diametral pitch system is the system of gear design which is generally applied to decimal-inch dimensional gears. The following table lists the symbols and formulas used in the diametral pitch system.

DECIMAL-INCH SPUR GEARS (AMERICAN NATIONAL STANDARD)

Term	Symbol	Formulas
Pitch (diametral pitch)	P	$P = \dfrac{N}{D}$ $P = \dfrac{3.1416}{P_C}$
Circular Pitch	P_C	$P_C = \dfrac{3.1416D}{N}$ $P_C = \dfrac{3.1416}{P}$
Pitch Diameter	D	$D = \dfrac{N}{P}$ $D = \dfrac{NP_C}{3.1416}$
Outside Diameter	D_O	$D_O = \dfrac{N+2}{P}$ $D_O = \dfrac{P_C(N+2)}{3.1416}$ $D_O = D + 2a$
Root Diameter	D_R	$D_R = D - 2d$
Addendum	a	$a = \dfrac{1}{P}$ $a = 0.3183P_C$
Dedendum	d	$d = \dfrac{1.157}{P}$ $d = 0.3683P_C$
Whole Depth	WD	$WD = \dfrac{2.157}{P}$ $WD = 0.6866P_C$ $WD = a + d$
Working Depth	W_D	$W_D = \dfrac{2.000}{P}$ $W_D = 0.6366P_C$
Clearance	c	$c = \dfrac{0.157}{P}$ $c = 0.050P_C$
Tooth Thickness	T	$T = \dfrac{1.5708}{P}$
Number of Teeth	N	$N = PD$ $N = \dfrac{3.1416D}{P_C}$

─GEAR CALCULATIONS_____

Most gear calculations are made by identifying the proper formula which is given in terms of the unknown, and substituting the known dimensions. It is sometimes necessary to re-arrange a formula in terms of a particular unknown. The solution of a problem may require the substitution of values in two or more formulas.

Examples: Refer to the Decimal-Inch Spur Gears Table on page 181.

1. Determine the pitch diameter of a 5-pitch gear which has 28 teeth.

Identify the formula whose parts consist of pitch diameter, pitch, and number of teeth.

$$D = \frac{N}{P}$$

Solve.

$$D = \frac{28}{5}$$

$$D = 5.6000 \text{ inches} \quad \text{Ans}$$

2. Determine the outside diameter of a gear which has 16 teeth and a circular pitch of 0.7854 inch.

Identify the formula whose parts consist of out-side diameter, number of teeth, and circular pitch.

$$D_O = \frac{P_C(N + 2)}{3.1416}$$

Solve.

$$D_O = \frac{0.7854(16 + 2)}{3.1416}$$

$$D_O = \frac{0.7854(18)}{3.1416}$$

$$D_O = 4.5000 \text{ inches} \quad \text{Ans}$$

3. Determine the circular pitch of a gear with a whole depth dimension of 0.3081 inch.

Identify the formula whose parts consist of circular pitch and whole depth.

$$WD = 0.6866 P_C$$

The formula must be rearranged in terms of circular pitch.

$$P_C = \frac{WD}{0.6866}$$

Solve.

$$P_C = \frac{0.3081}{0.6866}$$

$$P_C = 0.4487 \text{ inch} \quad \text{Ans}$$

4. Determine the addendum of a gear which has an outside diameter of 3.0000 inches and a pitch diameter of 2.7500 inches.

Identify the formula whose parts consist of ad-dendum, outside diameter, and pitch diameter.

$$D_O = D + 2a$$

The formula must be rearranged in terms of the addendum.

$$D_O - D = 2a$$

$$a = \frac{D_O - D}{2}$$

Solve.

$$a = \frac{3.0000 - 2.7500}{2}$$

$$a = 0.1250 \text{ inch} \quad \text{Ans}$$

5. Determine the working depth of a gear which has 46 teeth and a pitch diameter of 11.5000 inches.

There is no single formula in the table that consists of working depth, number of teeth, and pitch diameter. Therefore, it is necessary to substitute in two formulas in order to solve the problem.

Observe $W_D = \frac{2.0000}{P}$. The pitch must be found first.

$$P = \frac{N}{D}$$

$$P = \frac{46}{11.5000}$$

$$P = 4$$

Solve for W_D.

$$W_D = \frac{2.000}{P}$$

$$W_D = \frac{2.000}{4}$$

$$W_D = 0.5000 \text{ inch} \quad \text{Ans}$$

⌐GEARING — METRIC MODULE SYSTEM

The *module system* of gear design is generally the system which is used with metric system units of measure. The *module* of a gear equals the pitch diameter divided by the number of teeth. In the metric system, the module of a gear means the pitch diameter in millimeters is divided by the number of teeth. Module is an actual dimension in millimeters, not a ratio as with diametral pitch. For example, if a gear has 20 teeth and a 50 millimeter pitch diameter, the module is 2.5 millimeters (50 mm ÷ 20). A module of 2.5 millimeters means that there are 2.5 millimeters of pitch diameter per tooth.

A partial list of a standard series of modules (in millimeters) is listed as follows:

1	2	3	4	6	9
1.25	2.25	3.25	4.5	6.5	10
1.5	2.5	3.5	5	7	11
1.75	2.75	3.75	5.5	8	12

The relation between module and various gear parts using a metric module system is shown in the following table.

METRIC SPUR GEARS

Term	Symbol	Formulas
Module	m	$m = \dfrac{D}{N}$
Circular Pitch	P_C	$P_C = \dfrac{m}{0.3183}$
Pitch Diameter	D	$D = mN$
Outside Diameter	D_O	$D_O = m(N + 2)$
Addendum	a	$a = m$
Dedendum	d	$d = 1.157m*$ $d = 1.167m**$
Whole Depth	WD	$WD = 2.157m*$ $WD = 2.167m**$
Working Depth	W_D	$W_D = 2m$
Clearance	c	$c = 0.157m*$ $c = 0.1667m**$
Tooth Thickness	T	$T = 1.5708m$

*When clearance = 0.157 × module
**When clearance = 0.1667 × module

Examples: Refer to the Metric Spur Gears Table.

1. Determine the circular pitch of a 6 millimeter module gear.
    Circular Pitch   =   Module ÷ 0.3183
    Circular Pitch   = 6 mm ÷ 0.3183 = 18.850 mm   Ans

2. Determine the outside diameter of a 3.5 millimeter module gear with 20 teeth.
    Outside Diameter   =   (Number of teeth + 2) × *Module*
    Outside Diameter   = (20 + 2) × 3.5 mm = 22 × 3.5 mm = 77 mm   Ans

3. Compute the dedendum of a 4.5 millimeter module gear designed with a clearance of 0.157 × module.
    When *Clearance = 0.157* × *Module, the Dedendum = 1.157* × *Module.*
    Dedendum   = 1.157 × 4.5 mm = 5.207 mm   Ans

4. Compute the whole depth of 7 millimeter module gear designed with a clearance of 0.1667 × module.
    When *Clearance   =   0.1667* × *Module, the Whole Depth = 2.167* × *Module.*
    Whole Depth   = 2.167 × 7 mm = 15.169 mm   Ans

─APPLICATION─────────────────────────────

Gearing — Diametral Pitch System

Refer to the Decimal-Inch Spur Gears Table on page 181 for each of the following gearing problems.

	Given Values	Find	
1.	Circular Pitch = 1.5708″	Pitch	───────
2.	Pitch = 12	Circular Pitch	───────
3.	Pitch Diameter = 5.2000″ Number of Teeth = 26	Circular Pitch	───────
4.	Pitch Diameter = 12.5714″ Number of Teeth = 44	Pitch	───────
5.	Pitch = 7 Number of Teeth = 25	Pitch Diameter	───────
6.	Circular Pitch = 0.3142″ Number of Teeth = 12	Pitch Diameter	───────
7.	Pitch = 20	Circular Pitch	───────
8.	Pitch Diameter = 0.7273″ Number of Teeth = 16	Pitch	───────
9.	Pitch = 12 Pitch Diameter = 1.1667″	Number of Teeth	───────
10.	Circular Pitch = 0.6283″ Pitch Diameter = 8.4000″	Number of Teeth	───────
11.	Number of Teeth = 56 Pitch = 7	Outside Diameter	───────
12.	Pitch = 14	Addendum	───────
13.	Pitch Diameter = 1.3333″ Dedendum = 0.0643″	Root Diameter	───────
14.	Pitch = 3.5	Whole Depth	───────
15.	Circular Pitch = 0.2856″	Working Depth	───────
16.	Circular Pitch = 1.2566″	Clearance	───────
17.	Pitch = 20	Tooth Thickness	───────
18.	Pitch Diameter = 3.5000″ Addendum = 0.1818″	Outside Diameter	───────
19.	Circular Pitch = 0.0924″	Dedendum	───────
20.	Circular Pitch = 0.8976″	Addendum	───────
21.	Addendum = 0.1429″ Dedendum = 0.1653″	Whole Depth	───────
22.	Pitch = 4	Dedendum	───────
23.	Circular Pitch = 0.2094″	Whole Depth	───────
24.	Pitch = 17	Clearance	───────
25.	Pitch = 9	Working Depth	───────

Refer to the Decimal-Inch Spur Gears Table on page 181. The formula in terms of the unknown is not given. Choose the formula that consists of the given parts, rearrange in terms of the unknown, and solve.

	Given Values	Find	
26.	Addendum = 0.0769″	Circular Pitch	_____
27.	Addendum = 0.0666″	Pitch	_____
28.	Addendum = 0.2000″ Outside Diameter = 4.8000″	Pitch Diameter	_____
29.	Outside Diameter = 2.7144″ Number of Teeth = 17	Pitch	_____
30.	Outside Diameter = 4.3750″ Circular Pitch = 0.3927″	Number of Teeth	_____
31.	Working Depth = 0.0769″	Pitch	_____
32.	Working Depth = 0.5000″	Circular Pitch	_____
33.	Outside Diameter = 4.7144″ Pitch Diameter = 4.4286″	Addendum	_____

Refer to the Decimal-Inch Spur Gears Table on page 181. No single formula is given which consists of the given parts and the unknown. Two or more formulas, some in rearranged form, must be used in solving these problems.

	Given Values	Find	
34.	Number of Teeth = 72 Pitch Diameter = 6.0000″	Addendum	_____
35.	Number of Teeth = 44 Pitch Diameter = 3.6667″	Dedendum	_____
36.	Number of Teeth = 10 Pitch Diameter = 2.5000″	Whole Depth	_____
37.	Number of Teeth = 90 Pitch Diameter = 12.8571″	Working Depth	_____
38.	Pitch Diameter = 1.0625″ Pitch = 16	Outside Diameter	_____
39.	Pitch Diameter = 2.9167″ Pitch = 12	Root Diameter	_____
40.	Number of Teeth = 29 Pitch Diameter = 2.0714″	Root Diameter	_____
41.	Number of Teeth = 75 Pitch Diameter = 6.8182″	Clearance	_____
42.	Addendum = 0.1429″	Tooth Thickness	_____
43.	Pitch Diameter = 1.0455″ Addendum = 0.0455″	Number of Teeth	_____

Backlash is the amount that a tooth space is greater than the engaging tooth on the pitch circles of two gears. Determine the average backlash of each of the following using this formula.

$$\text{Average backlash} = \frac{0.030}{P}$$

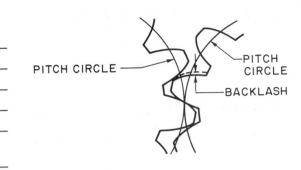

44. A 6-pitch gear   _____
45. A 13-pitch gear   _____
46. A 3.5-pitch gear   _____
47. A gear with a whole depth of 0.2606″   _____
48. A gear with a working depth of 0.1176″   _____
49. A gear with a pitch diameter of 4.800″
     and 24 teeth   _____

The center distance of a pinion and a gear is the distance between the centers of the pitch circles. Determine the center distance of each of the following using this formula.

$$\text{Center distance} = \frac{\text{pitch diameter of gear} + \text{pitch diameter of pinion}}{2}$$

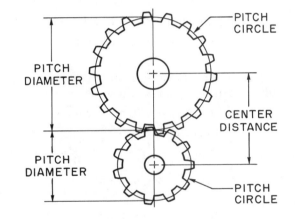

50. A pinion with a pitch diameter of 2.8333
     inches and a gear with a pitch diameter of
     4.1667 inches   _____

51. A pinion with a pitch diameter of 4.8889
     inches and a gear with a pitch diameter of
     8.6667 inches   _____

52. A 9-pitch pinion and gear; the pinion has
     23 teeth and the gear has 38 teeth   _____

53. A 16-pitch pinion and gear; the pinion has
     18 teeth and the gear has 44 teeth   _____

54. A gear and pinion with a circular pitch of
     0.1745 inch; the gear has 55 teeth and the
     pinion has 37 teeth   _____

Gearing — Metric Module System

Refer to the Metric Spur Gears Table on page 183 and determine the values in the following table.

	Module	Number of Teeth	a. Pitch Diameter	b. Circular Pitch	c. Outside Diameter	d. Addendum	e. Working Depth	f. Tooth Thickness
55.	6 mm	18						
56.	9 mm	24						
57.	2.5 mm	10						
58.	4 mm	16						
59.	10 mm	26						

Solve the following metric module system gearing problems. Certain problems require rearranging the data given in Metric Spur Gears Table on page 183.

60. Compute the whole depth of a 5 millimeter module gear designed with a
     clearance of 0.157 × module.   _____

61. What is the number of teeth on a 4 millimeter module gear with a pitch diameter of 120 millimeters?   _____

62. What is the module of a gear which has a working depth of 13 millimeters?   _____

63. Compute the dedendum of a 7 millimeter module gear designed with a clearance of 0.1667 × module.   _____

64. What is the module of a gear with 38 teeth and an outside diameter of 220 millimeters?   _____

⌐ UNIT 38 *ACHIEVEMENT REVIEW —*
SECTION 3 _____

⌐OBJECTIVE_____

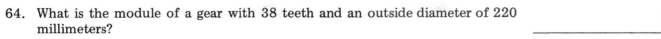

You should be able to solve the exercises and problems in this Achievement Review by applying the principles and methods covered in units 26–37.

1. Express each of the following problems as an algebraic expression.

   a. The sum of x and y reduced by c.   _____

   b. The product of a and b divided by c.   _____

   c. Twice M minus the square of P.   _____

2. Substitute the given numbers for letters and find the value for each of the following expressions.

   a. Find $(5a + 6b) \div 4b$ when $a = 4$ and $b = 2$.   _____

   b. Find $3xy - (2x + y)$ when $x = 6$ and $y = 3$.   _____

   c. Find $e^2 + (2f)^2 - \left(\frac{m}{f}\right)^2$ when $e = 5$, $f = 3$, and $m = 6$.   _____

   d. Find $\sqrt{hr + 5p}\left(\frac{5h}{p} + h\right)$ when $h = 10$, $p = 8$, and $r = 6$.   _____

3. Perform the operation or operations as indicated for each of the following exercises.

   a. $-25 + (-14)$ _____
   b. $24 + (-8)$ _____
   c. $-1.8 - (12.6)$ _____
   d. $18(-3)$ _____
   e. $(-0.3)(-2.6)$ _____
   f. $-18 \div 3$ _____
   g. $-12.8 \div (-0.2)$ _____

   h. $(-6)^2$ _____
   i. $(-4)^3$ _____
   j. $\sqrt[3]{-27}$ _____
   k. $\left(-\frac{1}{2}\right)^2$ _____
   l. $\sqrt[3]{\frac{-27}{8}}$ _____
   m. $(-4)^2 + \sqrt[3]{8} - (2)(-5)(-1)$ _____
   n. $[4(-10)(0.2)] \div [12 + 5(3 - 1) - 10]$ _____

4. The following expressions consist of literal terms. Perform the indicated operations.

   a. $-8P + 5P$ _____
   b. $-0.05H^2 - 1.13H^2$ _____
   c. $12d + 8d^2 - 7d + 14d^2$ _____
   d. $(3a - 2a^2) - (6a - 3a^2)$ _____
   e. $(-10x)(9x^2y)$ _____
   f. $(-5.9e^2f^2)(-f^2)$ _____

   g. $(16x^2 - 4x^3) \div 2x$ _____
   h. $(0.6f^2g^3 - fg - 2f^2) \div 0.2f$ _____
   i. $(6x^2 - y^3)(-3x^3 + y^2)$ _____
   j. $(-3a^2b^3c)^3$ _____
   k. $[(xy^2)^3 - (x^2y^2)]^2$ _____
   l. $\sqrt[3]{-27a^6b^3c^9}$ _____

m. $\sqrt{\frac{25}{64}\,e^4 g^2 h}$ _____   p. $-4a(2a)^2 + \sqrt{36a^4}$ _____

n. $-6(x^2 + y - 2x)$ _____   q. $9y - x[-8 + (xy)^2 - y] + 12\,x$ _____

o. $36 - (m^3 + m) + (m^3 - 12)$ _____   r. $b(b + 7m)^2 - b(b - 7m)^2$ _____

5. Solve for the unknown in each of the following equations using one of the six principles of equality. Check each answer.

     a. $x + 12 = 28$ _____   g. $7y = -63$ _____   m. $s^2 = 81$ _____

     b. $y - 15 = 23$ _____   h. $1.3E = 7.54$ _____   n. $x^3 = \frac{-8}{27}$ _____

     c. $-32 = B - 46$ _____   i. $11.22 = \frac{L}{6.6}$ _____   o. $\sqrt{M} = 7$ _____

     d. $H + 12.6 = 43.9$ _____   j. $\frac{x}{-0.8} = 8.48$ _____   p. $\sqrt[3]{V} = 5$ _____

     e. $14.3 = x + 53.6$ _____   k. $6.75 = -13.5x$ _____   q. $-0.027 = y^3$ _____

     f. $R - 7.8 = -9.2$ _____   l. $\frac{3}{4}x = 2\frac{1}{4}$ _____   r. $\sqrt[3]{B} = \frac{3}{4}$ _____

6. Solve for the unknown and check each of the following combined operations equations.

     a. $25 - (-P + 18) = 45$ _____   f. $12x^2 - 53 = (x - 3)(x + 3)$ _____

     b. $10(M - 4) = -5(M - 4)$ _____   g. $0.5G + 8(G - 3) = 4(0.75G - 7)$ _____

     c. $7.1E + 3(E - 6) = 0.5E + 22.8$ _____   h. $(T - 7)(T - 8) = T^2 + 41$ _____

     d. $\frac{H}{4} + 7.8 = 12.4$ _____   i. $7\sqrt{x} = 8(3\sqrt{x} + 1) - 25$ _____

     e. $\frac{1}{4}F + 6\left(F - \frac{1}{2}\right) = 15\frac{3}{4}$ _____   j. $H + 5\sqrt{H} = -9\sqrt{H} + H + 56$ _____

7. In each of the following formulas, substitute given numerical values for letter values and solve for the unknown and check. Calculate answers to 3 decimal places where necessary.

     a. $N = 0.707DP_n$.   Solve for D when $N = 36$ and $P_n = 12$. _____

     b. $I = \frac{nE}{R + nr}$.   Solve for E when $I = 0.3$, $n = 5$, $R = 6$, and $r = 4$. _____

     c. $S = \frac{0.290W}{t^2}$.   Solve for W when $S = 600$ and $t = 0.375$. _____

     d. $c = \sqrt{a^2 + b^2}$.   Solve for b when $a = 4.000$ and $c = 5.000$. _____

     e. $V = 1.570h(R^2 + r^2)$.   Solve for R when $V = 105$, $h = 5.5$ and $r = 3$. _____

8. Rearrange each of the following formulas in terms of the designated letter.

     a. $E = I(R + r)$.   Solve for R. _____

     b. $D_O = 2C - d + 2a$.   Solve for a. _____

     c. $M = D - 1.5155P + 3W$.   Solve for W. _____

     d. $r = \sqrt{x^2 + y^2}$.   Solve for x. _____

     e. $L = 3.14(0.5D + 0.5d) + 2x$.   Solve for x. _____

     f. $HP = \frac{D^2 N}{2.5}$.   Solve for D. _____

9. Solve for the unknown value in each of the following proportions and check.

     a. $\frac{P}{12.4} = \frac{3}{2}$ _____   d. $\frac{5}{6} = \frac{C + 7}{12}$ _____   g. $\frac{10}{5T} = \frac{1.8}{4.5}$ _____

     b. $\frac{3.6}{0.9} = \frac{E}{2.7}$ _____   e. $\frac{6.5}{M} = \frac{8.2}{41}$ _____   h. $\frac{10}{7.5} = \frac{2}{N - 2}$ _____

     c. $\frac{H}{\frac{1}{4}} = \frac{\frac{1}{2}}{\frac{3}{8}}$ _____   f. $\frac{22.5}{13.5} = \frac{1.2}{x}$ _____

10. Analyze each of the following problems to determine whether the problem is a direct or inverse proportion and solve.

   a. A reamer tapers 0.0975 inch along a 3.1875-inch length. What is the amount of taper along a 2.1250-inch length?   _____

   b. A machine produces 2550 parts in 8.5 hours. How many parts are produced by the machine in 10 hours?   _____

   c. Of two gears that mesh, one gear with 12 teeth revolves at 420 rpm. What is the revolutions per minute of the other gear which has 16 teeth?   _____

11. Solve the following cutting speed and gear problems.

   a. A steel shaft 2.250 inches in diameter is turned in a lathe at 250 rpm. Determine the cutting speed to the nearer whole number.   _____

   $$C = \frac{3.1416DN}{12}$$

   b. Determine the revolutions per minute to the nearer whole number of an aluminum cylinder 40.00 millimeters in diameter with a cutting speed of 150 meters per minute.   _____

   $$N = \frac{1000C}{3.1416D}$$

   c. Twenty 5.50-millimeter diameter holes each 62.00 millimeters deep are drilled in a workpiece. The drill turns at 6500 r/min with a feed of 0.05 millimeter per revolution. Determine the total cutting time to the nearer hundredth minute.   _____

   $$T = \frac{L}{FN}$$

   d. What is the pitch of a gear with 44 teeth and a pitch diameter of 12.5714 inches?   _____

   $$P = \frac{N}{D}$$

   e. Determine the whole depth to 4 decimal places of a gear with 20 teeth and a pitch diameter of 5.0000 inches.   _____

   $$P = \frac{N}{D} \quad \text{and} \quad WD = \frac{2.157}{P}$$

SECTION 4
FUNDAMENTALS OF PLANE GEOMETRY

¬UNIT 39 *INTRODUCTION TO GEOMETRIC FIGURES*

¬OBJECTIVES

After studying this unit you should be able to

- Add, subtract, multiply, and divide angles in terms of degrees, minutes, and seconds.
- Express decimal degrees as degrees, minutes, and seconds.
- Express degrees, minutes, and seconds as decimal degrees.

The fundamental principles of geometry generally applied to machine shop problems are those used to make the calculations required for machining parts from engineering drawings. An engineering drawing is an example of applied geometry.

¬PLANE GEOMETRY

Plane geometry is the branch of mathematics that deals with points, lines, and various figures that are made of combinations of points and lines. The figures lie on a flat surface, or *plane*. Examples of plane geometry are the views of a part as shown on an engineering drawing.

Since geometry is fundamental to machine technology, it is essential to understand the definitions and terms of geometry. It is equally important to be able to apply the geometric principles in problem solving. The methods and procedures used in problem solving are the same as those required for the planning, making, and checking of machined parts.

Procedure: To solve a geometry problem

- Study the figure.
- Relate it to the principle or principles that are needed for the solution.
- Base all conclusions on fact: given information and geometric principles.
- Do not assume that something is true because of its appearance or because of the way it is drawn.

Note: The same requirements are applied in reading engineering drawings.

¬AXIOMS

In the study of geometry certain basic statements called *axioms* are assumed to be true. The following statements are geometric axioms.

Things equal to the same thing, or to equal things, are equal to each other. Equals may be substituted for equals.

If equals are added to or subtracted from equals, the sums or remainders are equal.

If equals are multiplied or divided by equals, the products or quotients are equal.

The whole is equal to the sum of its parts.

Only one straight line can be drawn between two given points.

Through a given point, only one line can be drawn parallel to a given line.

Two straight lines can intersect at one point only.

¬POINTS AND LINES

A *point* has no size or form; it has only location. A point is shown as a dot. Each point is usually named by a capital letter as shown.

A•   •B
      C• D•

A *line*, as it is used in this book, always means a straight line. A line other than a straight line, such as a curved line, is identified. In this book, no distinction between a line and a line segment is made.

Parallel lines do not meet regardless of how far they are extended. They are the same distance apart (*equidistant*) at all points. The symbol ∥ means parallel. In the figure, line AB is parallel to line CD; therefore, AB and CD are equidistant (distance x) at all points.

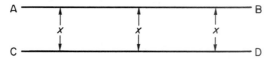

Perpendicular lines meet or intersect at a right, or 90°, angle. The symbol ⊥ means perpendicular. These figures are examples of perpendicular lines.

Oblique lines meet or intersect at an angle other than 90°. These figures are examples of oblique lines.

ANGLES

An *angle* is a figure which consists of two lines that meet at a point called the vertex. The symbol ∠ means angle. The size of an angle is determined by the number of degrees one side of the angle is rotated from the other. The length of the side does not determine the size of the angle. For example, ∠ 1 is equal to ∠ 2. The rotation of side AC from side AB is equal to the rotation of side DF from side DE although the lengths of the sides are not equal.

UNITS OF ANGULAR MEASURE

The degree is the basic unit of angular measure. The symbol (°) means degree. A radius which is rotated one revolution makes a complete circle or 360°. In the English system computations and measurements are in degrees and minutes. Degrees, minutes, and seconds are used for applications which require precise angular measure. In the metric system computations and measurements are in decimal degrees; however, degrees, minutes, and seconds are sometimes used.

UNITS OF ANGULAR MEASURE IN DEGREES, MINUTES, AND SECONDS

A degree is divided into 60 equal parts called minutes. The symbol for minute is ('). A minute is divided into 60 equal parts called seconds. The symbol for second is (").

1 Circle = 360 Degrees	1 Degree = $\frac{1}{360}$ Circle
1 Degree = 60 Minutes	1 Minute = $\frac{1}{60}$ Degree
1 Minute = 60 Seconds	1 Second = $\frac{1}{60}$ Minute

⌐ADDING ANGLES EXPRESSED IN DEGREES, MINUTES, AND SECONDS _____

Example 1: Determine $\angle 1$.

$\angle 1 = 15°18' + 63°37'$

$$
\begin{array}{r}
15°18' \\
+\ 63°37' \\
\hline
78°55' \quad \text{Ans}
\end{array}
$$

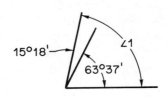

Example 2: Determine $\angle 2$.

$\angle 2 = 43°37' + 82°54'$

$$
\begin{array}{r}
43°37' \\
+\ 82°54' \\
\hline
125°91' = 126°31' \quad \text{Ans}
\end{array}
$$

Note: $91' = 60' + 31' = 1°31'$ therefore, $125°91' = 126°31'$.

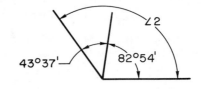

Example 3: Determine $\angle 3$.

$\angle 3 = 78°43'27'' + 29°38'52''$

$$
\begin{array}{r}
78°43'27'' \\
+\ 29°38'52'' \\
\hline
107°81'79'' = 107°82'19'' = 108°22'19'' \quad \text{Ans}
\end{array}
$$

Note: $79'' = 60'' + 19'' = 1'19''$ therefore, $107°81'79'' = 107°82'19''$.
$\qquad 82' = 60' + 22' = 1°22'$ therefore, $107°82'19'' = 108°22'19''$.

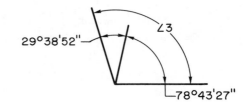

⌐SUBTRACTING ANGLES EXPRESSED IN DEGREES, MINUTES, AND SECONDS ____

Example 1: Determine $\angle 1$.

$\angle 1 = 123°47'32'' - 86°13'07''$

$$
\begin{array}{r}
123°47'32'' \\
-\ \ 86°13'07'' \\
\hline
37°34'25'' \quad \text{Ans}
\end{array}
$$

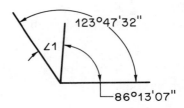

Example 2: Determine $\angle 2$.

$\angle 2 = 97°12' - 45°26'$

$$
\begin{array}{rl}
97°12' &= 96°72' \\
-\ 45°26' &= 45°26' \\
\hline
& 51°46' \quad \text{Ans}
\end{array}
$$

Note: Since $26'$ cannot be subtracted from $12'$, express $1°$ as $60'$.
$\qquad 97°12' = 96° + 1° + 12' = 96° + 60' + 12' = 96°72'$

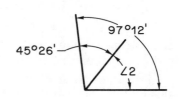

Example 3: Determine $\angle 3$.

$\angle 3 = 57°13'28'' - 44°19'42''$

$$
\begin{array}{rll}
57°13'28'' &= 56°73'28'' &=\ \ \ 56°72'88'' \\
-\ 44°19'42'' &= 44°19'42'' &= -\ 44°19'42'' \\
\hline
& & 12°53'46'' \quad \text{Ans}
\end{array}
$$

Note: Since $19'$ cannot be subtracted from $13'$, and $42''$ cannot be
subtracted from $28''$; express $1°$ as $60'$ and $1'$ as $60''$.
$\qquad 57°13'28'' = 56° + 1° + 13' + 28'' = 56°73'28'' = 56°72' + 1' + 28'' =$
$\qquad 56°72'88''$

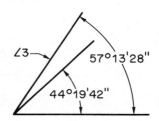

⎯MULTIPLYING ANGLES EXPRESSED IN DEGREES, MINUTES, AND SECONDS⎯

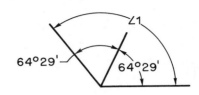

Example 1: Determine ∠1.

$$∠1 = 2(64°29')$$

$$\begin{array}{r} 64°29' \\ \times\ 2 \\ \hline 128°58' \quad \text{Ans} \end{array}$$

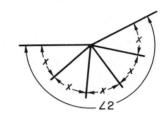

Example 2: Determine ∠2 when $x = 41°27'42''$.

$$∠2 = 5x = 5(41°27'42'')$$

$$\begin{array}{r} 41°\ \ 27'\ \ 42'' \\ \times\ \ 5 \\ \hline 205°135'210'' = 205°138'30'' = 207°18'30'' \quad \text{Ans} \end{array}$$

Note: $210'' = 3'30''$ and $138' = 2°18'$.

⎯DIVIDING ANGLES EXPRESSED IN DEGREES, MINUTES, AND SECONDS⎯

Example 1: Determine ∠1 and ∠2.

$$∠1 = ∠2 = 104°58' ÷ 2$$

$$\begin{array}{r} 52°29' \quad \text{Ans} \\ 2\,\overline{)\,104°58'} \end{array}$$

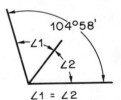

Example 2: Determine ∠1, ∠2, and ∠3.

$$∠1 = ∠2 = ∠3 = 128°37'21'' ÷ 3$$

Divide 128° by 3.
$128° ÷ 3 = 42°$ plus a remainder of $2°$.

$$\begin{array}{r} 42° \\ 3\,\overline{)\,128°} \\ 126 \\ \hline 2° \end{array}$$

Add the 2° (120') to the 37'.
$120' + 37' = 157'$
Divide 157' by 3.
$157' ÷ 3 = 52'$ plus a remainder of 1'.

$$\begin{array}{r} 52' \\ 3\,\overline{)\,157'} \\ 156 \\ \hline 1' \end{array}$$

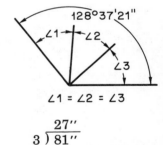

Add the 1' (60'') to the 21''.
$60'' + 21'' = 81''$
Divide 81'' by 3.
$81'' ÷ 3 = 27''$

$$\begin{array}{r} 27'' \\ 3\,\overline{)\,81''} \end{array}$$

Combine.

$$42° + 52' + 27'' = 42°52'27'' \quad \text{Ans}$$

⎯EXPRESSING DECIMAL DEGREES AS DEGREES, MINUTES, AND SECONDS⎯

The measure of an angle given in the form of decimal degrees, such as $47.1938°$, must often be expressed as degrees, minutes, and seconds.

Procedure: To express decimal degrees as degrees, minutes, and seconds

- Multiply the decimal part of the degrees by 60' in order to obtain minutes.
- If the number of minutes obtained is not a whole number, multiply the decimal part of the minutes by 60'' in order to obtain seconds. Round to the nearer whole second if necessary.
- Combine degrees, minutes, and seconds.

Example: Express 47.1938° as degrees, minutes, and seconds.

Multiply 0.1938 by 60′ to obtain minutes.	60′(0.1938) = 11.6280′
Multiply 0.6280 by 60″ to obtain seconds. Round to the nearer whole second.	60″(0.6280) = 38″
Combine degrees, minutes, and seconds.	47° + 11′ + 38″ = 47°11′38″ Ans

▬EXPRESSING DEGREES, MINUTES, AND SECONDS AS DECIMAL DEGREES▬▬

Often an angle given in degrees and minutes is to be expressed as decimal degrees.

Procedure: To express degrees and minutes as decimal degrees

- Divide the minutes by 60.
- Combine whole degrees and decimal degrees. Round the answer to 2 places.

Example: Express 76°29′ as decimal degrees.

Divide 29 by 60 to obtain decimal degrees.	29 ÷ 60 = 0.48°
Combine whole degrees with decimal degrees.	76° + 0.48° = 76.48° Ans

When working with English and metric units of measure, it may be necessary to express angles given in degrees, minutes, and seconds as angles in decimal degrees.

Procedure: To express degrees, minutes, and seconds as decimal degrees

- Divide the seconds by 60 in order to obtain decimal minutes.
- Combine whole minutes and decimal minutes.
- Divide the total minutes by 60 in order to obtain decimal degrees.
- Combine whole degrees and decimal degrees. Round the answer to 4 decimal places.

Example: Express 23°18′44″ as decimal degrees.

Divide 44 by 60 to obtain decimal minutes.	44 ÷ 60 = 0.7333′
Combine whole minutes and decimal minutes.	18′ + 0.7333′ = 18.7333′
Divide 18.7333 by 60 to obtain decimal degrees.	18.7333 ÷ 60 = 0.3122°
Combine whole degrees with decimal degrees.	23° + 0.3122° = 23.3122° Ans

▬APPLICATION▬▬▬

Definitions and Terms

1. Refer to the figure and identify each of the following as parallel, perpendicular, or oblique lines.

   a. Line AB and line CD   _____

   b. Line AB and EF   _____

   c. Line CD and GH   _____

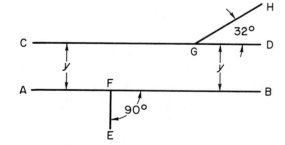

2. a. How many degrees are in a circle?   _____
   b. How many minutes are in 1 degree?   _____
   c. How many seconds are in 1 minute?   _____
   d. How many seconds are in 1 degree?   _____
   e. How many minutes are in a circle?   _____

3. Write the symbols for the following words.

    a. parallel   _____   c. degree   _____   e. second   _____

    b. perpendicular _____   d. minute   _____

Adding Angles Expressed in Degrees, Minutes, and Seconds

4. Determine $\angle 1$.   _____

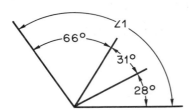

8. Determine $\angle 5$.   _____

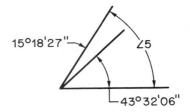

5. Determine $\angle 2$.   _____

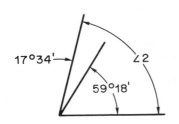

9. Determine $\angle 6$.   _____

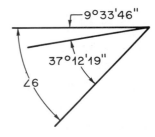

6. Determine $\angle 3$.   _____

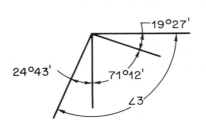

10. Determine $\angle 7 + \angle 8 + \angle 9$.   _____

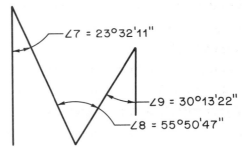

7. Determine $\angle 1 + \angle 2 + \angle 3$.   _____

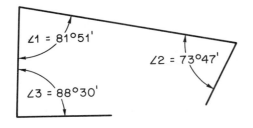

11. Determine $\angle 1 + \angle 2 + \angle 3 + \angle 4 + \angle 5$.   _____

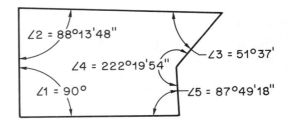

Subtracting Angles Expressed in Degrees, Minutes, and Seconds

Subtract the angles in each of the following exercises.

12. $114° - 89°$ _____

13. $92°35' - 73°16'$ _____

14. $63°23' - 32°58'$ _____

15. $122°36'17'' - 13°15'08''$ _____

16. $54°43'13'' - 19°13'42''$ _____

17. Determine ∠1.   _____

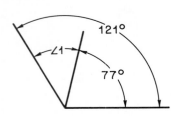

19. Determine ∠2.   _____

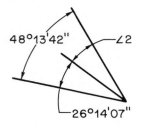

18. Determine ∠3 – ∠2.   _____

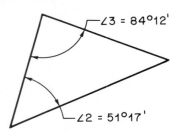

20. Determine ∠1 – ∠2.   _____

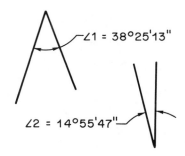

21. In the figure shown ∠6 = 720° – (∠1 + ∠2 + ∠3 + ∠4 + ∠5).
     Determine ∠6.   _____

     Note: 720° = 719°59′60″

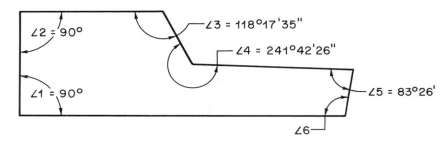

Multiplying Angles Expressed in Degrees, Minutes, and Seconds

Multiply the angles in each of the following exercises.

22. 7(15°)   _____

23. 3(27°18′)   _____

24. 2(43°43′)   _____

25. 5(22°10′13″)   _____

26. 8(34°24′30″)   _____

27. In the figure shown, ∠1 = ∠2 = 42°.
     Determine ∠3.   _____

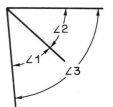

28. If x = 39°14′, find ∠4.   _____

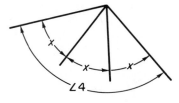

29. In the figure shown, ∠1 = ∠2 = ∠3 = ∠4 = ∠5 = 53°41′.
     Determine ∠6.   _____

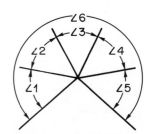

Dividing Angles Expressed in Degrees, Minutes, and Seconds

Divide the angles in each of the following exercises.

30. $94° \div 2$ _____

31. $93° \div 2$ _____

32. $105°20' \div 4$ _____

33. If $\angle 1 = \angle 2$, find $\angle 1$. _____

35. Determine y. _____

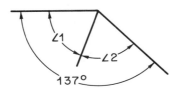

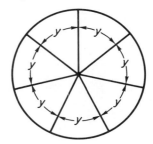

34. Determine x. _____

36. If $\angle 1 = \angle 2 = \angle 3 = \angle 4$, find $\angle 1$. _____

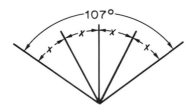

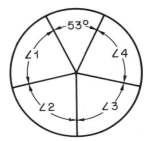

37. The sum of the angles in the figure shown equal $1440°$.
    If $\angle 1 = \angle 2 = \angle 3 = \angle 4 = \angle 5 = \angle 6$ and $\angle 7 = \angle 8 = \angle 9 = \angle 10 = 118°14'23''$.
    find $\angle 1$. _____

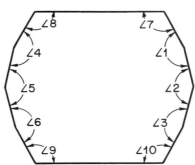

Expressing Decimal Degrees As Degrees and Minutes

Express the following decimal degrees as degrees and minutes. When necessary, round the answer to the nearer whole minute.

38. $13.50°$ _____

39. $76.95°$ _____

40. $48.10°$ _____

41. $117.70°$ _____

42. $20.30°$ _____

43. $93.15°$ _____

44. $81.08°$ _____

45. $6.47°$ _____

46. $120.93°$ _____

47. $77.67°$ _____

Expressing Decimal Degrees As Degrees, Minutes, and Seconds

Express the following decimal degrees as degrees, minutes, and seconds. When necessary, round the answer to the nearer whole second.

48. $52.1380°$ _____

49. $214.0820°$ _____

50. $7.9250°$ _____

51. $44.4440°$ _____

52. $75.8430°$ _____

53. $103.0090°$ _____

54. $37.9365°$ _____

55. $89.9056°$ _____

56. $176.0782°$ _____

57. $19.8973°$ _____

Expressing Degrees and Minutes As Decimal Degrees

Express the following degrees and minutes as decimal degrees. Round the answer to 2 decimal places.

58. 40°20′ _____   62. 122°07′ _____   65. 2°19′ _____

59. 107°45′ _____   63. 56°48′ _____   66. 34°04′ _____

60. 6°10′ _____   64. 88°57′ _____   67. 79°59′ _____

61. 93°18′ _____

Expressing Degrees, Minutes, and Seconds As Decimal Degrees

Express the following degrees, minutes, and seconds as decimal degrees. Round the answer to 4 decimal places.

68. 28°18′30″ _____   71. 98°20′25″ _____   74. 19°49′59″ _____

69. 53°10′45″ _____   72. 176°27′18″ _____   75. 61°12′06″ _____

70. 130°50′10″ _____   73. 5°06′17″ _____

UNIT 40 *PROTRACTORS — SIMPLE AND CALIPER*

OBJECTIVES

After studying this unit you should be able to

- Measure angles with a simple protractor.
- Lay out angles with a simple protractor.
- Read settings on a vernier bevel protractor.
- Compute complements and supplements of angles.

Protractors are used for measuring and laying out angles. Although all protractors are basically the same, different types are available for different uses and degrees of precision required.

SIMPLE PROTRACTOR

A simple protractor has two scales, each graduated from 0° to 180° so that it can be read from either the left or right side. The vertex of the angle to be measured or drawn is located at the middle of the base of the protractor.

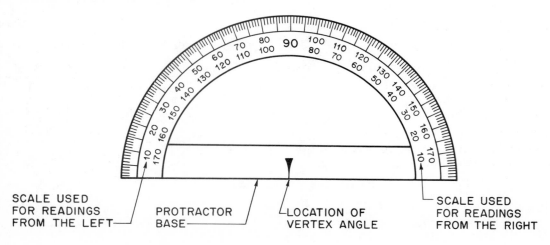

SCALE USED FOR READINGS FROM THE LEFT   PROTRACTOR BASE   LOCATION OF VERTEX ANGLE   SCALE USED FOR READINGS FROM THE RIGHT

Procedure: To lay out a given angle

- Draw a baseline.

- On the baseline mark a point as the vertex.

- Place the protractor base on the baseline with the protractor center on the vertex.

- If the angle rotates from the right, choose the scale which has a zero degree reading on the right side of the protractor. If the angle rotates from the left, choose the scale which has a zero degree reading on the left side of the protractor. At the scale reading for the angle being drawn, mark a point.

- Remove the protractor and connect the two points.

Example: Lay out an angle of 105°.

Draw baseline AB.

On AB mark point O as the vertex.

Place the protractor base on AB with the protractor center on point O.

The angle is rotated from the right. The inside scale has a zero degree reading on the right side of the protractor. Use the inside scale and mark a point at the scale reading of 105°.

Remove the protractor and connect points P and O.

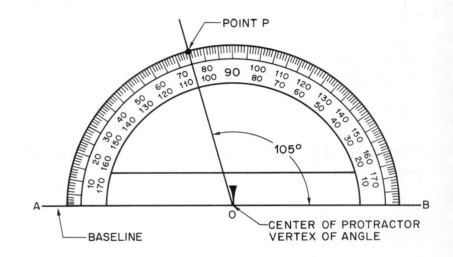

Procedure: To measure a given angle

- Place the protractor base on one side of the angle with the protractor center on the angle vertex.

- If the angle rotates from the right, choose the scale which has the zero degree reading on the right side of the protractor. If the angle rotates from the left, choose the scale which has the zero degree reading on the left side of the protractor. Read the measurement where the side crosses the protractor scale.

Example 1: Measure ∠1.

Extend the sides OA and OB of ∠1 as shown.

Place the protractor base on side OB with the protractor center on the angle vertex, point O.

Angle 1 is rotated from the right. The angle measurement is read from the inside scale since the inside scale has a zero degree (0°) reading on the right side of the protractor base. Read the measurement where the extension of side OA crosses the protractor scale. Angle 1 = 40°   Ans

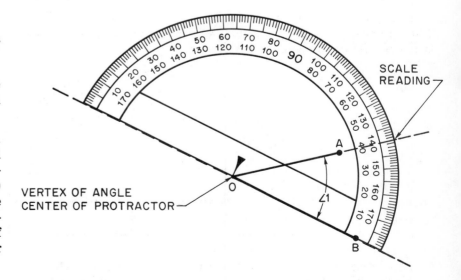

Example 2: Measure ∠2.

Extend the sides OD and OF of ∠2 as shown.

The protractor is positioned up-side down. Place the protractor base on the side OD with the protractor center on the angle vertex point O.

Angle 2 is rotated from the right. The angle measurement is read from the outside scale since the outside scale has a zero degree (0°) reading on the right side of the protractor base. Read the measurement where the extension of side OF crosses the protractor scale. Angle 2 = 125° Ans

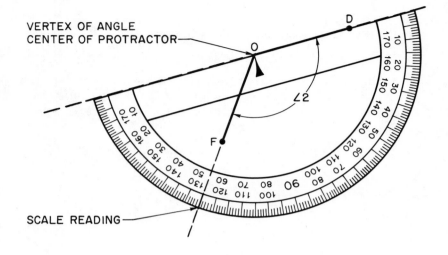

BEVEL PROTRACTOR WITH VERNIER SCALE

The bevel protractor is the most widely used vernier protractor in the machine shop. A vernier bevel protractor is shown.

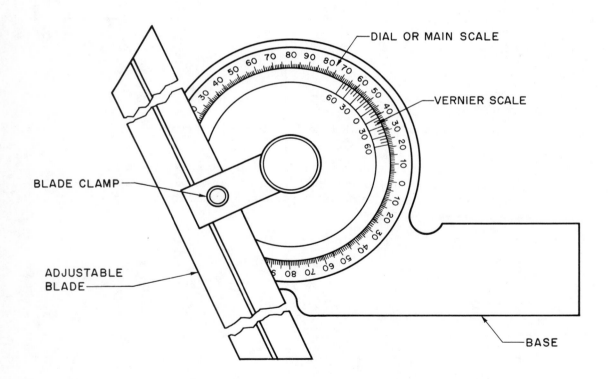

A vernier bevel protractor consists of a fixed dial or main scale. The main scale is divided into four sections, each from 0° to 90°. The vernier scale rotates within the main scale. A blade which can be adjusted to required positions is rotated to a desired angle.

The vernier scale permits accurate readings to 1/12 degree or 5 minutes. The vernier scale is divided into 24 units, with 12 units on each side of zero. The divisions on the vernier scale are in minutes. Each division is equal to 5 minutes.

The left vernier scale is used when the vernier zero is to the left of the dial zero. The right vernier scale is used when the vernier zero is to the right of the dial zero.

Example: Read the setting on the vernier protractor shown.

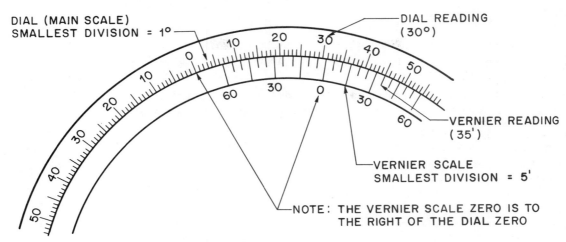

DIAL (MAIN SCALE) SMALLEST DIVISION = 1°

DIAL READING (30°)

VERNIER READING (35')

VERNIER SCALE SMALLEST DIVISION = 5'

NOTE: THE VERNIER SCALE ZERO IS TO THE RIGHT OF THE DIAL ZERO

The zero mark on the vernier scale is just to the right of the 30° division of the dial scale. The vernier zero is to the right of the dial zero; therefore, the right vernier scale is read. The 35′ vernier graduation coincides with a dial graduation. The protractor reading is 30° 35′.   Ans

COMPLEMENTS AND SUPPLEMENTS OF SCALE READINGS

When using the bevel protractor, the machinist must determine whether the desired angle of the part being measured is the actual reading on the protractor or the complement or the supplement of the protractor reading. Particular caution must be taken when measuring angles close to 45° and 90°.

Two angles are *complementary* when their sum is 90°. For example, 43° + 47° = 90°. Therefore, 43° is the complement of 47° and 47° is the complement of 43°.

Two angles are *supplementary* when their sum is 180°. For example, 92° + 88° = 180°. Therefore, 92° is the supplement of 88° and 88° is the supplement of 92°.

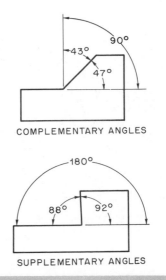

COMPLEMENTARY ANGLES

SUPPLEMENTARY ANGLES

APPLICATION

Simple Protractor

Write the values of angles A–J on the protractor scale shown.

A = _____
B = _____
C = _____
D = _____
E = _____
F = _____
G = _____
H = _____
I = _____
J = _____

2. Using a protractor, lay out the following angles.

   a. 17°   c. 80°   e. 4°   g. 97°   i. 150°
   b. 65°   d. 12°   f. 88°   h. 123°   j. 178°

3. Lay out a 3-sided closed figure (triangle) of any size containing angles of 47° and 105°. Measure the third angle. How many degrees are contained in the third angle?   _____

4. Lay out a 4-sided figure (quadrilateral) of any size containing angles of 89°, 76°, and 124°. Measure the fourth angle. How many degrees are contained in the fourth angle?   _____

5. Measure each of the following angles, ∠1–∠14, to the nearer degree. Extend the sides of the angles if necessary.

   ∠1 = _____
   ∠2 = _____
   ∠3 = _____
   ∠4 = _____
   ∠5 = _____
   ∠6 = _____
   ∠7 = _____
   ∠8 = _____
   ∠9 = _____
   ∠10 = _____
   ∠11 = _____
   ∠12 = _____
   ∠13 = _____
   ∠14 = _____

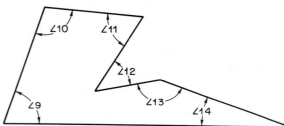

Vernier Protractor

Write the values of the settings on the following vernier protractor scales.

6. _____   8. _____   10. _____

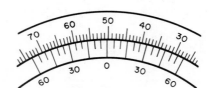
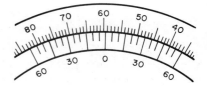

7. _____   9. _____   11. _____

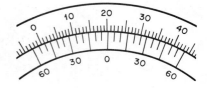

12. _____   13. _____   14. _____

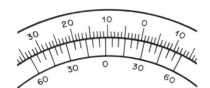   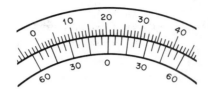   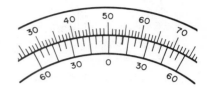

Complementary and Supplementary Angles

15. Write the complements of the following angles.

    a. 41° _____   d. 2° _____   g. 18°22′ _____

    b. 76° _____   e. 67°49′ _____   h. 78°19′27″ _____

    c. 17° _____   f. 45°19′ _____   i. 59°0′59″ _____

16. Write the supplements of the following angles.

    a. 7° _____   d. 179°59′ _____   g. 2°43′20″ _____

    b. 65° _____   e. 0°51′ _____   h. 67°18′27″ _____

    c. 91° _____   f. 89°59′ _____   i. 133°32′08″ _____

UNIT 41 ANGLES

OBJECTIVES

After studying this unit you should be able to

- Identify different types of angles.
- Determine unknown angles in geometric figures using the principles of opposite, alternate interior, corresponding, parallel, and perpendicular angles.

NAMING ANGLES

Angles are named by a number, a letter, or three letters. When an angle is named with three letters, the vertex must be the middle letter. For example, the angle shown can be called ∠1, ∠C, ∠ACB, or ∠BCA.

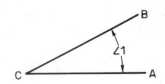

In cases where a point is the vertex of more than one angle, a single letter cannot be used to name an angle. For example, since point E is the vertex of three different angles, the single letter E cannot be used in naming the angle.

    ∠1 is called ∠GEH or ∠HEG.
    ∠2 is called ∠FEG or ∠GEF.
    ∠3 is called ∠FEH or ∠HEF.

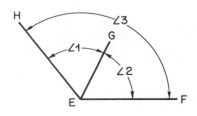

⌐TYPES OF ANGLES

An *acute angle* is an angle that is less than 90°. Angle 1 is acute.
A *right angle* is an angle of 90°. Angle A is a right angle.
An *obtuse angle* is an angle greater than 90° but less than 180°. Angle ABC is an obtuse angle.

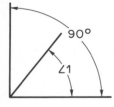 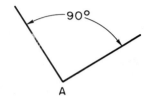 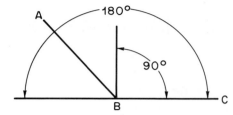

A *straight angle* is an angle of 180°. A straight line is a straight angle. Line EFG is a straight angle.
A *reflex angle* is an angle greater than 180° and less than 360°. Angle 3 is a reflex angle.

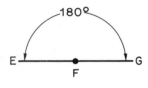 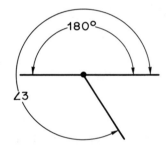

Adjacent Angles

Two angles are adjacent if they have a common side and a common vertex. Angle 1 and angle 2 shown are adjacent since they both contain the common side BC and the common vertex B. Angle 4 and angle 5 shown are not adjacent. The angles do not have a common vertex.

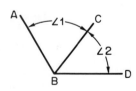 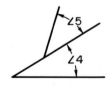

Angles Formed by a Transversal

A *transversal* is a line that intersects (cuts) two or more lines. Line EF is a transversal since it cuts lines AB and CD.

    Alternate interior angles are pairs of interior angles on opposite sides of the transversal. The angles have different vertices. For example, angles 3 and 5 and angles 4 and 6 are alternate interior angles.

    Corresponding angles are pairs of angles, one interior and one exterior with both angles on the same side of the transversal. The angles have different vertices. For example, angles 1 and 5, 2 and 6, 3 and 7, and 4 and 8 are corresponding angles.

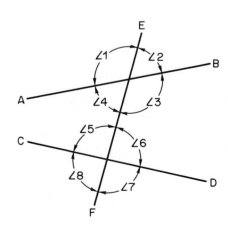

⌐GEOMETRIC PRINCIPLES

In this book, geometric postulates, theorems, and corollaries are grouped together and are called geometric principles. *Geometric principles* are statements of truth which are used as geometric rules. The principles will not be proved, but will be used as the basis for problem solving.

Principle 1

If two lines intersect, the opposite or vertical angles are equal.

Given: AB intersects CD.

Conclusion: $\angle 1 = \angle 3$ and $\angle 2 = \angle 4$.

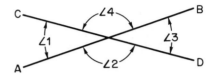

Principle 2

If two parallel lines are intersected by a transversal, the alternate interior angles are equal.

Given: AB || CD.

Conclusion: $\angle 3 = \angle 5$ and $\angle 4 = \angle 6$.

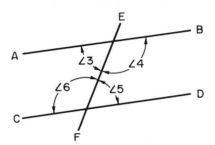

If two lines are intersected by a transversal and a pair of alternate interior angles are equal, the lines are parallel.

Given: $\angle 1 = \angle 2$.

Conclusion: AB || CD.

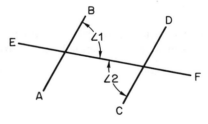

Principle 3

If two parallel lines are intersected by a transversal, the corresponding angles are equal.

Given: AB || CD.

Conclusion: $\angle 1 = \angle 5, \angle 2 = \angle 6, \angle 3 = \angle 7$ and $\angle 4 = \angle 8$.

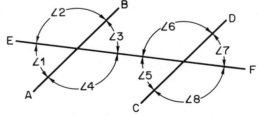

If two lines are intersected by a transversal and a pair of corresponding angles are equal, the lines are parallel.

Given: $\angle 1 = \angle 2$.

Conclusion: AB || CD.

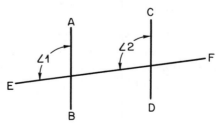

Principle 4

Two angles are either equal or supplementary if their corresponding sides are parallel.

Given: AB ∥ FG and BC ∥ DE.

Conclusion: ∠1 = ∠3 and ∠1 and ∠2 are supplementary.

(∠1 + ∠2 = 180°)

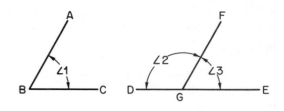

Principle 5

Two angles are either equal or supplementary if their corresponding sides are perpendicular.

Given: AB ⊥ DH and BC ⊥ EF.

Conclusion: ∠1 = ∠2; ∠1 and ∠3 are supplementary.

(∠1 + ∠3 = 180°)

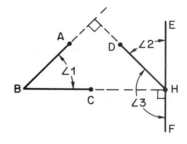

The following example illustrates the method of solving angular measure problems. Values of angles are determined by applying angular geometric principles and the fact that a straight angle (straight line) contains 180°.

Example: Given: AB ∥ CD, EF ∥ GH, ∠1 = 115°, and ∠2 = 82°. Determine the values of ∠3 through ∠9.

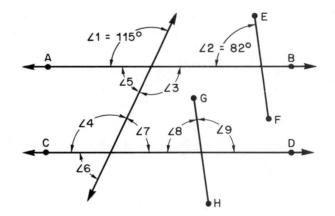

Solve for ∠3. Apply Principle 1. If two lines intersect, the opposite or vertical angles are equal.
∠3 = ∠1 = 115°   Ans

Solve for ∠4. Apply either Principle 2 or Principle 3.
   Applying Principle 2:
If two parallel lines are intersected by a transversal, the alternate interior angles are equal.
∠4 = ∠3 = 115°   Ans
   Or, applying Principle 3:
If two parallel lines are intersected by a transversal, the corresponding angles are equal.
∠4 = ∠1 = 115°   Ans

Solve for ∠5. Since a straight angle (straight line) contains 180°, ∠5 and ∠1 are supplementary.
∠5 = 180° − ∠1 = 180° − 115° = 65°   Ans

Solve for ∠6. Apply Principle 3. If two parallel lines are intersected by a transversal, the corresponding angles are equal. ∠6 = ∠5 = 65°   Ans

Solve for ∠7. Apply Principle 1. If two lines intersect, the opposite or vertical angles are equal. ∠7 = ∠6 = 65°   Ans

Solve for ∠8. Apply Principle 4. Two angles are either equal or supplementary if their corresponding sides are parallel. ∠8 = ∠2 = 82°   Ans

Solve for ∠9. Since a straight angle (straight line) contains 180°, ∠8 and ∠9 are supplementary. ∠9 = 180° − ∠8 = 180° − 82° = 98°   Ans

APPLICATION

Naming Angles

1. Name each of the following angles in three additional ways.

   a. ∠1 _____

   b. ∠2 _____

   c. ∠C _____

   d. ∠D _____

   e. ∠E _____

   f. ∠F _____

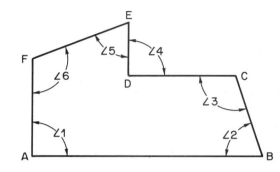

2. Name each of the following angles in two additional ways.

   a. ∠1 _____

   b. ∠CBF _____

   c. ∠3 _____

   d. ∠ECB _____

   e. ∠5 _____

   f. ∠BCD _____

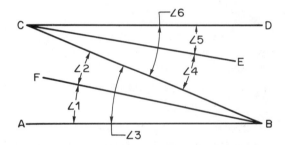

Types of Angles

3. Identify each of the following angles as acute, obtuse, right, straight, or reflex.

   a. ∠A _____   g. ∠ABC _____

   b. ∠ABF _____   h. ∠BCF _____

   c. ∠CBF _____   i. ∠AEC _____

   d. ∠DCE _____   j. ∠EFC _____

   e. ∠BFE _____   k. ∠1 _____

   f. ∠BFC _____

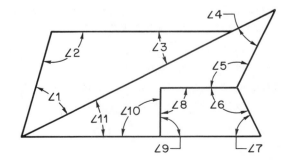

4. Name all pairs of adjacent angles shown in the figure.

   _____

   _____

   _____

5. Alternate interior angles and corresponding angles are shown in the figure.

   a. Name all pairs of alternate interior angles.   _____

   b. Name all pairs of corresponding angles.   _____

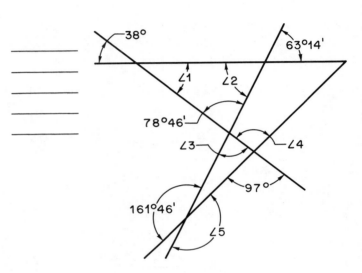

Applications of Geometric Principles

Solve the following problems.

6. Determine the values of ∠1 through ∠5.   _____
                                               _____
                                               _____
                                               _____
                                               _____

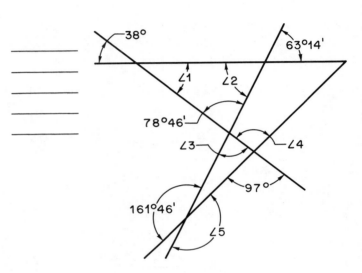

7. Determine the values of ∠2, ∠3, and ∠4 for these given values of ∠1.

   a. ∠1 = 28°   _____ _____ _____

   b. ∠1 = 33°17′   _____ _____ _____

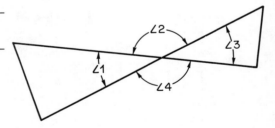

8. Given: AB ∥ CD. Determine the values of ∠2 through ∠8 for these given values of ∠1.

   a. ∠1 = 68°   _____ _____ _____
                   _____ _____ _____
                   _____

   b. ∠1 = 52°55′   _____ _____ _____
                      _____ _____ _____
                      _____

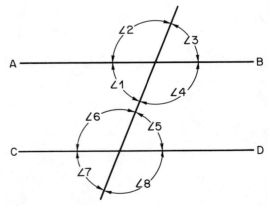

9. Given: Hole centerlines EF ‖ GH and MP ‖ KL. Determine the values of ∠1 through ∠15 for these values of ∠16.

a. ∠16 = 73° _____

_____

b. ∠16 = 87°08′ _____

_____

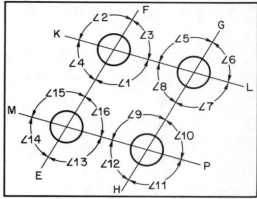

10. Given: Hole centerlines AB ‖ CD and EF ‖ GH. Determine the values of ∠1 through ∠22 for these given values of ∠23, ∠24, and ∠25.

a. ∠23 = 97°, ∠24 = 34°, and ∠25 = 102°

b. ∠23 = 112°23′, ∠24 = 27°53′, and ∠25 = 95°18′

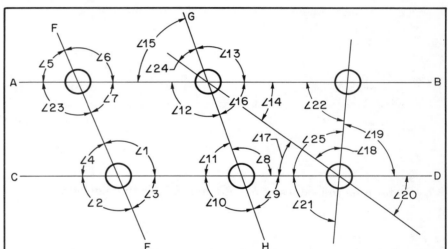

11. Given: AB ‖ CD, AC ‖ ED. Determine the value of ∠2 and ∠3 for these values of ∠1.

a. ∠1 = 68°   _____ _____

b. ∠1 = 77°26′   _____ _____

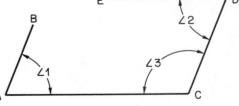

12. Given: FH ‖ GS ‖ KM and FG ‖ HK. Determine the values of ∠1, ∠2, and ∠3 for these values of ∠4.

a. ∠4 = 116°   _____ _____ _____

b. ∠4 = 107°43′   _____ _____ _____

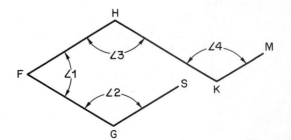

UNIT 42 *INTRODUCTION TO TRIANGLES*

OBJECTIVES

After studying this unit you should be able to

- Identify different types of triangles.
- Determine unknown angles based on the principle that all triangles contain 180°.
- Identify corresponding parts of triangles.

A *polygon* is a closed plane figure formed by three or more straight line segments. A *triangle* is a three-sided polygon; it is the simplest kind of polygon. The symbol △ means triangle.

TYPES OF TRIANGLES

A *scalene triangle* has three unequal sides. It also has three unequal angles. Triangle ABC is scalene. Sides AB, AC, and BC are unequal and angles A, B, and C are unequal.

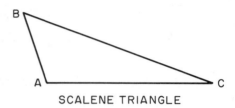
SCALENE TRIANGLE

An *isosceles triangle* has two equal sides. The equal sides are called *legs*. It also has two equal base angles. *Base angles* are the angles that are opposite the legs. In isosceles triangle RST, side RT = side ST and ∠R = ∠S.

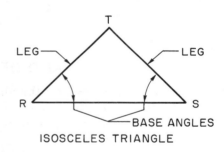
LEG   LEG
BASE ANGLES
ISOSCELES TRIANGLE

An *equilateral triangle* has three equal sides. It also has three equal angles. In equilateral triangle DEF, sides DE = DF = EF and ∠D = ∠E = ∠F.

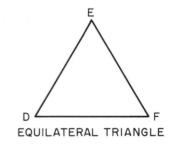
EQUILATERAL TRIANGLE

A *right triangle* has one right or 90° angle. The symbol ⌐ shown at the vertex of an angle means a right angle. The side opposite the right angle is called the *hypotenuse*. The other two sides are called *legs*. In right triangle HJK, ∠H = 90° and JK is the hypotenuse.

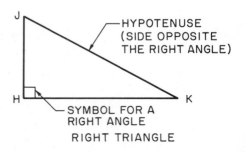
HYPOTENUSE (SIDE OPPOSITE THE RIGHT ANGLE)
SYMBOL FOR A RIGHT ANGLE
RIGHT TRIANGLE

⌐ANGLES OF A TRIANGLE

Principle 6

The sum of the angles of any triangle is equal to 180°.

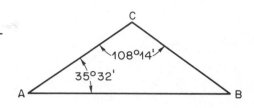

Example 1: In triangle ABC, $\angle A = 35°32'$ and $\angle C = 108°14'$. Determine $\angle B$.

$$\angle A + \angle B + \angle C = 180°$$
$$\angle B = 180° - (\angle A + \angle C)$$
$$\angle B = 179°60' - (35°32' + 108°14')$$
$$\angle B = 179°60' - 143°46'$$
$$\angle B = 36°14' \quad \text{Ans}$$

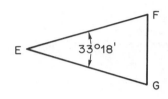

Example 2: In isosceles triangle EFG, EF = EG and $\angle E = 33°18'$. Determine $\angle F$ and $\angle G$.

$$\angle E + \angle F + \angle G = 180°$$
$$180° - \angle E = \angle F + \angle G$$
$$180° - 33°18' = \angle F + \angle G$$
$$146°42' = \angle F + \angle G$$

Since $\angle F = \angle G$, $\angle F$ and $\angle G$ each $= \dfrac{146°42'}{2} = 73°21'$ Ans

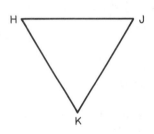

Example 3: Triangle HJK is equilateral. Determine $\angle H$, $\angle J$, and $\angle K$.

$$\angle H + \angle J + \angle K = 180°$$

Since $\angle H = \angle J = \angle K$, each angle $= \dfrac{180°}{3} = 60°$ Ans

⌐CORRESPONDING PARTS OF TRIANGLES

It is essential to develop the ability to identify corresponding angles and sides of two or more triangles. Corresponding sides and angles between triangles are not determined by the positions of the triangles. The smallest angle of a triangle lies opposite the shortest side and the largest angle of a triangle lies opposite the longest side. *Corresponding angles* between two triangles are determined by comparing the lengths of the sides which lie opposite the angles. *Corresponding sides* between two triangles are determined by comparing the sizes of the angles which lie opposite the sides.

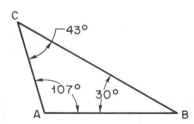

Example 1: In triangle ABC, determine the longest, next longest, and shortest sides.

The longest side is CB since it lies opposite the largest angle, 107°.   Ans

The next longest side is AB since it lies opposite the next largest angle, 43°.   Ans

The shortest side is AC since it lies opposite the smallest angle, 30°.   Ans

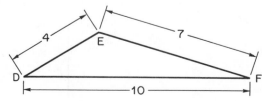

Example 2: In triangle DEF, determine the largest, next largest, and smallest angle. All dimensions are in inches.

The largest angle is $\angle E$ since it lies opposite the longest side, 10 inches.   Ans

The next largest angle is $\angle D$ since it lies opposite the next longest side, 7 inches.   Ans

The smallest angle is $\angle F$ since it lies opposite the shortest side, 4 inches.   Ans

Example 3: In triangles ABC and FED, determine the pairs of corresponding angles between the two triangles. All dimensions are in millimeters.

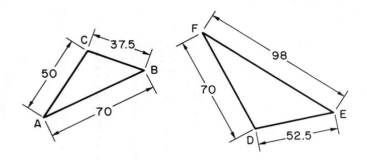

Angle C corresponds to ∠D since each angle lies opposite the longest side of each triangle.

Angle B corresponds to ∠E since each angle lies opposite the next longest side of each triangle.

Angle A corresponds to ∠F since each angle lies opposite the shortest side of each triangle.

APPLICATION

Types of Triangles

Identify each of the triangles 1–8 as scalene, isosceles, equilateral, or right.

1. _____

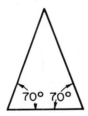

5. _____

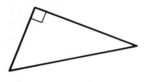

2. All dimensions are in inches.   _____

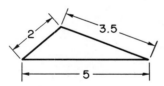

6. All dimensions are in millimeters.   _____

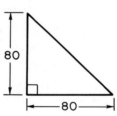

3. _____

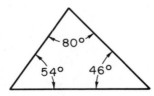

7. _____

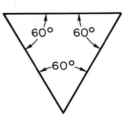

4. All dimensions are in millimeters.   _____

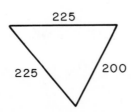

8. All dimensions are in inches.   _____

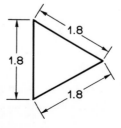

Angles of a Triangle

Solve the following problems.

9. Find the value of ∠A + ∠B + ∠C. _____

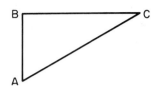

10. Find the value of the unknown angles for these given angle values.

    a. If ∠1 = 56° and ∠2 = 88°, find ∠3.
    _____

    b. If ∠2 = 79° and ∠3 = 46°, find ∠1.
    _____

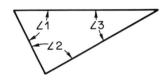

11. Find the value of the unknown angles for these given angle values.

    a. If ∠4 = 32°43′ and ∠5 = 119°17′, find ∠6.
    _____

    b. If ∠5 = 123°17′13″ and ∠6 = 27°, find ∠4.
    _____

12. Find the value of the unknown angles for these given angle values.

    a. If ∠A = 19°43′, find ∠B. _____
    b. If ∠B = 67°58′, find ∠A. _____

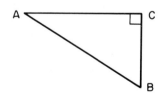

13. In triangle ABC, BC = 14.2 inches.

    a. Find AB. _____
    b. Find AC. _____

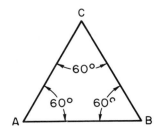

14. In triangle EFG, find the value of the unknown angles for these given angle values. All dimensions are in inches.

    a. If ∠E = 78°, find ∠G. _____
    b. If ∠G = 84°19′, find ∠F. _____

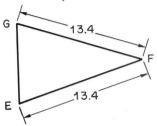

15. Find the value of the unknown angles for these given angle values. All dimensions are in millimeters.

    a. If ∠3 = 17°, find ∠1. _____
    b. If ∠3 = 25°19′, find ∠2. _____

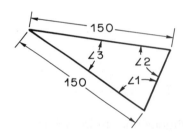

16. All dimensions are in inches.

    a. Find ∠3. _____
    b. Find ∠4. _____

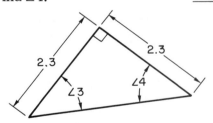

17. Find the value of the unknown angles for these given angle values.

    a. If ∠1 = 26° and ∠3 = 48°, find ∠2.
    _____

    b. If ∠1 = 28° and ∠2 = 15°, find ∠3.
    _____

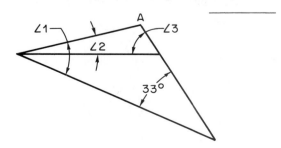

18. Hole centerlines AB ‖ CD.
    a. If ∠1 = 86°15′, find ∠2. _____
    b. If ∠2 = 67°17′, find ∠1. _____

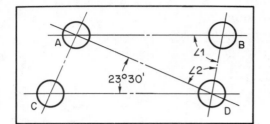

19. Find the value of the unknown angles listed.
    a. ∠3 _____
    b. ∠4 _____

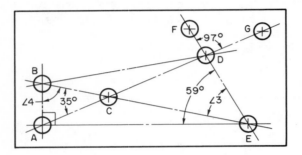

20. AB ‖ DE, BC is an extension of AB.
    a. If ∠E = 66°13′, find ∠A. _____
    b. If ∠A = 19°22′, find ∠E. _____

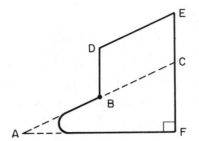

Corresponding Parts of Triangles

Determine the answers to the following problems which are based on corresponding parts.

21. All dimensions are in inches.
    a. Find the largest angle. _____
    b. Find the next largest angle. _____
    c. Find the smallest angle. _____

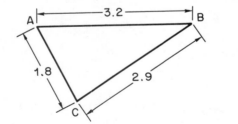

22. Refer to triangle EFG.
    a. Find the shortest side. _____
    b. Find the next shortest side. _____
    c. Find the longest side. _____

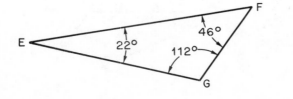

23. Identify the angle which corresponds with each angle listed. All dimensions are in millimeters.
    a. ∠A _____
    b. ∠B _____
    c. ∠1 _____

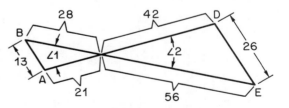

24. Identify the angle which corresponds with each angle listed. All dimensions are in inches.
    a. ∠F _____
    b. ∠G _____
    c. ∠H _____

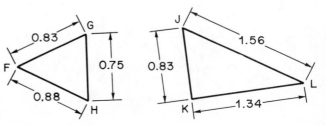

UNIT 43 GEOMETRIC PRINCIPLES FOR TRIANGLES AND OTHER COMMON POLYGONS

OBJECTIVES

After studying this unit you should be able to

- Identify similar triangles and compute unknown angles and sides.
- Compute angles and sides of isosceles, equilateral, and right triangles.
- Determine interior angles of any polygon.

CONGRUENT TRIANGLES

Two triangles are *congruent* if they are identical in size and shape. If one triangle is placed on top of the other, they fit together exactly. The symbol ≅ means congruent. Corresponding parts of congruent triangles are equal.

Example: △ABC ≅ △DEF.

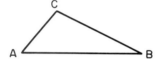 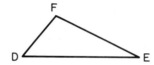

Corresponding parts of congruent triangles are equal.
∠A = ∠D, ∠B = ∠E and ∠C = ∠F.
AB = DE, AC = DF, and BC = EF.

SIMILAR TRIANGLES

Two triangles are *similar* if their corresponding angles are equal. The symbol ~ means similar. Corresponding sides of similar triangles are proportional.

Example 1: Triangles ABC and DEF have equal corresponding angles.

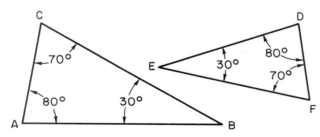

Two triangles are *similar* if their corresponding angles are equal.
△ABC ~ △DEF

Example 2: The lengths of the sides of triangles HJK and LMN are given in inches.

$$\frac{HJ}{LM} = \frac{JK}{MN} = \frac{HK}{LN}$$

$$\frac{2}{4} = \frac{4}{8} = \frac{5}{10}$$

$$\frac{1}{2} = \frac{1}{2} = \frac{1}{2}$$

 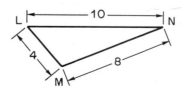

The corresponding sides are proportional.
△HJK ~ △LMN

Example 3: △PRS ~ △TWY.

All linear dimensions are in millimeters.

a. Determine the length of side PR.

b. Determine the length of side TY.

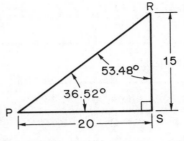
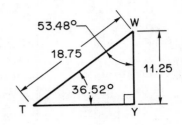

Set up proportions and solve for the unknown sides, PR and TY.

a.
$$\frac{WY}{RS} = \frac{WT}{PR}$$
$$\frac{11.25}{15} = \frac{18.75}{PR}$$
$$11.25PR = 15(18.75)$$
$$PR = \frac{15(18.75)}{11.25}$$
$$PR = 25 \text{ mm}\quad \text{Ans}$$

b.
$$\frac{WY}{RS} = \frac{TY}{PS}$$
$$\frac{11.25}{15} = \frac{TY}{20}$$
$$15TY = 20(11.25)$$
$$TY = \frac{20(11.25)}{15}$$
$$TY = 15 \text{ mm}\quad \text{Ans}$$

Principle 7

Two triangles are similar if their sides are respectively parallel.

Given: AB ∥ DE, AC ∥ DF, and BC ∥ EF.

Conclusion: △ABC ~ △DEF.

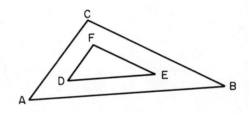

Two triangles are similar if their sides are respectively perpendicular.

Given: HJ ⊥ LM, HK ⊥ LP, and JK ⊥ MP.

Conclusion: △HJK ~ △LMP.

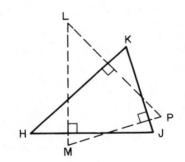

Within a triangle, if a line is drawn parallel to one side, the triangle formed is similar to the original triangle.

Given: DE ∥ BC.

Conclusion: △ADE ~ △ABC.

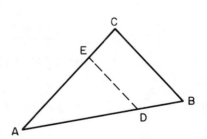

In a right triangle, if a line is drawn from the vertex of the right angle perpendicular to the opposite side, the two triangles formed and the original triangle are similar.

Given: In rt △HFG, FL ⊥ HG.

Conclusion: △FLH ~ △GLF ~ △GFH.

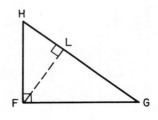

⌐ISOSCELES, EQUILATERAL, AND RIGHT TRIANGLES _____

Principle 8

In an isosceles triangle, an altitude to the base bisects the base and the vertex angle.
> An *altitude* is a line drawn from a vertex perpendicular to the opposite side.
> To *bisect* means to divide into two equal parts.

Given: Isosceles △ABC with AC = CB and line CD the altitude to base AB.

Conclusion: AD = BD and ∠1 = ∠2.

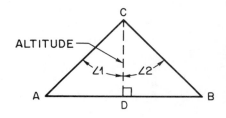

In an equilateral triangle, an altitude to any side bisects the side and the vertex angle.

Given: Equilateral △EFG with EH the altitude to FG.

Conclusion: FH = GH and ∠3 = ∠4.

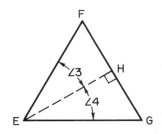

Principle 9

In a right triangle, the square of the hypotenuse is equal to the sum of the squares of the other two sides.
> This principle, called the *Pythagorean Theorem*, is often used for solving machine problems.

Example 1: In right △ABC, side a = 6 inches and side b = 8 inches. Determine side c.

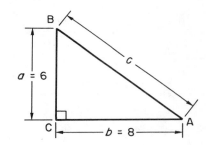

Side c is the hypotenuse.
Substitute the given values and solve for side c.

$$c^2 = a^2 + b^2$$
$$c^2 = (6\text{ in})^2 + (8\text{ in})^2$$
$$c^2 = 36\text{ sq in} + 64\text{ sq in}$$
$$c^2 = 100\text{ sq in}$$
$$c = \sqrt{100\text{ sq in}}$$
$$c = 10\text{ in}\quad\text{Ans}$$

Example 2: In right △EFG, f = 5.800 inches and hypotenuse g = 7.200 inches. Determine side e.

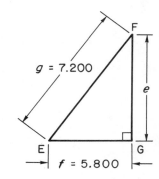

Side g is the hypotenuse.
Substitute the given values, rearrange the equation, and solve for e.

$$g^2 = e^2 + f^2$$
$$(7.200\text{ in})^2 = e^2 + (5.800\text{ in})^2$$
$$51.840\text{ sq in} = e^2 + 33.640\text{ sq in}$$
$$18.200\text{ sq in} = e^2$$
$$\sqrt{18.200\text{ sq in}} = e$$
$$e = 4.266\text{ in}\quad\text{Ans}$$

⌐POLYGONS _____

The types of polygons most common to machine trade applications in addition to triangles are squares, rectangles, parallelograms, and regular hexagons. A *regular polygon* is one which has equal sides and equal angles.

> A *square* is a regular four-sided polygon. Each angle equals 90°. In the square ABCD shown, AB = BC = CD = AD and ∠A = ∠B = ∠C = ∠D = 90°.

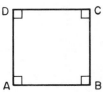

A *rectangle* is a four-sided polygon with opposite sides parallel and equal. Each angle equals 90°. In the rectangle EFGH shown, EF ∥ GH, EH ∥ FG; EF = GH, EH = FG; ∠E = ∠F = ∠G = ∠H = 90°.

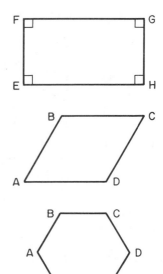

A *parallelogram* is a four-sided polygon with opposite sides parallel and equal. Opposite angles are equal. In the parallelogram ABCD shown, AB ∥ CD, AD ∥ BC; AB = CD, AD = BC; ∠A = ∠C, ∠B = ∠D.

A *regular hexagon* is a six-sided figure with all sides equal and all angles equal. In the regular hexagon ABCDEF shown, AB = BC = CD = DE = EF = AF, and ∠A = ∠B = ∠C = ∠D = ∠E = ∠F.

Principle 10

The sum of the interior angles of a polygon of N sides is equal to (N − 2) times 180°.

Example 1: In figure EFGH, ∠E = 72°, ∠F = 95°, ∠G = 108°. Determine ∠H.

Since EFGH has 4 sides, N = 4.
The sum of the 4 angles = (4 − 2)180° = 2(180°) = 360°.

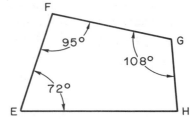

Add the 3 given angles and subtract from 360° to find ∠H.
∠H = 360° − (∠E + ∠F + ∠G)
∠H = 360° − (72° + 95° + 108°)
∠H = 360° − 275°
∠H = 85°   Ans

Example 2: Refer to figure ABCDEF and determine ∠1.

Since ABCDEF has 6 sides, N = 6.
The sum of the 6 angles = (6 − 2)180° = 4(180°) = 720°.

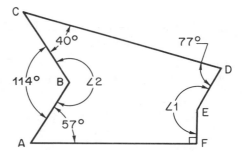

Find ∠2.
∠2 = 360° − 114° = 246°

Add the 5 known interior angles and subtract from 720° to find ∠1.
∠1 = 720° − (57° + 246° + 40° + 77° + 90°)
∠1 = 720° − 510°
∠1 = 210°   Ans

APPLICATION

Similar Triangles

1. Determine which of the following pairs of triangles (A–F) are similar. All linear dimensions are in inches.

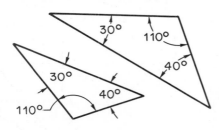

PAIR A

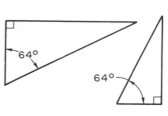

PAIR B

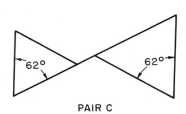

PAIR C

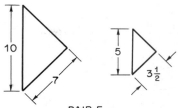

PAIR E

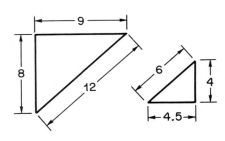

PAIR D

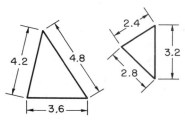

PAIR F

The similar pairs of triangles are _____ .

Solve the following problems.

2. In △ABC and △DEF, ∠A = ∠D, ∠B = ∠E, ∠C = ∠F. All dimensions are in inches.

    a. Find AC.   _____

    b. Find DE.   _____

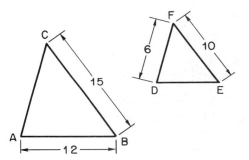

3. In the figure, ∠H = ∠P, ∠J = ∠M, ∠K = ∠L. All dimensions are in millimeters.

    a. Find HK.   _____

    b. Find LM.   _____

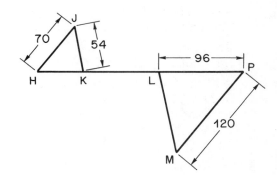

4. In △ABC and △DEF, AB ∥ DE, AC ∥ DF, BC ∥ EF.

    a. Find ∠A.   _____

    b. Find ∠F.   _____

    c. Find ∠B.   _____

    d. Find ∠E.   _____

5. Use the figure to find the value of the following angles.

    a. Find ∠1.   _____

    b. Find ∠2.   _____

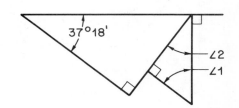

6. In △HJK, PM ∥ JK.
    a. Find ∠HPM. _____
    b. Find ∠PMK. _____

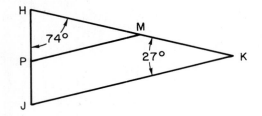

7. Refer to the figure to find these angles.
    a. ∠1 _____
    b. ∠2 _____
    c. ∠3 _____

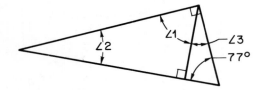

8. Refer to the figure to find these dimensions.
    All dimensions are in millimeters.
    a. Find dimension A. _____
    b. Find dimension B. _____

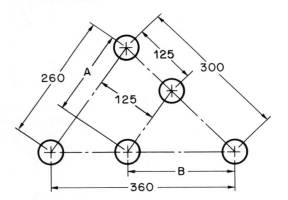

9. In this figure, AB ∥ DE and CB ∥ EF. All dimensions are in inches.
    a. Find x. _____
    b. Find y. _____

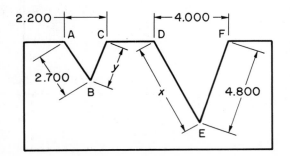

Isosceles, Equilateral, and Right Triangles

Solve the following problems.

10. All dimensions are in inches.
     a. Find x. _____
     b. Find ∠1. _____

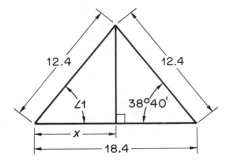

11. All dimensions are in millimeters.
     a. Find x. _____
     b. Find y. _____

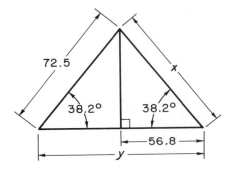

12. All dimensions are in inches.
     a. Find ∠1. _____
     b. Find ∠2. _____

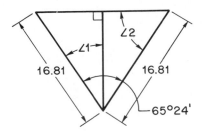

13. All dimensions are in millimeters.
     a. Find x. _____
     b. Find y. _____

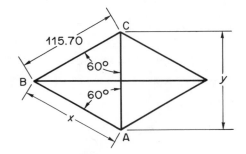

14. All dimensions are in inches.
    a. Find ∠1. _____
    b. Find x. _____

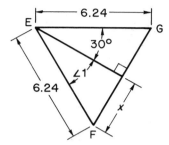

15. Refer to this figure. Using the given values, find the values of x.
    a. If $d = 9''$ and $e = 12''$, find x. _____
    b. If $d = 3''$ and $e = 4''$, find x. _____

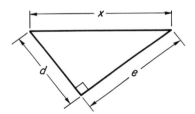

16. Using the figure and these given values, find the values of y.
    a. If $g = 111$ mm and $m = 125$ mm, find y.
       _____
    b. If $g = 155.40$ mm and $m = 175$ mm, find y.
       _____

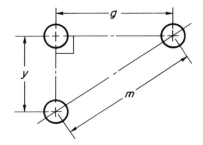

17. Using the figure and these given values, find the values of y.
    a. If Radius A = 360 mm and $x = 480$ mm, find y. _____
    b. If Radius A = 216 mm and $x = 288$ mm, find y. _____

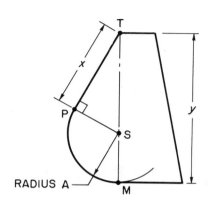

18. All dimensions are in inches.
    a. If $y = 2.800''$, find x. _____
    b. If $y = 3.000''$, find x. _____

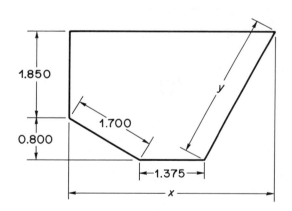

19. All dimensions are in inches.
    a. If $y = 2.145''$, find x. _____
    b. If $y = 2.265''$, find x. _____

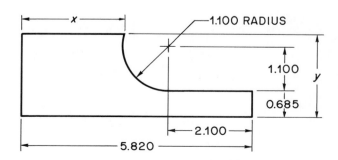

Polygons

Solve the following problems.

20. Refer to polygon ABCD.

    a. If ∠2 = 87°, find ∠1.   _____

    b. If ∠1 = 114°, find ∠2.   _____

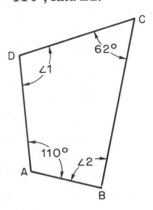

21. Use the angle values given.

    a. If ∠1 = 116°, find ∠2.   _____

    b. If ∠2 = 85°, find ∠1.   _____

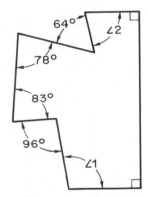

22. Use the angle values given to find ∠2.

    a. If ∠1 = 37°, find ∠2.   _____

    b. If ∠1 = 29°, find ∠2.   _____

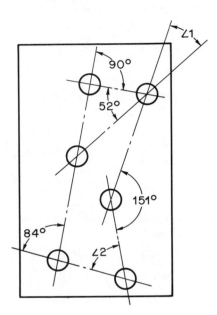

─ UNIT 44 *INTRODUCTION TO CIRCLES* ──

─OBJECTIVES_____

After studying this unit you should be able to

- Identify parts of a circle.
- Solve problems by using geometric principles which involve chords, arcs, central angles, perpendiculars, and tangents.

DEFINITIONS

A *circle* is a closed curve of which every point on the curve is equally distant from a fixed point called the center.

Refer to (1) for the following definitions:
The *circumference* is the length of the curved line which forms the circle.

A *chord* is a straight line segment that joins two points on the circle. AB is a chord.

A *diameter* is a chord that passes through the center of a circle. CD is a diameter.

A *radius* (plural radii) is a straight line segment that connects the center of the circle with a point on the circle. The radius is one-half the diameter. OE is a radius.

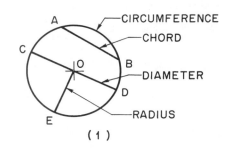

(1)

Refer to (2) for the following definitions:
An *arc* is that part of a circle between any two points on the circle. The symbol ⌢ written above the letters means arc. ÂB is an arc.

A *tangent* to a circle is a straight line that touches the circle at one point only. The point on the circle touched by the tangent is called the *point of tangency* or *tangent point*. CD is a tangent and point P is a tangent point.

A *secant* is a straight line that passes through a circle and intersects the circle at two points. EF is a secant.

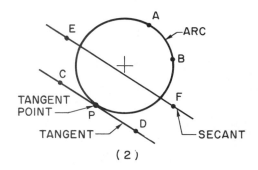

(2)

Refer to (3) for the following definitions:
A *segment* is that part of a circle which is bounded by a chord and its arc. The shaded figure ABC is a segment.

A *sector* is that part of a circle which is bounded by two radii and the arc intercepted by the radii. The shaded figure EOF is a sector.

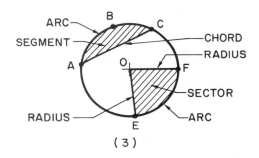

(3)

Refer to (4) for the following definitions:
A *central angle* is an angle whose vertex is at the center of a circle and whose sides are radii. Angle MON is a central angle.

An *inscribed angle* is an angle whose vertex is on the circle and whose sides are chords. Angle SRT is an inscribed angle.

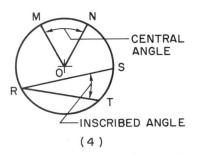

(4)

CIRCUMFERENCE FORMULA

The circumference of a circle is equal to pi (π) times the diameter. Generally, for the degree of precision required in machining applications, a value of 3.1416 is used for π.

$$C = \pi d$$
or
$$C = 2\pi r$$

where C = circumference
π = pi
d = diameter
r = radius

Example 1: Compute the circumference of a circle with a 50.70-mm diameter.
$C = \pi d = 3.1416(50.70 \text{ mm}) = 159.28 \text{ mm}$ Ans

Example 2: Determine the radius of a circle which has a circumference of 14.860 inches.
$C = 2\pi r$
$14.860 \text{ in} = 2(3.1416)(r)$
$r = 2.365 \text{ in}$ Ans

―GEOMETRIC PRINCIPLES――――――――――――――――――――――――

Principle 11

In the same circle or in equal circles, equal chords cut off equal arcs.

Given: Circle A = Circle B and chords CD = EF = GH = MS.

Conclusion: $\overset{\frown}{CD} = \overset{\frown}{EF} = \overset{\frown}{GH} = \overset{\frown}{MS}$.

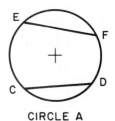 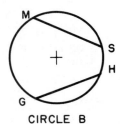

CIRCLE A   CIRCLE B

Principle 12

In the same circle or in equal circles, equal central angles cut off equal arcs.

Given: Circle D = Circle E and $\angle 1 = \angle 2 = \angle 3 = \angle 4$.

Conclusion: $\overset{\frown}{AB} = \overset{\frown}{FG} = \overset{\frown}{HK} = \overset{\frown}{MP}$.

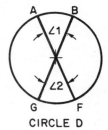 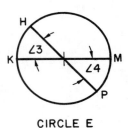

CIRCLE D   CIRCLE E

Principle 13

In the same circle or in equal circles, two central angles have the same ratio as the arcs which are cut off by the angles.

Example: Circle A = Circle B. If $\angle COD = 90°$, $\angle EOF = 50°$, $\overset{\frown}{CD} = 1.400''$, and $\overset{\frown}{GH} = 2.100''$, determine (a) the length of $\overset{\frown}{EF}$ and (b) the size of $\angle GOH$. All dimensions are in inches.

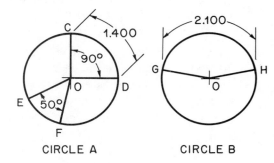

CIRCLE A   CIRCLE B

a. Set up a proportion between $\overset{\frown}{CD}$ and $\overset{\frown}{EF}$ with their respective central angles. Solve for $\overset{\frown}{EF}$.

$$\frac{\angle COD}{\angle EOF} = \frac{\overset{\frown}{CD}}{\overset{\frown}{EF}}$$

$$\frac{90°}{50°} = \frac{1.400''}{\overset{\frown}{EF}}$$

$$9\overset{\frown}{EF} = 5(1.400'')$$

$$\overset{\frown}{EF} = \frac{5(1.400'')}{9}$$

$$\overset{\frown}{EF} = 0.778''\quad \text{Ans}$$

b. Set up a proportion between $\overset{\frown}{CD}$ and $\overset{\frown}{GH}$ with their central angles. Solve for $\angle GOH$.

$$\frac{\angle COD}{\angle GOH} = \frac{\overset{\frown}{CD}}{\overset{\frown}{GH}}$$

$$\frac{90°}{\angle GOH} = \frac{1.400''}{2.100''}$$

$$1.400(\angle GOH) = 90°(2.100)$$

$$\angle GOH = \frac{90°(2.100)}{1.400}$$

$$\angle GOH = 135°\quad \text{Ans}$$

Principle 14

A line drawn from the center of a circle perpendicular to a chord bisects the chord and the arc cut off by the chord. The perpendicular bisector of a chord passes through the center of a circle.

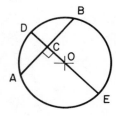

Given: Diameter DE ⊥ chord AB.

Conclusion: AC = BC and $\overset{\frown}{AD} = \overset{\frown}{BD}$ and $\overset{\frown}{AE} = \overset{\frown}{BE}$.

The use of Principle 14 with the Pythagorean Theorem (Principle 9) has wide practical application in the machine trades.

Example: Holes A, B, and C are to be drilled in the plate shown. The centers of holes A and C lie on a 280-mm diameter circle. The center of hole B lies on the intersection of chord AC and segment OB which is perpendicular to AC. Compute working dimensions F, G, and H. All dimensions are in millimeters.

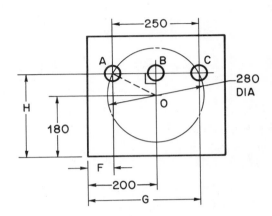

Compute dimension F: Applying Principle 14, AC is bisected by OB.

$\quad$ AB = BC = 250 mm ÷ 2 = 125 mm

$\quad$ F = 200 mm − 125 mm = 75 mm $\quad$ Ans

Compute dimension G.

$\quad$ G = 200 mm + 125 mm = 325 mm $\quad$ Ans

Compute dimension H. In right △ABO, AB = 125 mm, AO = 280 mm ÷ 2 = 140 mm

Compute OB by applying the Pythagorean Theorem (Principle 9).

$$AO^2 = OB^2 + AB^2$$
$$(140 \text{ mm})^2 = OB^2 + (125 \text{ mm})^2$$
$$OB = 63.05 \text{ mm}$$

H $=$ 180 mm + 63.05 mm = 243.05 mm $\quad$ Ans

Principle 15

A line perpendicular to a radius at its extremity is tangent to the circle. A tangent is perpendicular to a radius at its tangent point.

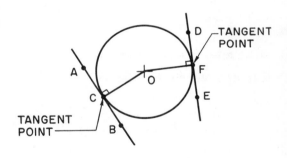

Example 1:

$\quad$ Given: Line AB ⊥ to a radius CO at point C.

$\quad$ Conclusion: Line AB is a tangent.

Example 2:

$\quad$ Given: Tangent DE passes through point F of radius FO.

$\quad$ Conclusion: Tangent DE ⊥ radius FO.

Principle 16

Two tangents drawn to a circle from a point outside the circle are equal. The angle at the outside point is bisected by a line drawn from the point to the center of the circle.

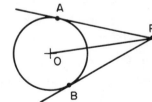

Example 1:

$\quad$ Given: Tangents AP and BP are drawn to the circle from point P.

$\quad$ Conclusion: AP = BP.

Example 2:

$\quad$ Given: Line OP which extends from outside point P to center O.

$\quad$ Conclusion: ∠APO = ∠BPO.

Principle 17

If two chords intersect inside a circle, the product of the two segments of one chord is equal to the product of the two segments of the other chord.

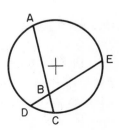

Example 1:

$\quad$ Given: Chords AC and DE intersect at point B.

$\quad$ Conclusion: AB(BC) = BD(BE).

Example 2: If AB = 7.5 inches, BC = 2.8 inches, and BD = 2.1 inches, determine the length of BE.

$$AB(BC) = BD(BE)$$
$$7.5(2.8) = 2.1(BE)$$
$$21.0 = 2.1BE$$
$$BE = 10.0 \text{ inches} \quad \text{Ans}$$

APPLICATION

Definitions

Name each of the parts of circles for the following problems.

1. a. AB   _____
   b. CD   _____
   c. EO   _____
   d. Point O _____

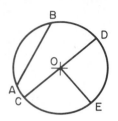

3. a. M   _____
   b. P   _____
   c. SO   _____
   d. TO   _____
   e. RW   _____
   f. $\overset{\frown}{RW}$ _____

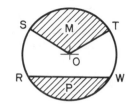

2. a. $\overset{\frown}{GF}$ _____
   b. HK   _____
   c. LM   _____
   d. GF   _____
   e. Point P _____

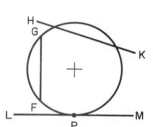

4. a. $\angle 1$ _____
   b. $\angle 2$ _____
   c. AO   _____
   d. CD   _____
   e. CE   _____
   f. $\overset{\frown}{AB}$ _____

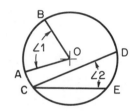

Circumference Formula

Use $C = \pi d$ or $C = 2\pi r$ where C = circumference
π = 3.1416
d = diameter
r = radius

5. Determine the unknown value for each of the following problems. Round the answers to 3 decimal places.

   a. If d = 8.000″, find C.   _____
   b. If d = 30.000 mm, find C.   _____
   c. If r = 18.600 mm, find C.   _____
   d. If r = 2.850″, find C.   _____
   e. If C = 35.000″, find d.   _____
   f. If C = 218.000 mm, find d.   _____
   g. If C = 312.000 mm, find r.   _____
   h. If C = 7.680″, find r.   _____

6. Determine the length of wire, in feet, in a coil of 60 turns. The average diameter of the coil is 30 inches. Round the answer to 1 decimal place.   _____

7. A pipe with a wall thickness of 5.00 millimeters has an outside diameter of 87.50 millimeters. Compute the inside circumference of the pipe. Round the answer to 2 decimal places.   _____

8. The flywheel of a machine has a 0.80-meter diameter and revolves 240 times per minute. How many meters does a point on the outside of the flywheel rim travel in 5 minutes? Round the answer to 1 decimal place.   _____

Geometric Principles

Solve the following problems. These problems are based on principles 11–14, although a problem may require the application of two or more of any of the principles.

9. △ABC is equilateral. All dimensions are in inches.

   a. Find $\overset{\frown}{AB}$. _____

   b. Find $\overset{\frown}{BC}$. _____

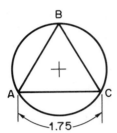

10. All dimensions are in inches.

    a. Find $\overset{\frown}{AB}$. _____

    b. Find $\overset{\frown}{BC}$. _____

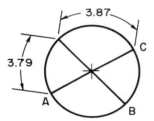

11. a. If $\overset{\frown}{EF}$ = 160 mm, find $\overset{\frown}{HP}$. _____

    b. If $\overset{\frown}{HP}$ = 284 mm, find $\overset{\frown}{EF}$. _____

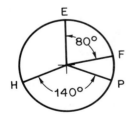

12. a. If $\overset{\frown}{SW}$ = 3.800″ and $\overset{\frown}{TM}$ = 5.700″, find ∠1.

    _____

    b. If $\overset{\frown}{TM}$ = 4.128″ and $\overset{\frown}{SW}$ = 2.064″, find ∠1.

    _____

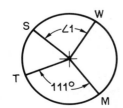

13. a. If AB = 5.378″ and $\overset{\frown}{AC}$ = 3.782″, find
    (1) DB and (2) $\overset{\frown}{ACB}$. _____

    b. If DB = 3.017″ and $\overset{\frown}{ACB}$ = 7.308″, find
    (1) AB and (2) $\overset{\frown}{CB}$. _____

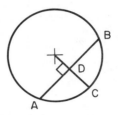

14. Find $\overset{\frown}{HK}$ when $\overset{\frown}{EF}$ = 21.23 mm. _____

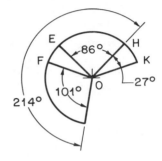

15. All dimensions are in inches.

    a. If ∠1 = 240°, find $\overset{\frown}{ABC}$. _____

    b. If $\overset{\frown}{ABC}$ = 2.500″, find ∠1. _____

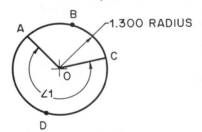

16. All dimensions are in inches.

    a. If x = 5.100″, find ∠1. _____

    b. If x = 4.750″, find ∠1. _____

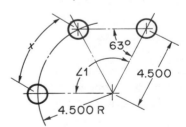

17. a. If radius x = 7.500″ and y = 4.500″, find PM. _____

    b. If radius x = 8.000″ and y = 4.800″, find PM. _____

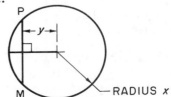

18. The circumference of this circle is 15.500".
     a. If x = 3.100", find $\angle 1$.   _____
     b. If $\angle 1$ = 40°, find x.   _____

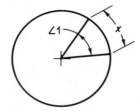

19. Determine the centerline distance between
     hole A and hole B for these values.
     a. Radius x = 8.000" and DO = 2.100"

        _____

     b. Radius x = 1.200" and DO = 0.700"

        _____

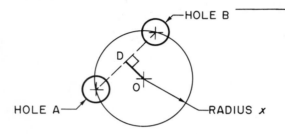

Solve the following problems. These problems are
based on principles 15–17, although a problem may
require the application of two or more of any of
the principles.

20. Point P is a tangent point and $\angle 1$ = 107°18'.
     a. If $\angle 2$ = 42°12', find (1) $\angle$E and (2) $\angle$F.

        _____   _____

     b. If $\angle 2$ = 49°53', find (1) $\angle$E and (2) $\angle$F.

        _____   _____

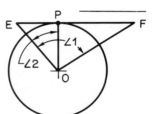

21. AB and CB are tangents.
     a. If y = 137.20 mm and $\angle$ABC = 67°, find
        (1) $\angle 1$ and (2) x.   _____   _____
     b. If x = 207.70 mm and $\angle 1$ = 33°49', find
        (1) $\angle$ABC and (2) y.   _____   _____

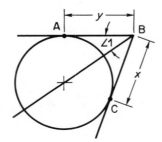

22. Point A is a tangent point. All dimensions are
     in inches.
     a. If y = 1.400", find x.   _____
     b. If y = 1.800", find x.   _____

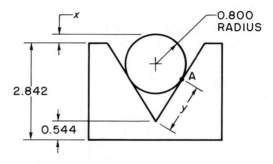

23. Points E, G, and F are tangent points.
     a. If $\angle 1$ = 117°, find $\angle 2$.   _____
     b. If $\angle 1$ = 122°15', find $\angle 2$.   _____

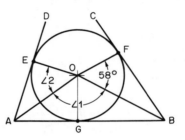

24. All dimensions are in millimeters.
     a. If EK = 150.00 mm, find GK.   _____
     b. If GK = 120.00 mm, find EK.   _____

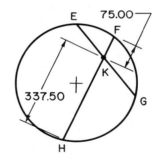

25. All dimensions are in inches.
     a. If PT = 1.800", find x.   _____
     b. If PT = 2.000", find x.   _____

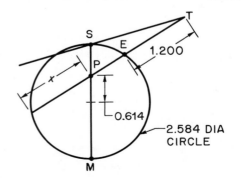

UNIT 45 *ARCS AND ANGLES OF CIRCLES*

OBJECTIVES

After studying this unit you should be able to

- Solve problems by using geometric principles which involve angles formed inside, on, and outside a circle.
- Solve problems by using geometric principles which involve internally and externally tangent circles.

ANGLES FORMED INSIDE A CIRCLE

Principle 18

A central angle is equal to its intercepted arc.

(An *intercepted arc* is an arc which is cut off by a central angle.)

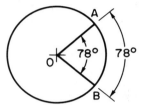

Given: $\widehat{AB} = 78°$.

Conclusion: $\angle AOB = 78°$.

An angle formed by two chords which intersect inside a circle is equal to one-half the sum of its two intercepted arcs.

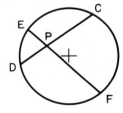

Example 1:

Given: Chords CD and EF intersect at point P.

Conclusion: $\angle EPD = \frac{1}{2} (\widehat{CF} + \widehat{DE})$.

Example 2: If $\widehat{CF} = 106°$ and $\widehat{ED} = 42°$, determine $\angle EPD$.

$\angle EPD = \frac{1}{2} (106° + 42°) = 74°$ Ans

Example 3: If $\angle EPD = 64°$ and $\widehat{CF} = 96°$, determine DE.

$\angle EPD = \frac{1}{2} (\widehat{CF} + \widehat{DE})$

$64° = \frac{1}{2} (96° + \widehat{DE})$

$64° = 48° + \frac{1}{2} \widehat{DE}$

$16° = \frac{1}{2} \widehat{DE}$

$\widehat{DE} = 32°$ Ans

An inscribed angle is equal to one-half of its intercepted arc.

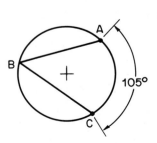

Given: $\widehat{AC} = 105°$.

Conclusion: $\angle ABC = \frac{1}{2} \widehat{AC} = \frac{1}{2} (105°) = 52°30'$.

ARC LENGTH FORMULA

Consider a complete circle as an arc of 360°. The ratio of the number of degrees of an arc to 360° is the fractional part of the circumference that is used to find the length of an arc. **The length of an arc equals the ratio of the number of degrees of the arc to 360° times the circumference.**

$$\text{Arc Length} = \frac{\text{Arc Degrees}}{360°} (2\pi r)$$

or

$$\text{Arc Length} = \frac{\text{Central Angle}}{360°} (2\pi r)$$

Example 1: $\overset{\frown}{ABC}$ = 130° and the radius is 120 mm. Determine the arc length $\overset{\frown}{ABC}$.

$$\text{Arc Length} = \frac{\text{Arc Degrees}}{360°} (2\pi r)$$

$$\text{Arc Length} = \frac{130°}{360°} [2(3.1416)(120 \text{ mm})]$$

$$\text{Arc Length} = 272.272 \text{ mm}   \text{Ans}$$

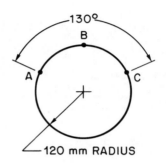

130°

120 mm RADIUS

Example 2: The arc length of $\overset{\frown}{DEF}$ is 8.400″ and the radius is 5.000″. Determine ∠1. All dimensions are in inches.

$$\text{Arc Length} = \frac{\text{Central Angle}}{360°} (2\pi r)$$

$$8.400″ = \frac{\angle 1}{360°} [2(3.1416)(5.000″)]$$

$$\angle 1 = 96.257° \text{ or } 96°15′   \text{Ans}$$

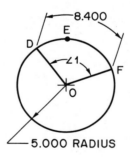

8.400

5.000 RADIUS

ANGLES FORMED ON A CIRCLE

Principle 19

An angle formed by a tangent and a chord at the tangent point is equal to one-half of its intercepted arc.

Example 1: Tangent CD meets chord AB at tangent point A and $\overset{\frown}{AEB}$ = 110°. Determine ∠CAB.

$$\angle CAB = \tfrac{1}{2} \overset{\frown}{AEB} = \tfrac{1}{2} (110°) = 55°   \text{Ans}$$

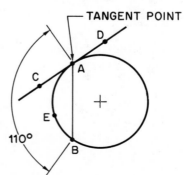

TANGENT POINT

110°

Example 2: The centers of 3 holes lie on line ABC. Line ABC is tangent to circle O at hole-center B. The hole-center D, of a fourth hole, lies on the circle. Determine ∠ABD.

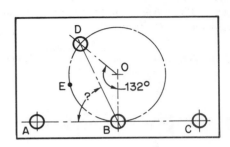

A central angle is equal to its intercepted arc (Principle 18).
$\overset{\frown}{DEB} = \angle DOB = 132°$

Apply Principle 19.
$\angle ABD = \frac{1}{2}\overset{\frown}{DEB} = \frac{1}{2}(132°) = 66°$ Ans

---

ANGLES FORMED OUTSIDE A CIRCLE

Principle 20

An angle formed at a point outside a circle by two secants, two tangents, or a secant and a tangent is equal to one-half the difference of the intercepted arcs.

TWO SECANTS

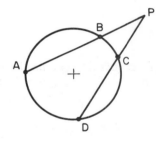

Example 1: Given: Secants AP and DP meet at point P and intercept $\overset{\frown}{BC}$ and $\overset{\frown}{AD}$.
Conclusion: $\angle P = \frac{1}{2}(\overset{\frown}{AD} - \overset{\frown}{BC})$.

Example 2: If $\overset{\frown}{AD} = 85°40'$ and $\overset{\frown}{BC} = 39°17'$, find $\angle P$.
$\angle P = \frac{1}{2}(\overset{\frown}{AD} - \overset{\frown}{BC}) = \frac{1}{2}(85°40' - 39°17') = \frac{1}{2}(46°23') = 23°11'30''$ Ans

Example 3: If $\angle P = 28°$ and $\overset{\frown}{BC} = 40°$, determine $\overset{\frown}{AD}$.
$\angle P = \frac{1}{2}(\overset{\frown}{AD} - \overset{\frown}{BC})$
$28° = \frac{1}{2}(\overset{\frown}{AD} - 40°)$
$\overset{\frown}{AD} = 96°$ Ans

TWO TANGENTS

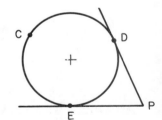

Example 1: Given: Tangents DP and EP meet at point P and intercept $\overset{\frown}{DE}$ and $\overset{\frown}{DCE}$.
Conclusion: $\angle P = \frac{1}{2}(\overset{\frown}{DCE} - \overset{\frown}{DE})$.

Example 2: If $\overset{\frown}{DCE} = 253°37'$ and $\overset{\frown}{DE} = 106°23'$, determine $\angle P$.
$\angle P = \frac{1}{2}(\overset{\frown}{DCE} - \overset{\frown}{DE}) = \frac{1}{2}(253°37' - 106°23') = \frac{1}{2}(147°14') = 73°37'$ Ans

A TANGENT AND A SECANT

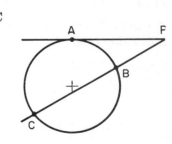

Example 1: Given: Tangent AP and secant CP meet at point P and intercept $\overset{\frown}{AC}$ and $\overset{\frown}{AB}$.
Conclusion: $\angle P = \frac{1}{2}(\overset{\frown}{AC} - \overset{\frown}{AB})$.

Example 2: If $\overset{\frown}{AC} = 126°38'$ and $\angle P = 28°50'$, determine $\overset{\frown}{AB}$.
$\angle P = \frac{1}{2}(\overset{\frown}{AC} - \overset{\frown}{AB})$
$28°50' = \frac{1}{2}(126°38' - \overset{\frown}{AB})$
$28°50' = 63°19' - \frac{1}{2}\overset{\frown}{AB}$
$\frac{1}{2}\overset{\frown}{AB} = 63°19' - 28°50'$
$\overset{\frown}{AB} = 68°58'$ Ans

⌐INTERNALLY AND EXTERNALLY TANGENT CIRCLES

Two circles that are tangent to the same line at the same point are tangent to each other. Circles are either internally or externally tangent.

Internally tangent — two circles are internally tangent if both circles are on the same side of the common tangent line.

Externally tangent — Two circles are externally tangent if the circles are on opposite sides of the common tangent line.

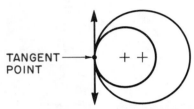

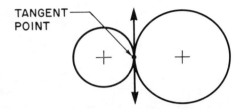

INTERNALLY TANGENT CIRCLES   EXTERNALLY TANGENT CIRCLES

Principle 21

If two circles are either internally or externally tangent, a line connecting the centers of the circles passes through the point of tangency and is perpendicular to the tangent line.

⌐INTERNALLY TANGENT CIRCLES

Example: Given: Circle D and Circle E are internally tangent at point C. D is the center of Circle D and E is the center of Circle E. Line AB is tangent to both circles at point C.

Conclusion: An extension of line DE passes through tangent point C and line CDE ⊥ tangent line AB.

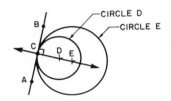

Principle 21 is often used as the basis for computing dimensions of parts on which two or more radii blend to give a smooth curved surface. This type of application is illustrated by the following example.

Example: A part is to be machined as shown. The proper locations of the two radii will result in a smooth curve from point A to point B. Note that the curve from A to B is not an arc of one circle; it is made up of two different size circles. In order to make the part, the location to the center of the 12.000-inch radius (dimension x) must be determined. Compute x. All dimensions are in inches.

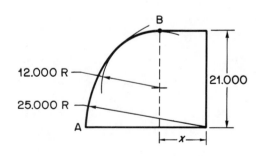

The 12.000″ radius arc and the 25.000″ radius arc are internally tangent. Apply Principle 21. A line connecting arc centers F and H passes through tangent point C.

Tangent point C is the endpoint of the 25.000″ radius, CFH = 25.000″.

Tangent point C is the endpoint of the 12.000″ radius, CF = 12.000″.

FH = 25.000″ – 12.000″ = 13.000″

Since BFE is vertical and AEH is horizontal, ∠FEH = 90°.

In right △FEH, FH = 13.000″, FE = 21.000″ – BF = 9.000″.

Apply the Pythagorean Theorem (Principle 9) to compute EH.

$$FH^2 = EH^2 + FE^2$$
$$(13.000 \text{ in})^2 = EH^2 + (9.000 \text{ in})^2$$
$$EH = 9.381 \text{ in}$$
$$x = EH = 9.381 \text{ in} \quad \text{Ans}$$

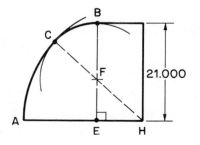

⌐EXTERNALLY TANGENT CIRCLES⎯⎯⎯⎯⎯⎯⎯⎯⎯⎯⎯⎯⎯⎯⎯⎯⎯

Example 1: Given: Circle D and Circle E are externally tangent at point C. D is the center of Circle D and E is the center of Circle E. Line AB is tangent to both circles at point C.

Conclusion: Line DE passes through tangent point C and line DE ⊥ tangent line AB at point C.

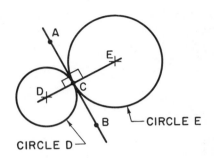

Example 2: Three holes are to be bored in a steel plate as shown. The 42.00-mm and 61.40-mm diameter holes are tangent at point D. CD is the common tangent line. Determine the distances between hole centers (AB, AC, and BC). All dimensions are in millimeters.

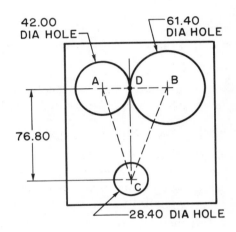

Compute AB. Apply Principle 21. Since AB connects the centers of two tangent circles, AB passes through tangent point D.

AB = AD + DB = 21.00 mm + 30.70 mm = 51.70 mm   Ans

Compute AC and BC.

Since AB connects the centers of two tangent circles, AB ⊥ tangent line DC. Triangle ADC and triangle BDC are right triangles. Apply the Pythagorean Theorem (Principle 9).

In right △ADC, AD = 21.00 mm and DC = 76.80 mm.

AC^2 = $AD^2 + DC^2$
AC^2 = $(21.00 \text{ mm})^2 + (76.80 \text{ mm})^2$
 AC = 79.62 mm   Ans

In right △BDC, DB = 30.70 mm and DC = 76.80 mm.

BC^2 = $DB^2 + DC^2$
BC^2 = $(30.70 \text{ mm})^2 + 76.80 \text{ mm})^2$
 BC = 82.71 mm   Ans

⌐APPLICATION⎯⎯⎯⎯⎯⎯⎯⎯⎯⎯⎯⎯⎯⎯⎯⎯⎯⎯⎯⎯⎯⎯⎯⎯⎯⎯⎯⎯⎯⎯

Arc Length Formula

Determine the unknown value for each of the following problems. Round the answers to 3 decimal places.

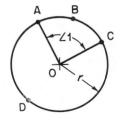

1. $\overset{\frown}{ABC} = 90°$ and $r = 4.000$ in. Find arc length $\overset{\frown}{ABC}$. ⎯⎯⎯⎯⎯

2. $\overset{\frown}{ABC} = 85°$ and $r = 60.000$ mm. Find arc length $\overset{\frown}{ADC}$. ⎯⎯⎯⎯⎯

3. Arc length $\overset{\frown}{ABC} = 510.000$ mm and $r = 120.000$ mm. Find $\angle 1$. ⎯⎯⎯⎯⎯

4. Arc length $\overset{\frown}{ADC} = 21.600$ in and $r = 5.200$ in. Find $\angle 1$. ⎯⎯⎯⎯⎯

5. Arc length $\overset{\frown}{ABC} = 18.750$ in and $\angle 1 = 72°$. Find r. ⎯⎯⎯⎯⎯

6. Arc length $\overset{\frown}{ABC} = 620.700$ mm and $\angle 1 = 69.30°$. Find r. ⎯⎯⎯⎯⎯

Geometric Principles

Solve the following problems. These problems are
based on principles 18–21, although a problem may
require the application of two or more of any of
the principles.

7. a. If $\angle 1 = 73°$, find
    (1) $\widehat{DC}$ _____
    (2) $\angle EOD$ _____
    (3) $\widehat{ABC}$ _____

  b. If $\angle 1 = 68°$, find
    (1) $\widehat{DC}$ _____
    (2) $\angle EOD$ _____
    (3) $\widehat{BCD}$ _____

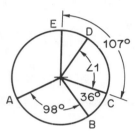

8. a. If $\angle 1 = 63°$, find
    (1) $\widehat{HK}$ _____
    (2) $\widehat{HM}$ _____

  b. If $\angle 1 = 59°47'$, find
    (1) $\widehat{HK}$ _____
    (2) $\widehat{HM}$ _____

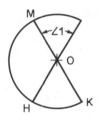

9. a. If $\widehat{PS} = 46°$, find
    (1) $\angle 1$ _____
    (2) $\angle 2$ _____

  b. If $\widehat{PS} = 39°$, find
    (1) $\angle 1$ _____
    (2) $\angle 2$ _____

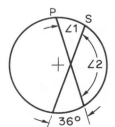

10. a. If $\widehat{DC} = 28°$, find $\widehat{AB}$.   _____
  b. If $\widehat{AB} = 131°$, find $\widehat{DC}$.   _____

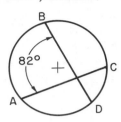

11. a. If $\angle 3 = 47°$ and $\widehat{GH} = 32°$, find
    (1) $\widehat{EF}$ _____
    (2) $\angle 4$ _____

  b. If $\angle 4 = 17°53'$ and $\widehat{EF} = 103°$, find
    (1) $\angle 3$ _____
    (2) $\widehat{GH}$ _____

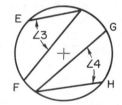

12. a. If $\angle 1 = 25°$ and $\widehat{MPT} = 95°$, find
    (1) $\widehat{KTP}$ _____
    (2) $\widehat{PT}$ _____
    (3) $\widehat{MP}$ _____

  b. If $\angle 1 = 17°30'$ and $\widehat{MPT} = 103°$, find
    (1) $\widehat{KTP}$ _____
    (2) $\widehat{PT}$ _____
    (3) $\widehat{MP}$ _____

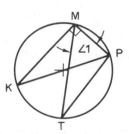

13. a. If $\widehat{AB} = 114°$, find
    (1) $\angle 1$ _____
    (2) $\angle 2$ _____

  b. If $\widehat{AB} = 110°42'$, find
    (1) $\angle 1$ _____
    (2) $\angle 2$ _____

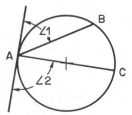

14. a. If $\overparen{EF} = 84°$, find
   (1) $\angle EFD$ _____
   (2) $\overparen{HF}$ _____
   (3) $\angle 1$ _____
 b. If $EF = 79°$, find
   (1) $\angle EFD$ _____
   (2) $\overparen{HF}$ _____
   (3) $\angle 1$ _____

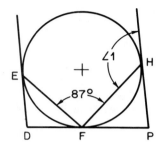

15. a. If $\overparen{ST} = 18°$ and $\overparen{SM} = 40°$, find
   (1) $\angle 1$ _____
   (2) $\angle 2$ _____
 b. If $\overparen{ST} = 23°$ and $\overparen{SM} = 39°$, find
   (1) $\angle 1$ _____
   (2) $\angle 2$ _____

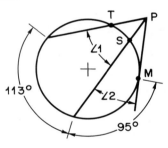

16. a. If $\overparen{AB} = 72°$ and $\overparen{CD} = 50°$, find
   (1) $\angle 1$ _____
   (2) $\angle 2$ _____
   (3) $\angle 3$ _____
 b. If $\overparen{CD} = 43°$ and $\overparen{AD} = 106°$, find
   (1) $\angle 1$ _____
   (2) $\angle 2$ _____
   (3) $\angle 3$ _____

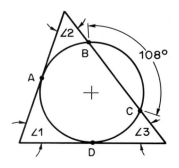

17. a. If $\angle 1 = 28°$ and $\angle 2 = 62°$, find
   (1) $\overparen{DH}$ _____
   (2) $\overparen{EDH}$ _____
 b. If $\angle 1 = 25°$ and $\angle 2 = 67°$, find
   (1) $\overparen{DH}$ _____
   (2) $\overparen{EDH}$ _____

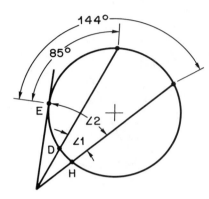

18. a. If Dia A = 3.756″ and Dia B = 1.622″, find x. _____
 b. If $x = 0.975″$ and Dia B = 1.026″, find Dia A. _____

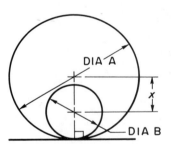

19. a. If $x = 25.07$ mm and $y = 29.85$ mm, find Dia A. _____
 b. If $x = 79.58$ mm and $y = 117.96$ mm, find Dia A. _____

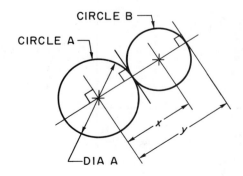

20. a. If ∠1 = 67° and ∠2 = 93°, find
     (1) $\widehat{AB}$ _____
     (2) $\widehat{DE}$ _____
   b. If ∠1 = 75° and ∠2 = 85°, find
     (1) $\widehat{AB}$ _____
     (2) $\widehat{DE}$ _____

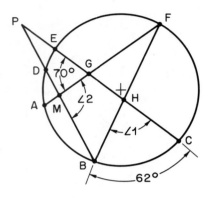

21. All dimensions are in inches.
   a. If Dia A = 1.000″, find x. _____
   b. If Dia A = 0.800″, find x. _____

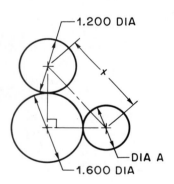

22. All dimensions are in inches.
   a. If y = 0.350″, find x. _____
   b. If y = 0.410″, find x. _____

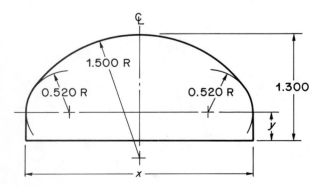

23. AC is a diameter.
   a. If ∠2 = 24°, find ∠1. _____
   b. If ∠2 = 31°14′, find ∠1. _____

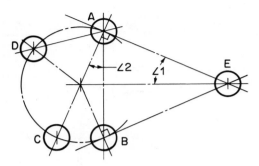

24. All dimensions are in millimeters.
   a. If radius R = 15.24 mm, find y.

     _____

   b. If radius R = 16.51 mm, find y.

     _____

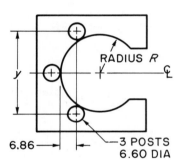

25. Points A, B, C, D, and E are tangent points.
   a. If $\widehat{AB}$ = 46° and $\widehat{DE}$ = 66°, find ∠1.

     _____

   b. If $\widehat{AB}$ = 53° and $\widehat{DE}$ = 70°, find ∠1.

     _____

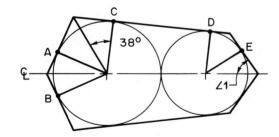

⌐UNIT 46 *FUNDAMENTAL GEOMETRIC CONSTRUCTIONS* ⎯⎯⎯⎯⎯

⌐OBJECTIVES⎯⎯⎯⎯⎯⎯⎯⎯⎯⎯⎯⎯⎯⎯⎯⎯⎯⎯⎯⎯⎯⎯⎯⎯

After studying this unit you should be able to

- Make constructions which are basic to the machine trades.
- Lay out typical machine shop problems using the methods of construction.

A knowledge of basic geometric constructions done with a compass or dividers and a steel rule is required of a machinist in laying out work. The constructions are used in determining stock allowances and reference locations on castings, forgings, and sheet stock.

For certain jobs where wide dimensional tolerances are permissible, the most practical and efficient way of producing a part may be by scribing and centerpunching locations. Layout dimensions are sometimes used as a reference for machining complex parts which require a high degree of precision. Locations lightly scribed on a part are used as a precaution to insure that the part or table movement is in the proper direction. It is particularly useful in operations which require part rotation or repositioning.

There are many geometric constructions, some of which are relatively complex. The constructions presented in this book are those which are most basic and common to a wide range of practical applications.

⌐CONSTRUCTION 1. TO CONSTRUCT A PERPENDICULAR BISECTOR OF A LINE SEGMENT⎯⎯⎯⎯⎯⎯⎯⎯⎯⎯⎯⎯⎯⎯⎯⎯⎯⎯⎯⎯⎯⎯⎯⎯⎯⎯⎯⎯⎯⎯

Required: Construct a perpendicular bisector to line segment AB.

Procedure:

- With endpoint A as a center and using a radius equal to more than half AB, draw arcs above and below AB.
- With endpoint B as a center and with the same radius used at A, draw arcs above and below AB which intersect the first pair of arcs.
- Draw a connecting line between the intersection of the arcs above and below AB. Line CD is perpendicular to AB and point O is the midpoint of AB.

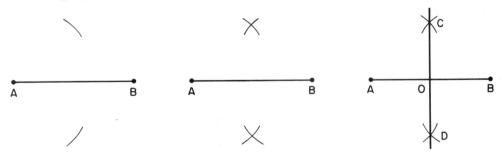

Practical Application

Locate the center of a circle.

Solution

The perpendicular bisector of a chord passes through the center of the circle. The center of a circle is located by drawing two chords and constructing a perpendicular bisector to each chord. The intersection of the two perpendicular bisectors locates the center of the circle. The construction lines are shown in this figure.

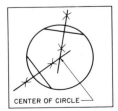

CENTER OF CIRCLE

CONSTRUCTION 2. TO CONSTRUCT A PERPENDICULAR TO A LINE SEGMENT AT A GIVEN POINT ON THE LINE SEGMENT

Required: Construct a perpendicular at point O on line segment AB.

Procedure:

- With given point O as a center, and with a radius of any convenient length, draw arcs intersecting AB at points C and D.
- With C as a center, and with a radius greater than OC, draw an arc. With D as a center, and with the same radius used at C, draw an arc which intersects the first arc at E.
- Draw a line connecting point E and point O. Line EO is perpendicular to line AB at point O.

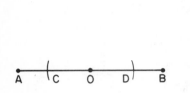 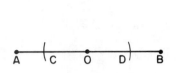 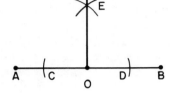

Practical Application

A triangular piece is to be scribed and cut. The piece is laid out as follows:

The 22-inch base is measured and marked off.

The $10\frac{7}{64}$-inch distance is measured and marked off at point A on the baseline.

From point A, a perpendicular to the baseline is constructed.

The construction lines are shown.

The $7\frac{1}{2}$-inch distance is measured and marked off at point B on the constructed perpendicular.

Lines are scribed connecting vertex B with the endpoints of the baseline.

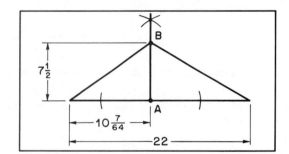

CONSTRUCTION 3. TO CONSTRUCT A LINE PARALLEL TO A GIVEN LINE AT A GIVEN DISTANCE

Required: Construct a line parallel to line AB at a given distance of 1 inch.

Procedure:

- Set the compass to the required distance (1 inch).
- With any points C and D as centers on AB, draw arcs with the given distance (1 inch) as the radius.
- Draw a line, EF, that touches each arc at one point (the tangent point). Line EF is parallel to line AB and EF is 1 inch from AB.

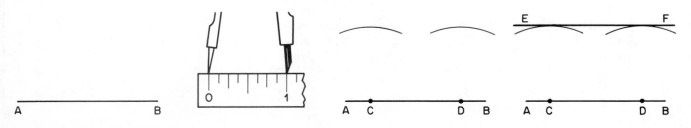

Practical Application

The cutout shown in the drawing is laid out on a sheet as follows: All dimensions are in millimeters.

The centerline (₵) is scribed and the 310-mm distance is marked off. Points A and B are the endpoints of the 310-mm segment.

From points A and B perpendiculars are constructed. The perpendiculars are extended more than 70 mm (140 mm ÷ 2) above and below AB.

With points C and D as centers on AB, 70-mm radius arcs are drawn above and below AB. A line is scribed above and a line is scribed below AB touching the pairs of arcs. The lines are extended to intersect the perpendiculars constructed. The points of intersection are E, F, G, and H.

From point A and from point B on AB, 40-mm distances are marked off. Point J and point K are the endpoints.

Lines are scribed connecting J to E and G and connecting K to F and H. Scribed figure JEFKHG is the required cutout.

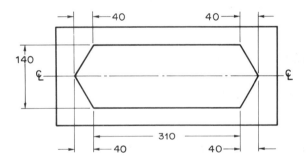 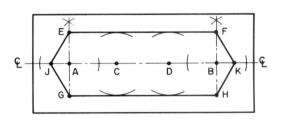

CONSTRUCTION 4. TO BISECT A GIVEN ANGLE

Required: Bisect ∠ABC.

Procedure:

- With point B as the center draw an arc intersecting sides BA and BC at points D and E.
- With D as the center, and with a radius equal to more than half the distance DE, draw an arc. With E as the center, and with the same radius, draw an arc. The intersection of the two arcs is point F.
- Draw a line from point B to point F. Line BF is the bisector of ∠ABC.

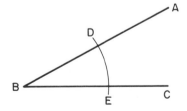 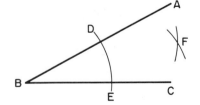 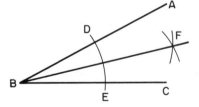

Practical Application

The centers of the three $\frac{1}{2}$-inch diameter holes in the mounting plate shown are located and center punched. Two $\frac{1}{4}$-inch diameter holes are located by constructing the bisector of ∠ABC as shown and marking and center punching the $1\frac{3}{8}$-inch and $4\frac{1}{4}$-inch hole center locations on the bisector.

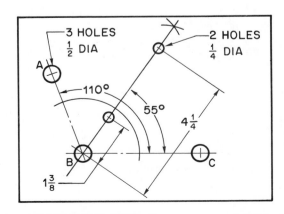

CONSTRUCTION 5. TO CONSTRUCT TANGENTS TO A CIRCLE FROM AN OUTSIDE POINT

Required: Construct tangents to given circle O from given outside point P.

Procedure:

- Draw a line segment connecting center O and point P. Bisect OP. Point A is the midpoint of OP.
- With point A as the center and AP as a radius, draw arcs intersecting circle 0 at points B and C. Points B and C are tangent points.
- Connect points B and P, and C and P. Line segments BP and CP are tangents.

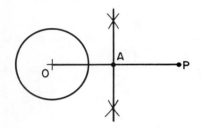 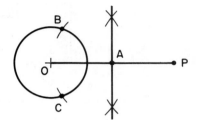 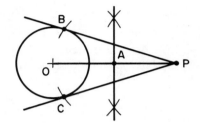

Practical Application

A piece is to be made as shown in the drawing. All dimensions are in millimeters. The piece is laid out as follows:

A baseline is scribed and AB (170 mm) is marked off.

Distance OA (152 mm) is set on dividers and with OA as the radius, an arc is scribed. Distance OB (104 mm) is set on dividers and with OB as the radius, an arc is scribed to intersect with the OA radius arc. The intersection of the arcs locates center O of the 42-mm radius circle.

Dividers are set to the 42-mm radius dimension, and the circle is scribed from center O.

Tangents to the circle from points A and B are constructed resulting in tangent points C and D and tangent line segments AC and BD. The piece is now laid out and ready to be cut to the scribed lines.

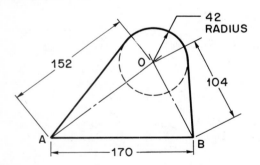 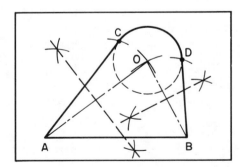

CONSTRUCTION 6. TO DIVIDE A LINE SEGMENT INTO A GIVEN NUMBER OF EQUAL PARTS

Required: Divide line segment AB into three equal parts.

Procedure:

- From point A, draw line AC forming any convenient angle with AB.
- On AC, with a compass, lay off any three equal segments, AD, DE, and EF.
- Connect point F with point B. With centers at points F, E, and D, draw arcs of equal radii. The arc with a center at point F intersects AC at point G and BF at point H. Set distance GH on the compass and mark off this distance on the other two arcs. The points of intersection are K and M.

- Connect points E and K, and D and M, extending the lines past AB. Line AB is divided into three equal segments; AP = PS = SB. Note: line segment AB can be divided into any required number of equal segments by laying off the required number of equal segments on AC and following the procedure given.

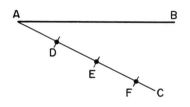 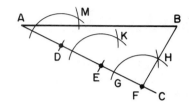 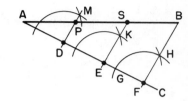

Practical Application

Six holes are to be equally spaced within a distance of $2\frac{11}{16}$ inches. Since six holes are required, there will be five equal spaces between holes. Dividing $2\frac{11}{16}$ inches by 5 results in fractional distances which are difficult to accurately measure or transfer, such as $\frac{8.6}{16}$ inch, $\frac{17.2}{32}$ inch, or $\frac{34.4}{64}$ inch. By careful construction, the hole centers are accurately located as shown in the figure.

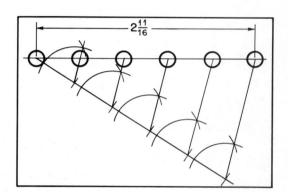

APPLICATION

Construction 1 and 2 Applications

Show construction lines and arcs for each of these problems.

1. Trace each line segment in problems a–d and construct perpendicular bisectors to each segment.

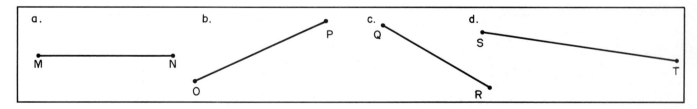

2. Trace each line in problems a–c, and construct perpendiculars to each line at the given points on the lines.

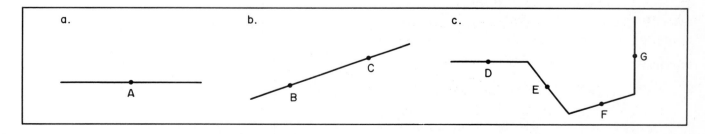

3. With a compass, draw a circle 2 inches in diameter. By construction, locate the center of the circle.

4.   Lay out a figure as follows:

   a. Draw a horizontal line and mark off a distance of 2 1/2 inches. Label the left endpoint of the 2 1/2-inch line segment point A and label the right endpoint point D.

   b. From point A and above point A, construct a perpendicular to AD. Mark off a distance of 1 7/8 inches on the perpendicular from point A. Label the top endpoint point B.

   c. From point B and to the right of point B, construct a perpendicular to AB. Mark off a distance of 2 1/2 inches on the perpendicular from point B. Label the right endpoint point C.

   d. From point C and below point C, construct a perpendicular to BC. Mark off a distance of 1 7/8 inches on the perpendicular from point C. If your constructions are accurate the 1 7/8-inch distance marked off coincides with point D. What kind of a figure is formed by this construction? _____

Construction 3 and 4 Applications

Show construction lines and arcs for each of these problems.

5.   Trace each of the lines in problems a–d and construct a line parallel to the line at a distance of 1 1/2 inches.

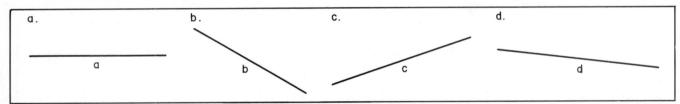

6.   Trace each of the angles a–c and construct a bisector to each.

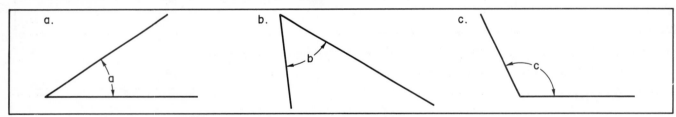

7.   Lay out the following angles by construction. Check the angle with a protractor but do not lay out angles with a protractor.

   a. $45°$ c. $67°30'$ e. $168\frac{3}{4}°$

   b. $22°30'$ d. $157\frac{1}{2}°$

8.   Lay out the plate shown. Make the layout full size using construction methods. Use a protractor only for checking. All dimensions are in inches.

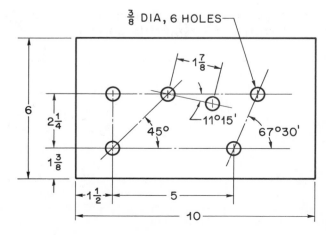

9. Lay out the gage shown. Make the layout full size using construction methods. Use a protractor only for checking. All dimensions are in inches.

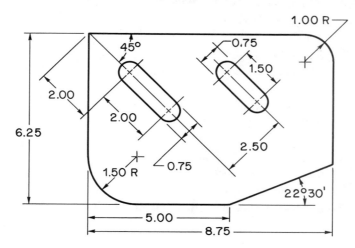

Construction 5 and 6 Applications

Show construction lines and arcs for each of these problems.

10. Trace each circle and point in problems a–c and construct tangents to the circles from the given points.

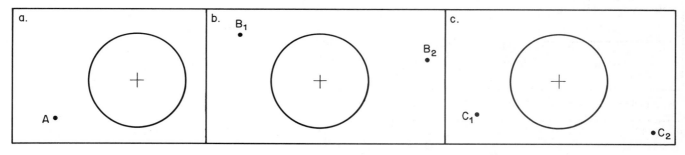

11. Trace each line segment of problems a, b, and c. Divide the given lines into the designated number of segments by means of construction.

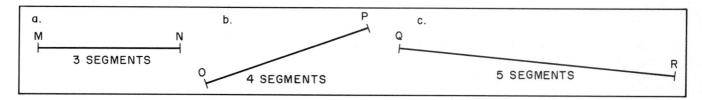

12. Lay out the template shown. Make the layout full size using construction methods. All dimensions are in inches.

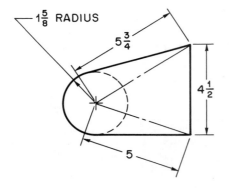

13. Lay out the cutout shown. Make the layout full size using construction methods. All dimensions are in inches.

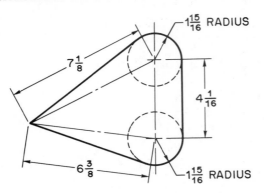

14. Trace the plate shown. Lay out three sets of holes by construction. Follow the given directions.

    Directions:

    - Bisect ∠A and construct 4 equally spaced 3/16-inch diameter holes. Make the first hole 7/8 inch from point A and the last hole 2 7/16 inches from point A.

    - Bisect ∠B and construct 8 equally spaced 1/4-inch diameter holes. Make the first hole 3/4 inch from point B and the last hole 3 11/16 inches from the first hole.

    - Bisect ∠C and construct 4 equally spaced 3/16-inch diameter holes. Make the first hole 9/16 inch from point C and the last hole 2 11/16 inches from point C.

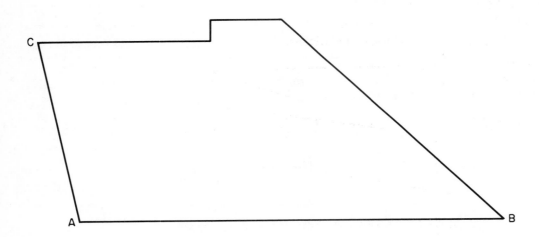

¯UNIT 47 *ACHIEVEMENT REVIEW —*
SECTION 4 _____

¯OBJECTIVE_____

    You should be able to solve the exercises and problems in this Achievement Review by applying the principles and methods covered in units 39–46.

1. Add, subtract, multiply, or divide each of the following exercises as indicated.

   a. $37°18' + 86°23'$ _____   e. $4(18°21')$ _____

   b. $38°46' + 21°32'$ _____   f. $3(7°23'43'')$ _____

   c. $136°36'28'' - 94°17'15''$ _____   g. $87° \div 2$ _____

   d. $58°14' - 44°58'$ _____   h. $103°20' \div 4$ _____

2. Determine $\angle A$. _____

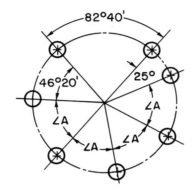

3. Given: The sum of all angles = $720°$.
   $\angle 3 = \angle 4 = \angle 5 = \angle 6$.
   $\angle 1 = \angle 2 = 68°42'18''$.

   Determine $\angle 3$. _____

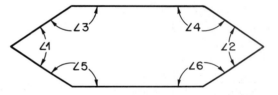

4. Express $73.65°$ as degrees and minutes. _____

5. Express $64.1420°$ as degrees, minutes, and seconds. _____

6. Express $37°23'$ as decimal degrees to 2 decimal places. _____

7. Express $104°18'47''$ as decimal degrees to 4 decimal places. _____

8. Using a simple protractor, measure each of the angles, 1–7 to the nearer degree. It may be necessary to extend sides of angles.

   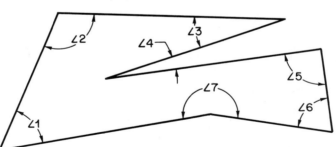

   $\angle 1 =$ _____
   $\angle 2 =$ _____
   $\angle 3 =$ _____
   $\angle 4 =$ _____
   $\angle 5 =$ _____
   $\angle 6 =$ _____
   $\angle 7 =$ _____

9. Write the values of the settings shown in the following vernier protractor scales.

   a. _____   b. _____   c. _____

   a.
   b.
   c.

10. Write the complement of each of the following angles.

    a. $53°$ _____   b. $19°47'$ _____   c. $58°17'26''$ _____

11. Write the supplement of each of the following angles.

    a. $47°$ _____   b. $93°18'$ _____   c. $102°43'33''$ _____

12. Given: AB || CD and EF || GH.
    Determine the value of each angle, $\angle 1 - \angle 10$.

    $\angle 1$ = _____

    $\angle 2$ = _____

    $\angle 3$ = _____

    $\angle 4$ = _____

    $\angle 5$ = _____

    $\angle 6$ = _____

    $\angle 7$ = _____

    $\angle 8$ = _____

    $\angle 9$ = _____

    $\angle 10$ = _____

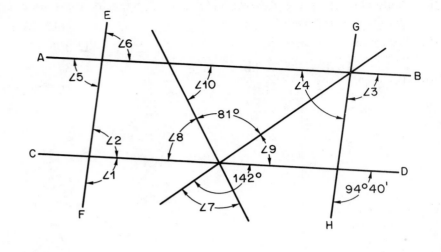

13.

a. Determine:	b. Determine:	c. Determine:

a. Determine:
   (1) $\angle 1$ _____
   (2) Side a _____

b. Determine:
   (1) $\angle 1$ _____
   (2) Side b _____
   (3) Side c _____

c. Determine:
   (1) $\angle 1$ _____
   (2) $\angle 2$ _____

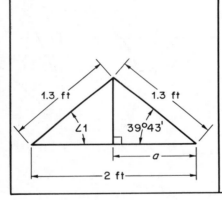
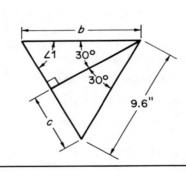
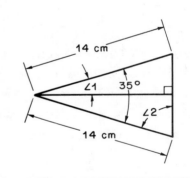

14. a. Given: $a = 7.300''$ and $b = 8.100''$. Find c. _____

    b. Given: $b = 70.00$ mm and $c = 120.00$ mm. Find a. _____

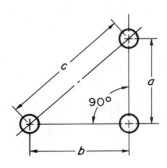

15. Compute $\angle 1$. _____

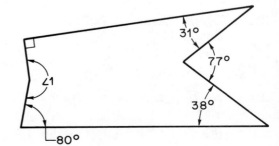

16. Determine the circumference of a circle which has a 4.210-inch radius. Round the answer to 3 decimal places.

_____

17. Determine the diameter of a circle which has a 360.00-millimeter circumference. Round the answer to 2 decimal places.

_____

18. a. Given: CD = 184 mm and $\overset{\frown}{CE}$ = 118 mm.
    Determine CF and $\overset{\frown}{CED}$.

    b. Given: FD = 26 mm and $\overset{\frown}{CED}$ = 78 mm.
    Determine CD and $\overset{\frown}{ED}$.

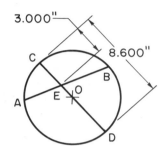

CF = _____

$\overset{\frown}{CED}$ = _____

CD = _____

$\overset{\frown}{ED}$ = _____

19. a. Given: EB = 5.200''.
    Determine AE.

    b. Given: AE = 4.300''.
    Determine AB.

_____

_____

20. Given: Points A and E are tangent points. EB is a diameter. $\overset{\frown}{AFE}$ = 156°, $\overset{\frown}{CDE}$ = 140°, and $\overset{\frown}{ED}$ = 60°. Determine angles 1–10.

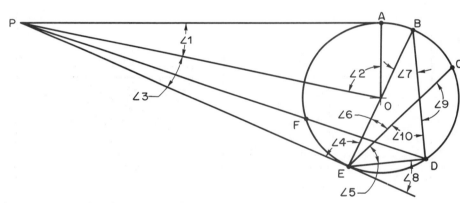

∠1 = _____   ∠3 = _____   ∠5 = _____   ∠7 = _____   ∠9 = _____

∠2 = _____   ∠4 = _____   ∠6 = _____   ∠8 = _____   ∠10 = _____

21. a. Given: $\overset{\frown}{ABC}$ = 110° and r = 5.200''.
    Compute arc length $\overset{\frown}{ABC}$.

    b. Given: Arc length $\overset{\frown}{ABC}$ = 480.20 mm and r = 105.00 mm.
    Compute ∠1.

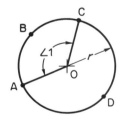

_____

_____

22. a. Given: Dia H = 14.520'' and d = 8.300''.
    Compute Dia M.

    b. Given: Dia M = 37.260'', e = 15.840'', and d = 12.560''.
    Compute f.

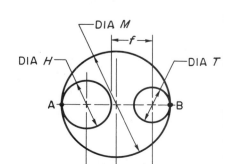

_____

_____

23. Lay out the template shown. Make the layout full size using construction methods. Do not use a protractor. All dimensions are in inches.

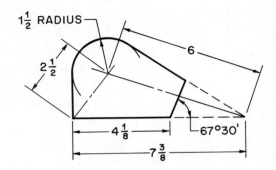

SECTION 5
TRIGONOMETRY

UNIT 48 *INTRODUCTION TO TRIGONOMETRIC FUNCTIONS* _____

OBJECTIVES _____

After studying this unit you should be able to

- Identify the sides of a right triangle with reference to any angle.
- Determine functions of given angles using the table of trigonometric functions.
- Determine angles which correspond to given functions using the table of trigonometric functions.

Trigonometry is the branch of mathematics which is used to compute unknown angles and sides of triangles. Many problems that cannot be solved by the use of geometry alone are easily solved by trigonometry.

Practical machine shop problems are often solved by using a combination of elements of algebra, geometry, and trigonometry. Therefore, it is essential to develop the ability to analyze a problem in order to relate and determine the mathematical principles which are involved in its solution. Then the problem must be worked in clear orderly steps, based on mathematical facts.

When solving a problem, it is important to understand the trigonometric operations involved rather than to mechanically "plug in" values. Attempting to solve trigonometry problems without understanding the principles involved will prove to be unsuccessful, particularly in practical shop applications such as those found later in the text.

RATIO OF RIGHT TRIANGLE SIDES _____

In a right triangle, the ratio of two sides of the triangle determine the sizes of the angles, and the angles determine the ratio of two sides. Refer to the triangles shown. The size of angle A is determined by the ratio of side a to side b. When side $a = 1$ inch and side $b = 2$ inches, the ratio of a to b is 1:2 or 1/2. If side a is increased to 2 inches and side b remains 2 inches, the ratio of a to b is 1:1 or 1/1. Observe the increase in angle A as the ratio changed from 1/2 to 1/1.

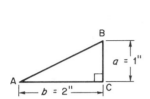 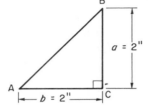 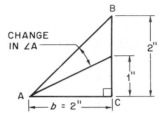

IDENTIFYING RIGHT TRIANGLE SIDES BY NAME _____

The sides of a right triangle are named opposite side, adjacent side, and hypotenuse. The *hypotenuse* (hyp) is always the side opposite the right angle. It is always the longest side of a right triangle. The positions of the opposite and adjacent sides depend on the reference angle. The *opposite side* (opp side) is opposite the reference angle and the *adjacent side* (adj side) is next to the reference angle.

In the triangle showing ∠A as the reference angle, side b is the adjacent side and side a is the opposite side. In the triangle showing ∠B as the reference angle, side b is the opposite side and side a is the adjacent side. It is important to be able to identify the opposite and adjacent sides of right triangles in reference to any angle regardless of the positions of the triangles.

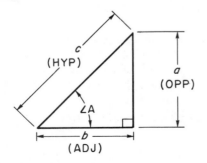   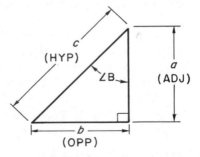

TRIGONOMETRIC FUNCTIONS: RATIO METHOD

There are two methods of defining trigonometric functions: the unity or unit circle method and the ratio method. Only the ratio method is presented in this book.

Since a triangle has three sides and a ratio is the comparison of any two sides, there are six different ratios. The names of the ratios are the sine, cosine, tangent, cotangent, secant, and cosecant.

The six trigonometric functions are defined in this table in relation to the triangle shown. The reference angle is A, the adjacent side is b, the opposite side is a, and the hypotenuse is c.

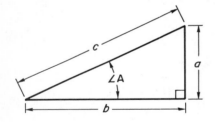

Function	Symbol	Definition of Function
sine of Angle A	sin A	$\sin A = \dfrac{\text{opp side}}{\text{hyp}} = \dfrac{a}{c}$
cosine of Angle A	cos A	$\cos A = \dfrac{\text{adj side}}{\text{hyp}} = \dfrac{b}{c}$
tangent of Angle A	tan A	$\tan A = \dfrac{\text{opp side}}{\text{adj side}} = \dfrac{a}{b}$
cotangent of Angle A	cot A	$\cot A = \dfrac{\text{adj side}}{\text{opp side}} = \dfrac{b}{a}$
secant of Angle A	sec A	$\sec A = \dfrac{\text{hyp}}{\text{adj side}} = \dfrac{c}{b}$
cosecant of Angle A	csc A	$\csc A = \dfrac{\text{hyp}}{\text{opp side}} = \dfrac{c}{a}$

To properly use trigonometric functions, it is essential to know that the function of an angle depends upon the ratio of the sides and <u>not</u> the size of the triangle. The functions of similar triangles are the same regardless of the sizes of the triangles since the sides of similar triangles are proportional. For example, in the similar triangles shown, the functions of angle A are the same for the three triangles. The equality of the tangent function is shown. Each of the other five functions have equal values for the three similar triangles.

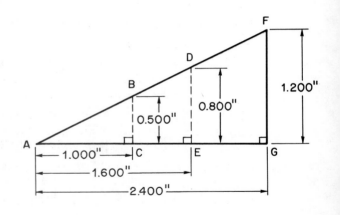

In △ABC, $\tan \angle A = \dfrac{0.500}{1.000} = 0.500$

In △ADE, $\tan \angle A = \dfrac{0.800}{1.600} = 0.500$

In △AFG, $\tan \angle A = \dfrac{1.200}{2.400} = 0.500$

⌐TRIGONOMETRIC FUNCTION TABLE

The Trigonometric Function Table found in the Appendix lists functions of angles from $0°$ to $90°$ in 10-minute increments. Since decimal-degree units can easily be expressed as degrees and minutes, this table can be used for both English and metric units of measure. Angular measure answers for English unit exercises are expressed in degrees and minutes. Angular measure answers for metric unit exercises are expressed as decimal degrees correct to 2 decimal places.

In the Trigonometric Function Table, angles from $0°$ to $45°$ are located in the left column and increase in value reading from the top to the bottom of a page. Angles from $45°$ to $90°$ are located in the right column and increase in value from the bottom to the top of the page. Observe that a column which is labeled *sin* on the top of a page is labeled *cos* on the bottom; the same is also true for the other functions. The top function names are used when locating functions of angles from $0°$ to $45°$, the bottom function names are used when locating functions of angles from $45°$ to $90°$.

Examples of Locating Functions of Given Angles

1. Find the sine of $17°40'$.

   Locate $17°$ in the left column and move down to the $40'$ row.

   Locate the *sin* function on the top of the page and move down the *sin* column to the $17°40'$ row. The value found is 0.30348.

   The sine of $17°40'$ is 0.30348.   Ans

2. Find the tangent of $57°50'$.

   Locate $57°$ in the right column and move up to $50'$ row.

   Locate the *tan* function on the bottom of the page and move up the *tan* column to the $57°50'$ row. The value found is 1.5900.

   The tangent of $57°50'$ is 1.5900.   Ans

3. Find the cosine of $16.50°$.

   Express $16.50°$ as degrees and minutes.
     Multiply the decimal part of degrees by $60'$ to obtain minutes. $0.50(60') = 30'$
   $16.50° = 16°30'$

   Locate $16°$ in the left column and move down to the $30'$ row.

   Locate the *cos* function on the top of the page and move down the *cos* column to the $16°30'$ row. The value found is 0.95882.

   The cosine of $16.50°$ is 0.95882   Ans

Examples of Locating Angles of Given Functions

1. Find the angle whose cosine is 0.88020.

   Locate the *cos* column in the table and read down the column until 0.88020 is located. Since the top *cos* function was used, the corresponding angle ($28°20'$) is located directly across from 0.88020.

   The angle whose cosine is 0.88020 is $28°20'$.   Ans

2. Find the angle whose secant is 2.9957.

   Locate the *sec* column in the table. Observe that 2.9957 cannot be found in the table using the top *sec* function. Therefore, the angle must be greater than $45°$ and the bottom *sec* function is used. Read up the column until 2.9957 is located.

   Since the bottom *sec* function was used, the corresponding angle ($70°30'$) is located directly across from 2.9957.
   The angle whose secant is 2.9957 is $70°30'$.   Ans

3. Find the angle in decimal degrees whose tangent is 0.62892. Round the answer to 2 decimal places.

   Locate the *tan* column in the table and read down the column until 0.62892 is located.

Since the top *tan* function was used, the corresponding angle (32°10′) is located directly across from 0.62892.

Express 32°10′ as decimal degrees.
   Divide 10′ by 60′ to obtain decimal degrees. 10 ÷ 60 = 0.17
32°10′ = 32.17°

The angle whose tangent is 0.62892 is 32.17°.   Ans

INTERPOLATION

To determine the function of an angle or the angle of a function not listed in the trigonometric function tables, a method called interpolation is used. Interpolation is a method of finding values between two known values.

When interpolating values, it is important to consider whether the function of an angle increases or decreases as an angle increases. Functions that do not begin with "co" (sine, tangent, secant) increase as the angle increases. Functions that begin with "co" (cosine, cotangent, cosecant) decrease as the angle increases. When interpolating functions of given angles, whether the function increases or decreases must be kept in mind. The type of function determines whether an obtained value is to be added or subtracted in the final interpolation computation. This fact is illustrated in the interpolation examples which follow.

Observe the following when using the tables in this book in solving problems which require interpolation. When interpolating functions from given angles or angles from given functions, do not use the cotangent, secant, or cosecant functions for angles less than 15°. Do not use the tangent, secant or cosecant functions for angles greater than 75°. The trigonometric functions which are listed in 10′ in the table, produce changes which are either too small or too large to always obtain accurate interpolated values.

In interpolation, when expressing decimal degrees as degrees and minutes, compute minute values to the nearer tenth minute.

Example 1: Determine the tangent of 42°13′.

   The angle 42°13′ lies between 42°10′ and 42°20′. Therefore, the tangent function of 42°13′ lies between the tangent of 42°10′ and the tangent of 42°20′.

   Determine a ratio using the given angles. The difference between 42°10′ and 42°20′ is 10′ and the difference between 42°10′ and 42°13′ is 3′. The resulting ratio is 3/10 or 0.3.

   Look up the tangent of 42°10′ and the tangent of 42°20′. The tangent of 42°10′ is 0.90569. The tangent of 42°20′ is 0.91099. The difference between 0.91099 and 0.90569 is 0.00530.

   Multiply the function difference by the ratio.
   0.3 × 0.00530 = 0.00159

   Add 0.00159 and 0.90569, the tangent of the smaller angle (42°10′).
   0.90569 + 0.00159 = 0.90728

   The tangent of 42°13′ = 0.90728.   Ans

$$
10'\left\{ \begin{array}{c} 3'\left\{ \begin{array}{l} \tan 42°10' = 0.90569 \\ \tan 42°13' = 0.90728 \end{array} \right\} 0.00159 \\ \tan 42°20' = 0.91099 \end{array} \right\} 0.00530
$$

Note: The value 0.00159 was added to 0.90569 because the tangent is an increasing function.

Example 2: Determine the cosine of 56°47′.

   The cosine function of 56°47′ lies between the cosine of 56°40′ and the cosine of 56°50′.

   The difference between 56°40′ and 56°50′ is 10′. The difference between 56°40′ and 56°47′ is 7′. The resulting ratio is 7/10 or 0.7.

   The cosine of 56°40′ is 0.54951. The cosine of 56°50′ is 0.54708. The difference between 0.54951 and 0.54708 is 0.00243.

Multiply.
0.7 × 0.00243 = 0.00170

Subtract 0.00170 from 0.54951, the cosine of the smaller angle (56°40′).
0.54951 − 0.00170 = 0.54781

The cosine of 56°47′ = 0.54781.   Ans

$$10' \begin{cases} 7' \begin{cases} \cos 56°40' = 0.54951 \\ \cos 56°47' = 0.54781 \end{cases} 0.00170 \\ \cos 56°50' = 0.54708 \end{cases} 0.00243$$

Note: The value 0.00170 was subtracted from 0.54951 because the cosine is a decreasing function.

Example 3: Determine the sine of 18.27°.

Express 18.27° as degrees and minutes. In interpolation, when expressing decimal degrees as degrees and minutes, compute minute values to the nearer tenth minute.
0.27° = 0.27(60′) = 16.2′
18.27° = 18°16.2′

The sine function of 18°16.2′ lies between the sine of 18°10′ and the sine of 18°20′.

The difference between 18°10′ and 18°20′ is 10′. The difference between 18°10′ and 18°16.2′ is 6.2′. The resulting ratio is 6.2/10 or 0.62.

The sine of 18°10′ is 0.31178. The sine of 18°20′ is 0.31454. The difference between 0.31454 and 0.31178 is 0.00276.

Multiply.
0.62 × 0.00276 = 0.00171

Add 0.00171 to 0.31178, the sine of the smaller angle (18°10′).
0.31178 + 0.00171 = 0.31349

The sine of 18.27° = 0.313 49.   Ans

$$10' \begin{cases} 6.2' \begin{cases} \sin 18°10' \;\;= 0.31178 \\ \sin 18°16.2' = 0.31349 \end{cases} 0.00171 \\ \sin 18°20' \;\;= 0.31454 \end{cases} 0.00276$$

Note: The value 0.00171 was added to 0.31178 because the sine is an increasing function.

Example 4: Determine the angle whose sine is 0.52349.

Look up the two nearer sine functions that 0.52349 lies between. 0.52349 lies between the sine function 0.52250 whose angle is 31°30′ and the sine function 0.52498 whose angle is 31°40′.

The difference between 0.52498 and 0.52250 is 0.00248, and the difference between 0.52349 and 0.52250 is 0.00099. The resulting ratio is 0.00099/0.00248 or 99/248.

The difference between 31°30′ and 31°40′ is 10′.

Multiply the angle difference by the ratio.
$$\frac{99}{248} \times 10' = \frac{990}{248} = 3.99 = 4'$$
Add 4′ to the smaller angle.
31°30′ + 4′ = 31°34′

The angle whose sine is 0.52349 is 31°34′.   Ans

$$10' \begin{cases} 4' \begin{cases} \sin 31°30' = 0.52250 \\ \sin 31°34' = 0.52349 \end{cases} 0.00099 \\ \sin 31°40' = 0.52498 \end{cases} 0.00248$$

Note: In this example, the numerator 0.00099 of the ratio 0.00099/0.00248 was determined in reference to the sine function of the smaller angle 31°30′. Always compute the numerator of the ratio in reference to the function of the smaller angle whether the function is increasing or decreasing. In so doing, the final computation always involves adding the computed minute value to the smaller angle.

APPLICATION

Identifying Right Triangle Sides by Name

With reference to ∠1, name the sides of each of the following triangles as opposite, adjacent, or hypotenuse.

1. Name sides r, x, and y.

r = _____

x = _____

y = _____

2. Name sides r, x, and y.

r = _____

x = _____

y = _____

3. Name sides a, b, and c.

a = _____

b = _____

c = _____

4. Name sides a, b, and c.

a = _____

b = _____

c = _____

5. Name sides a, b, and c.

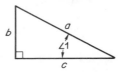

a = _____

b = _____

c = _____

6. Name sides d, m, and p.

d = _____

m = _____

p = _____

7. Name sides d, m, and p.

d = _____

m = _____

p = _____

8. Name sides e, f, and g.

e = _____

f = _____

g = _____

9. Name sides h, k, and l.

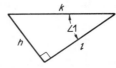

h = _____

k = _____

l = _____

10. Name sides h, k, and l.

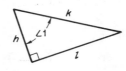

h = _____

k = _____

l = _____

11. Name sides m, p, and s.

m = _____

p = _____

s = _____

12. Name sides m, p, and s.

m = _____

p = _____

s = _____

13. Name sides m, r, and t.

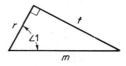

m = _____

r = _____

t = _____

14. Name sides m, r, and t.

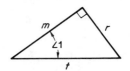

m = _____

r = _____

t = _____

15. Name sides f, g, and h.

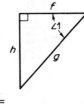

f = _____

g = _____

h = _____

16. Name sides f, g, and h.

f = _____

g = _____

h = _____

Trigonometric Functions

The sides of each of the following triangles are labeled with different letters. State the ratio of each of the 6 functions in relation to $\angle 1$ for each of the triangles. For example, for the triangle in exercise number 17, $\sin \angle 1 = \frac{y}{r}$, $\cos \angle 1 = \frac{x}{r}$, $\tan \angle 1 = \frac{y}{x}$, $\cot \angle 1 = \frac{x}{y}$, $\sec \angle 1 = \frac{r}{x}$, and $\csc \angle 1 = \frac{r}{y}$.

17.

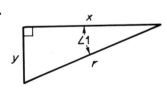

——— ———
——— ———
——— ———

19.

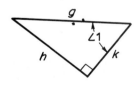

——— ———
——— ———
——— ———

21.

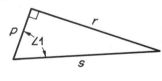

——— ———
——— ———
——— ———

18.

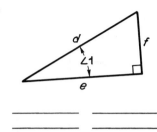

——— ———
——— ———
——— ———

20.

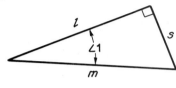

——— ———
——— ———
——— ———

22.

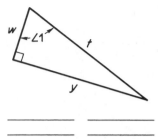

——— ———
——— ———
——— ———

23. Three groups of triangles are given below. Each group consists of four triangles. Within each group, name the triangles — a, b, c, or d — in which angles A are equal.

Group 1 _____

a.   b.   c.   d.

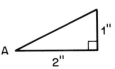

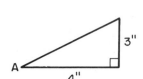
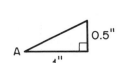

Group 2 _____

a.   b.   c.   d.

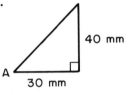
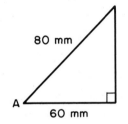
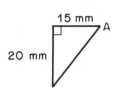
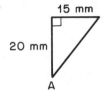

Group 3 _____

a.   b.   c.   d.

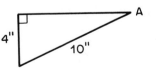
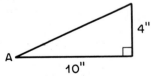
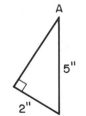

Trigonometric Function Table

Use the Trigonometric Function Table to determine the functions of the following angles. These exercises do not require interpolation.

24. sin 30° _____	32. cos 10°20′ _____	40. csc 53°20′ _____
25. cos 16° _____	33. cot 26°40′ _____	41. sec 12°40′ _____
26. tan 25° _____	34. sin 66°30′ _____	42. sin 89°10′ _____
27. cot 50° _____	35. cos 79°10′ _____	43. tan 0°50′ _____
28. sin 73° _____	36. tan 5°50′ _____	44. cos 0°10′ _____
29. tan 80° _____	37. cot 18°30′ _____	45. cot 89°50′ _____
30. csc 14° _____	38. sec 80°10′ _____	46. sec 1°20′ _____
31. sec 72° _____	39. cos 47°50′ _____	47. csc 45°10′ _____

Use the Trigonometric Function Table to determine the values of angle A that correspond to the following functions. These exercises do not require interpolation.

48. tan A = 0.12278 _____	56. tan A = 0.60483 _____	64. tan A = 0.09923 _____
49. cot A = 19.081 _____	57. sec A = 1.2309 _____	65. sin A = 0.98531 _____
50. sin A = 0.15643 _____	58. cos A = 0.69675 _____	66. csc A = 1.0314 _____
51. cos A = 0.90631 _____	59. csc A = 1.5294 _____	67. cot A = 0.33783 _____
52. sec A = 2.0627 _____	60. sin A = 0.87882 _____	68. tan A = 0.49858 _____
53. csc A = 1.0946 _____	61. cos A = 0.93565 _____	69. sin A = 0.37730 _____
54. cot A = 2.7475 _____	62. cot A = 0.50587 _____	70. cot A = 0.01164 _____
55. sin A = 0.60182 _____	63. sec A = 1.0891 _____	71. cos A = 0.99989 _____

Interpolation

Use the Trigonometric Function Table to determine the functions of the following angles. These exercises require interpolation.

72. sin 25°13′ _____	79. cos 4°24′ _____	86. sin 63°28′ _____
73. tan 36°26′ _____	80. cot 37°18′ _____	87. cos 31°41′ _____
74. cos 8°44′ _____	81. tan 49°51′ _____	88. tan 19°15′ _____
75. cot 60°12′ _____	82. sin 2°34′ _____	89. cot 50°7′ _____
76. sec 29°55′ _____	83. csc 26°33′ _____	90. sec 38°53′ _____
77. csc 52°47′ _____	84. cos 88°46′ _____	91. csc 56°44′ _____
78. sin 76°9′ _____	85. sec 46°2′ _____	

Use the Trigonometric Function Table to determine the values of angle A to the nearer minute that correspond to the following functions. These exercises require interpolation.

92. tan A = 0.58384 _____	99. tan A = 0.51065 _____	106. sec A = 1.6395 _____
93. cot A = 0.81752 _____	100. cos A = 0.90862 _____	107. cot A = 0.66364 _____
94. sin A = 0.64470 _____	101. sin A = 0.94682 _____	108. tan A = 0.42660 _____
95. cos A = 0.48750 _____	102. tan A = 0.08930 _____	109. sin A = 0.32110 _____
96. csc A = 2.0354 _____	103. cos A = 0.16440 _____	110. sec A = 1.5753 _____
97. sec A = 1.6666 _____	104. csc A = 1.3260 _____	111. csc A = 1.9762 _____
98. cot A = 1.1340 _____	105. sin A = 0.78177 _____	

Use the Trigonometric Function Table to determine the functions of the following angles given in decimal degrees. These exercises require interpolation.

112. sin 38.75° _____
115. sin 7.06° _____
118. sec 32.91° _____

113. tan 14.80° _____
116. cot 88.30° _____
119. cos 66.43° _____

114. cos 59.67° _____
117. tan 23.72° _____
120. csc 41.58° _____

Use the Trigonometric Function Table to determine the values of angle A in decimal-degrees that correspond to the following functions. Round the answers to 2 decimal places. These exercises require interpolation.

121. tan A = 0.333 62 _____
124. cot A = 1.1681 _____
127. tan A = 1.3925 _____

122. cos A = 0.340 38 _____
125. sin A = 0.746 64 _____
128. sec A = 1.1964 _____

123. sin A = 0.684 55 _____
126. cos A = 0.873 91 _____
129. csc A = 1.0606 _____

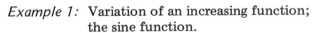

UNIT 49 *ANALYSIS OF TRIGONOMETRIC FUNCTIONS* _____

OBJECTIVES _____

After studying this unit you should be able to

- Determine the variations of functions as angles change.
- Compute cofunctions of complementary angles.

VARIATION OF FUNCTIONS _____

As the size of an angle increases the sine, tangent, and secant functions increase while the cofunctions (cosine, cotangent, cosecant) decrease. As the reference angles approach 0° or 90°, the function variation can be shown. These examples illustrate variations of an increasing function and a decreasing function for a reference angle which is increasing in size.

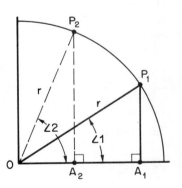

Note: Use this figure for Examples 1–2.

OP_1 and OP_2 are radii of the arc of a circle. $OP_1 = OP_2 = r$

Example 1: Variation of an increasing function; the sine function.

The sine of an angle $= \dfrac{\text{opposite side}}{\text{hypotenuse}}$

$$\sin \angle 1 = \frac{A_1 P_1}{r}$$

$$\sin \angle 2 = \frac{A_2 P_2}{r}$$

$A_2 P_2$ is greater than $A_1 P_1$; therefore, $\sin \angle 2$ is greater than $\sin \angle 1$. Observe that if $\angle 1$ decreases to 0°, side $A_1 P_1 = 0$.

$$\sin 0° = \frac{0}{r} = 0$$

If $\angle 2$ increases to 90°, side $A_2 P_2 = r$.

$$\sin 90° = \frac{r}{r} = 1$$

Conclusion: As an angle increases from 0° to 90°, the sine of the angle increases from 0 to 1.

Example 2: Variation of a decreasing function; the cosine function.

The cosine of an angle $= \dfrac{\text{adjacent side}}{\text{hypotenuse}}$

$$\cos \angle 1 = \frac{OA_1}{r}$$

$$\cos \angle 2 = \frac{OA_2}{r}$$

OA_2 is less than OA_1; therefore, $\cos \angle 2$ is less than $\cos \angle 1$. Observe that if $\angle 1$ decreases to 0°, side $OA_1 = r$.

$$\cos 0° = \frac{r}{r} = 1$$

If $\angle 2$ increases to 90°, side $OA_2 = 0$.

$$\cos 90° = \frac{0}{r} = 0$$

Conclusion: As an angle increases from 0° to 90°, the cosine of the angle decreases from 1 to 0.

Study the table of trigonometric functions found in the Appendix and observe the changes in functions as angles increase or decrease. It is also helpful to sketch figures for all functions in order to further develop an understanding of the relationship of angles and their functions. Particular attention should be given to functions of angles close to 0° and 90°.

A summary of the variations taken from the table of trigonometric functions is shown for an angle increasing from 0° to 90°.

As an angle increases from 0° to 90°	
sin increases from 0 to 1	cos decreases from 1 to 0
tan increases from 0 to ∞	cot decreases from ∞ to 0
sec increases from 1 to ∞	csc decreases from ∞ to 1

The symbol ∞ means infinity. Infinity is the quality of existing beyond or being greater than any countable value. It cannot be used for computations at this level of mathematics.

Rather than to attempt to treat ∞ as a value, think of the tangent and secant functions not at an angle of 90°, but at angles very close to 90°. Observe that as an angle approaches 90°, the tangent and secant functions get very large. Think of the cotangent and cosecant functions not at an angle of 0°, but as very small angles close to 0°. Observe that as an angle approaches 0° the cotangent and cosecant functions get very large.

⌐FUNCTIONS OF COMPLEMENTARY ANGLES_____

Two angles are complementary when their sum is 90°. For example, 20° is the complement of 70° and 70° is the complement of 20°. In the triangle shown, ∠A is the complement of ∠B and ∠B is the complement of ∠A. The six functions of the angle and the cofunctions of the complementary angle are shown.

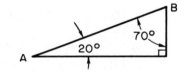

sin 20° = cos 70° = 0.342 02	cos 20° = sin 70° = 0.939 69
tan 20° = cot 70° = 0.363 97	cot 20° = tan 70° = 2.747 5
sec 20° = csc 70° = 1.064 2	csc 20° = sec 70° = 2.923 8

A function of an angle is equal to the cofunction of the complement of the angle.

The complement of an angle equals 90° minus the angle. The relationships of the six functions of angles and the cofunctions of the complementary angles are shown.

sin A = cos (90° − A)	cos A = sin (90° − A)
tan A = cot (90° − A)	cot A = tan (90° − A)
sec A = csc (90° − A)	csc A = sec (90° − A)

Examples: For each function of an angle, write the cofunction of the complement of the angle.

1. $\sin 30° = \cos (90° − 30°) = \cos 60°$ Ans
2. $\cot 10° = \tan (90° − 10°) = \tan 80°$ Ans
3. $\tan 72.53° = \cot (90° − 72.53°) = \cot 17.47°$ Ans
4. $\sec 40°20' = \csc (90° − 40°20') = \csc (89°60' − 40°20') = \csc 49°40'$ Ans
5. $\cos 90° = \sin (90° − 90°) = \sin 0°$ Ans

APPLICATION

Variation of Functions

Refer to this figure in answering exercises 1–7. It may be helpful to sketch figures.

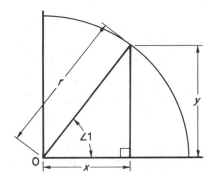

1. When ∠1 is almost 90°:
    a. how does side y compare to side r?   _____
    b. how does side x compare to side r?   _____
    c. how does side x compare to side y?   _____

2. When ∠1 is 90°:
    a. what is the value of side x?   _____
    b. how does side y compare to side r?   _____

3. When ∠1 is slightly greater than 0°:
    a. how does side y compare to side r?   _____
    b. how does side x compare to side r?   _____
    c. how does side x compare to side y?   _____

4. When ∠1 is 0°:
    a. what is the value of side y?   _____
    b. how does side x compare to side r?   _____

5. When side x = side y:
    a. what is the value of ∠1?   _____
    b. what is the value of the tangent function? _____
    c. what is the value of the cotangent function? _____

6. When side x = side r:
    a. what is the value of the cosine function?   _____
    b. what is the value of the secant function?   _____
    c. what is the value of the sine function?   _____
    d. what is the value of the tangent function? _____

7. When side y = side r:
    a. what is the value of the sine function?   _____
    b. what is the value of the cosecant function? _____
    c. what is the value of the cosine function?   _____
    d. what is the value of the cotangent function? _____

For each exercise, functions of two angles are given. Which of the functions of the two angles is greater? Do <u>not</u> use the tables of trigonometric functions.

8. sin 38°; sin 43° _____

9. tan 17°; tan 18° _____

10. cos 76°; cos 80° _____

11. cot 40°; cot 36° _____

12. sec 5°; sec 8° _____

13. csc 22°; csc 25° _____

14. tan 19°20'; tan 16°40' _____

15. cos 81°19'; cos 81°20' _____

16. sin 0.42°; sin 0.37° _____

17. csc 39.30°; csc 39.25° _____

18. cot 27°23'; cot 87°0' _____

19. sec 55°; sec 54°50' _____

Functions of Complementary Angles

For each function of an angle, write the cofunction of the complement of the angle.

20. tan 17° _____

21. sin 49° _____

22. cos 26° _____

23. sec 87° _____

24. cot 35° _____

25. csc 51° _____

26. cos 90° _____

27. sin 0° _____

28. tan 66.5° _____

29. cos 12.2° _____

30. cot 7°10' _____

31. sec 31°26' _____

32. csc 0°38' _____

33. sin 5.89° _____

34. cos 5.89° _____

35. cot 0° _____

36. tan 90° _____

37. sec 44°29' _____

38. cos 0.01° _____

39. sin 89°59' _____

For each exercise, functions and cofunctions of two angles are given. Which of the functions or cofunctions of the two angles is greater? Do <u>not</u> use the tables of trigonometric functions.

40. cos 55°; sin 20° _____

41. cos 55°; sin 40° _____

42. tan 30°; cot 65° _____

43. tan 30°; cot 45° _____

44. sec 43°; csc 56° _____

45. sec 43°; csc 58° _____

46. sin 12°; cos 80° _____

47. sin 12°; cos 75° _____

48. cot 89°10'; tan 1°20' _____

49. cot 89°10'; tan 0°40' _____

50. sec 0.2°; csc 89.9° _____

51. sec 0.2°; csc 89.0° _____

UNIT 50 *BASIC CALCULATIONS OF ANGLES AND SIDES OF RIGHT TRIANGLES* _____

OBJECTIVES

After studying this unit you should be able to
- Compute an unknown angle of a right triangle when two sides are known.
- Compute an unknown side of a right triangle when an angle and a side are known.

DETERMINING AN UNKNOWN ANGLE WHEN TWO SIDES OF A RIGHT TRIANGLE ARE KNOWN

In order to solve for an unknown angle of a right triangle where neither acute angle is known, at least two sides must be known. An understanding of the procedures required for solving for unknown angles is essential to the machinist.

Procedure: To determine an unknown angle when two sides of a right triangle are known

- Identify two given sides as adjacent, opposite, or hypotenuse in relation to the desired angle.
- Determine the functions that are ratios of the sides identified in relation to the desired angle.

  Note: Two of the six trigonometric functions are ratios of the two known sides. Either of the two functions can be used. Both produce the same value for the unknown, except for cotangents, secants, and cosecants of angles less than 15° and tangents, secants, and cosecants of angles greater than 75°.

- Choose one of the two functions, substitute the given sides in the ratio, and divide.
- Using the Table of Trigonometric Functions, determine the angle that corresponds to the quotient obtained. It is often necessary to interpolate. When sides are given in inches (English units), compute the angle to the nearer minute. When sides are given in millimeters (metric units), compute the angle to the nearer hundredth degree.

Example 1: Determine ∠A of the triangle shown. All dimensions are in inches.

In relation to ∠A, the 10.774-inch side is the adjacent side and the 7.500-inch side is the opposite side.

Determine the two functions whose ratios consist of the adjacent and opposite sides. The tangent $\angle A = \frac{\text{opposite side}}{\text{adjacent side}}$, and the cotangent $\angle A = \frac{\text{adjacent side}}{\text{opposite side}}$. Either the tangent or cotangent function can be used.

Choosing the cotangent function:

$$\cot \angle A = \frac{\text{adj side}}{\text{opp side}}$$
$$\cot \angle A = \frac{10.774}{7.500}$$
$$\cot \angle A = 1.4365$$
$$A = 34°51' \quad \text{Ans}$$

Interpolate from the function table to determine the angle whose cotangent function is nearer 1.4365.

Example 2: Determine ∠B of the triangle shown. All dimensions are in millimeters.

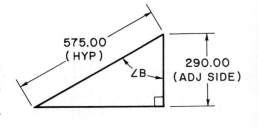

In relation to ∠B, the 290.00-mm side is the adjacent side and the 575.00-mm side is the hypotenuse.

Determine the two functions whose ratios consist of the adjacent side and the hypotenuse. The secant $\angle B = \frac{\text{hypotenuse}}{\text{adjacent side}}$, and the cosine $\angle B = \frac{\text{adjacent side}}{\text{hypotenuse}}$. Either the secant or cosine function can be used.

Choosing the secant function:

$$\sec \angle B = \frac{\text{hyp}}{\text{adj side}}$$
$$\sec \angle B = \frac{575.00}{290.00}$$
$$\sec \angle B = 1.9828$$
$$\angle B = 59.72° \quad \text{Ans}$$

Interpolate from the function table to determine the angle whose secant function is nearer 1.9828. The angle interpolated to the nearer minute is 59°43'. Express 59°43' as decimal degrees.

43 ÷ 60 = 0.72
59°43' = 59.72°

Example 3: Determine ∠1 and ∠2 of the triangle shown. All dimensions are in inches.

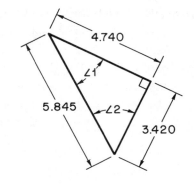

Calculate either ∠1 or ∠2. Choose any two of the three given sides.   In relation to ∠1, the 3.420-inch side is the opposite side. The 5.845-inch side is the hypotenuse.

Determine the two functions whose ratios consist of the opposite side and the hypotenuse. The sine ∠1 = $\frac{\text{opposite side}}{\text{hypotenuse}}$, and the cosecant ∠1 = $\frac{\text{hypotenuse}}{\text{opposite side}}$. Either the sine or cosecant function can be used.

Choosing the sine function:

$$\sin \angle 1 = \frac{\text{opp side}}{\text{hyp}}$$

$$\sin \angle 1 = \frac{3.420}{5.845}$$

$$\sin \angle 1 = 0.58512$$

Interpolate from the function table to determine the angle nearer 0.58512. The angle interpolated to the nearer minute is 35°49′.

$$\angle 1 = 35°49' \text{Ans}$$

Since ∠1 + ∠2 = 90°,

$$\angle 2 = 90° - 35°49'$$
$$\angle 2 = 54°11' \text{Ans}$$

DETERMINING AN UNKNOWN SIDE WHEN AN ACUTE ANGLE AND ONE SIDE OF A RIGHT TRIANGLE ARE KNOWN

In order to solve for an unknown side of a right triangle, at least an acute angle and one side must be known.

Procedure: To determine an unknown side when an acute angle and one side of a right triangle are known

- Identify the given side and the unknown side as adjacent, opposite, or hypotenuse in relation to the given angle.

- Determine the trigonometric functions that are ratios of the sides identified in relation to the given angle.

  Note: Two of the six functions will be found as ratios of the two identified sides. Either of the two functions can be used. Both produce the same value for the unknown, except for cotangents, secants, and cosecants of angles less than 15° and tangents, secants, and cosecants of angles greater than 75°. If the unknown side is made the numerator of the ratio, the problem is solved by multiplication. If the unknown side is made the denominator of the ratio, the problem is solved by division.

- Choose one of the two functions and substitute the given side and given angle. Express an angle given in decimal degrees as degrees and minutes.

- Using the trigonometric function table, look up the function of the given angle and substitute this value. If the angle is not given in the table, interpolate the function of the angle.

- Solve as a proportion for the unknown side.

Example 1: Determine side x of the triangle shown. All dimensions are in inches.

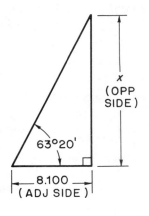

In relation to the 63°20′ angle, the 8.100-inch side is the adjacent side and side x is the opposite side.

Determine the two functions whose ratios consist of the adjacent and opposite sides. The tangent $63°20' = \frac{\text{opposite side}}{\text{adjacent side}}$ and the cotangent $63°20' = \frac{\text{adjacent side}}{\text{opposite side}}$. Either the tangent or cotangent function can be used.

Choosing the tangent function:

$$\tan 63°20' = \frac{\text{opp side}}{\text{adj side}}$$

$$\tan 63°20' = \frac{x}{8.100}$$

Look up the tangent of 63°20′ in the function table and substitute. tan 63°20′ = 1.9912

$$1.9912 = \frac{x}{8.100}$$

Solve as a proportion.

$$\frac{1.9912}{1} = \frac{x}{8.100}$$

$$x = 1.9912(8.100)$$

$$x = 16.129 \text{ in} \quad \text{Ans}$$

Example 2: Determine side z of the triangle shown. All dimensions are in millimeters.

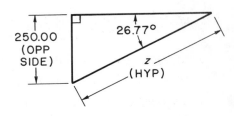

In relation to the 26.77° angle, the 250.00-mm side is the opposite side. Side z is the hypotenuse.

Determine the two functions whose ratios consist of the opposite side and the hypotenuse. sine $26.77° = \frac{\text{opposite side}}{\text{hypotenuse}}$ and cosecant $26.77° = \frac{\text{hypotenuse}}{\text{opposite side}}$. Either the sine or cosecant function can be used.

Choosing the sine function:

$$\sin 26.77° = \frac{\text{opp side}}{\text{hyp}}$$

$$\sin 26.77° = \frac{250.00}{z}$$

Express 26.77° as degrees and minutes. $\sin 26°46' = \frac{250.00}{z}$
0.77 (60′) = 46′
26.77° = 26°46′

Interpolate the sine 26°46′ and substitute.
sin 26°46′ = 0.450 36

$$0.450\ 36 = \frac{250.00}{z}$$

Solve as a proportion.

$$\frac{0.450\ 36}{1} = \frac{250.00}{z}$$

$$0.450\ 36z = 250.00$$

$$z = 555.11 \text{ mm} \quad \text{Ans}$$

Example 3: Determine side x, side y, and $\angle 1$ of the triangle shown.
All dimensions are in inches.

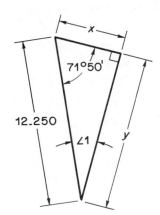

Calculate either side x or side y. In relation to the 71°50′ angle, side x is the adjacent side. The 12.250-inch side is the hypotenuse.

Determine the two functions whose ratios consist of the adjacent side and the hypotenuse in order to solve for side x. Either the cosine or secant function can be used.

Choosing the cosine function:
$$\cos 71°50' = \frac{\text{adj side}}{\text{hyp}}$$

$$\cos 71°50' = \frac{x}{12.250}$$

Look up the cosine of 71°50′ in the function table and substitute.
$\cos 71°50' = 0.31178$

$$0.31178 = \frac{x}{12.250}$$

Solve as a proportion.
$$\frac{0.31178}{1} = \frac{x}{12.250}$$

$$x = 0.31178(12.250)$$

$$x = 3.819 \text{ in}\quad \text{Ans}$$

Solve for side y by using either a trigonometric function or the Pythagorean Theorem. It is generally more convenient to solve for y using a trigonometric function. In relation to the 71°50′ angle, side y is the opposite side. The 12.250-inch side is the hypotenuse.

Determine the two functions whose ratios consist of the opposite side and the hypotenuse. Either the sine or cosecant function can be used.

Choosing the sine function:
$$\sin 71°50' = \frac{\text{opp side}}{\text{hyp}}$$

$$\sin 71°50' = \frac{y}{12.250}$$

$$0.95015 = \frac{y}{12.250}$$

$$\frac{0.95015}{1} = \frac{y}{12.250}$$

$$y = 0.95015(12.250)$$

$$y = 11.639 \text{ in}\quad \text{Ans}$$

Determine $\angle 1$.
$$\angle 1 = 90° - 71°50' = 18°10'\quad \text{Ans}$$

APPLICATION

A Trigonometric Function Table can be found in the Appendix.

Determining an Unknown Angle When Two Sides of a Right Triangle Are Known

Solve the following problems. Compute angles to the nearer minute in triangles with English unit sides. Compute angles to the nearer hundredth degree in triangles with metric unit sides.

1. Determine ∠A.

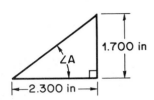

4. Determine ∠x.

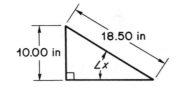

7. Determine ∠y.

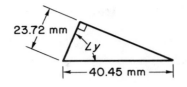

10. a. Determine ∠A.

b. Determine ∠B.

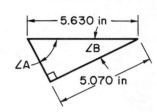

2. Determine ∠B.

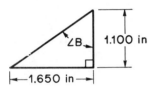

5. Determine ∠1.

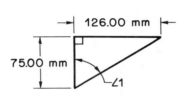

8. Determine ∠B.

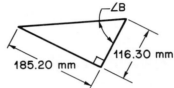

11. a. Determine ∠x.

b. Determine ∠y.

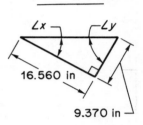

3. Determine ∠1.

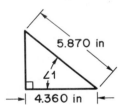

6. Determine ∠A.

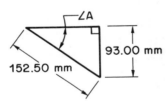

9. a. Determine ∠1.

b. Determine ∠2.

12. a. Determine ∠C.

b. Determine ∠D.

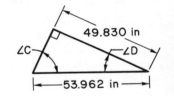

Determining an Unknown Side When an Acute Angle and One Side of a Right Triangle Are Known

Solve the following problems. Compute the sides to 3 decimal places in triangles dimensioned in English units. Compute the sides to 2 decimal places in triangles dimensioned in metric units.

13. Determine side b.

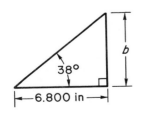

14. Determine side c.

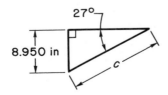

15. Determine side x.

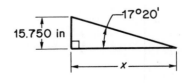

16. Determine side d.

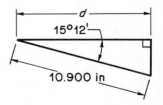

17. Determine side *y*. _____

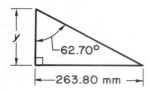

18. Determine side *f*. _____

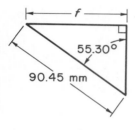

19. Determine side *p*. _____

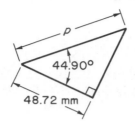

20. Determine side *y*. _____

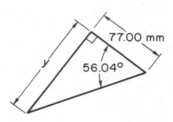

21. a. Determine side *d*. _____

    b. Determine side *e*. _____

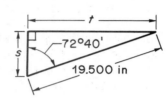

22. a. Determine side *s*. _____

    b. Determine side *t*. _____

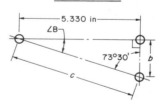

23. a. Determine side *x*. _____

    b. Determine side *y*. _____

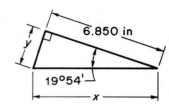

24. a. Determine side *p*. _____

    b. Determine side *n*. _____

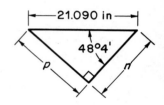

Determining Unknown Sides and Angles

Solve the following problems. For triangles dimensioned in English units, compute the sides to 3 decimal places and the angles to the nearer minute. For triangles dimensioned in metric units, compute the sides to 2 decimal places and the angles to the nearer hundredth degree.

25. a. Determine $\angle B$. _____

    b. Determine side *x*. _____

    c. Determine side *y*. _____

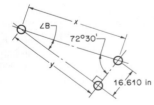

26. a. Determine $\angle 1$. _____

    b. Determine $\angle 2$. _____

    c. Determine side *a*. _____

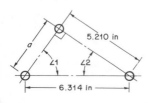

27. a. Determine side *a*. _____

    b. Determine side *b*. _____

    c. Determine $\angle 2$. _____

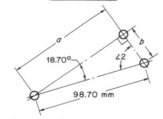

28. a. Determine $\angle A$. _____

    b. Determine $\angle B$. _____

    c. Determine side *r*. _____

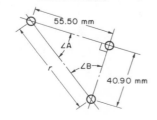

29. a. Determine $\angle B$. _____

    b. Determine side *b*. _____

    c. Determine side *c*. _____

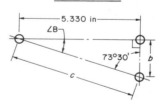

30. a. Determine $\angle D$. _____

    b. Determine $\angle E$. _____

    c. Determine side *m*. _____

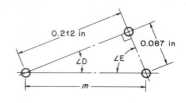

31. a. Determine $\angle 1$. _____

    b. Determine side *g*. _____

    c. Determine side *h*. _____

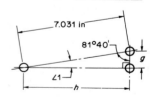

─UNIT 51 *SIMPLE PRACTICAL MACHINE* APPLICATIONS ──────────

─OBJECTIVE────────────────────

After studying this unit you should be able to
* Solve simple machine shop problems which require the projection of auxiliary lines and the use of geometric principles and trigonometric functions.

─METHOD OF SOLUTION──────────────

The examples discussed in this unit are simple practical shop applications of right angle trigonometry, although they may not be given directly in the form of right triangles. To solve most of the examples, it is necessary to project auxiliary lines to produce a right triangle. The unknown, or a dimension required to compute the unknown, is part of the triangle. The auxiliary lines may be projected between given points, or from given points. The lines may be projected parallel or perpendicular to centerlines, tangents, or other reference lines.

It is important to study carefully the procedures and the use of auxiliary lines as they are applied to the examples which follow. The same basic method is used in solving many similar machine shop problems. A knowledge of both geometric principles and trigonometric functions and the ability to relate and apply them to specific situations are required in solving many machine shop problems.

─SINE BAR AND SINE PLATE────────────────

Sine bars and sine plates are used to measure angles which have been cut in parts and to position parts which are to be cut at specified angles. One end of the sine bar or plate is raised with gage blocks in order to set a desired angle. The most common sizes of bars and plates are 5 inches and 10 inches between rolls. In setting angles, the sine bar or the top plate of the sine plate is the hypotenuse of a right triangle, and the gage blocks are the opposite side in reference to the desired angle.

Example 1: Determine the gage block height x which is required to set an angle of $24°20'$ with a 5-inch sine bar as shown.

$$\sin 24°20' = \frac{\text{gage block height } x}{\text{sine bar length}}$$

$$\sin 24°20' = \frac{x}{5}$$

$$0.41204 = \frac{x}{5}$$

$$x = 5(0.41204)$$
$$x = 2.0602 \text{ in} \quad \text{Ans}$$

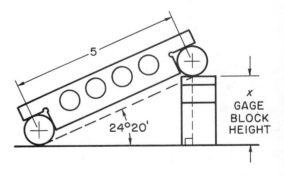

Note: In order to determine gage block heights, look up the sine of the angle in the function table and multiply it by the length of the sine bar or sine plate. When using a 10-inch bar or plate, look up the sine of the angle and move the decimal point one place to the right (the same as multiplying by 10).

Example 2: Determine the angle set on a 10-inch sine plate using a gage block height of 3.0625 inches.

$$\sin \angle x = \frac{3.0625}{10} = 0.30625$$
$$\angle x = 17°50' \quad \text{Ans}$$

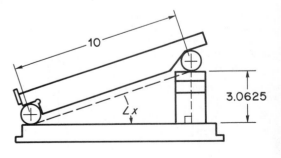

⌐TAPERS AND BEVELS

Example 1: Determine the included taper angle of the shaft shown. All dimensions are in inches.

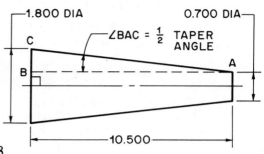

The problem must be solved by using a figure in the form of a right triangle. Therefore, project line AB from point A parallel to the centerline. Right △ABC is formed in which ∠BAC is one-half the included taper angle. Side AB = 10.500″.

Side BC = $\frac{1.800″ - 0.700″}{2}$ = 0.550″

Using sides AB and BC, solve for ∠BAC. $\tan \angle BAC = \frac{BC}{AB} = \frac{0.550}{10.500} = 0.05238$

$\angle BAC = 3°0'$

The included taper angle = 2(3°0′) = 6°0′   Ans

Example 2: Determine diameter x of the part shown. All dimensions are in millimeters.

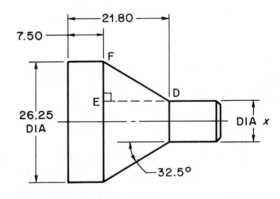

Project line DE from point D parallel to the centerline, in order to form right △DEF.

Side DE = 21.80 mm − 7.50 mm = 14.30 mm

∠EDF = 32.5° = 32°30′

Using side DE and ∠EDF, solve for side EF.

$$\tan \angle EDF = \frac{EF}{DE}$$

$$\tan 32°30' = \frac{EF}{14.30}$$

$$0.637\,07 = \frac{EF}{14.30}$$

$$EF = 0.637\,07(14.30)$$

$$EF = 9.11 \text{ mm}$$

Dia x = 26.25 mm − 2(9.11 mm) = 8.03 mm   Ans

⌐ISOSCELES TRIANGLE APPLICATIONS: DISTANCE BETWEEN HOLES AND V-SLOTS

The solutions to many practical trigonometry problems are based on recognizing figures as isosceles triangles. In an isosceles triangle, an altitude to the base bisects the base and the vertex angle.

Example 1: In this figure, five holes are equally spaced on a 5.200-inch diameter circle. Determine the straight line distance between two consecutive holes.

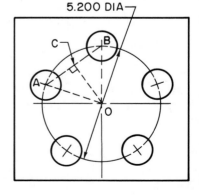

Project radii from center O to hole centers A and B.

Project a line from A to B. $\angle AOB = \frac{360°}{5} = 72°$

Since OA = OB, △AOB is isosceles. Project line OC ⊥ to AB from center O. Line OC bisects ∠AOB and side AB.

In right △AOC, $\angle AOC = \frac{72°}{2} = 36°$

$$AO = \frac{5.200 \text{ in}}{2} = 2.600 \text{ in}$$

Solve for side AC.

$$\sin \angle AOC = \frac{AC}{AO}$$

$$\sin 36° = \frac{AC}{2.600}$$

$$0.58779 = \frac{AC}{2.600}$$

$$AC = 0.58779(2.600)$$

$$AC = 1.528 \text{ in}$$

AB = 2(1.528 in) = 3.056 in   Ans

Example 2: Determine the depth of cut x required to machine the V-slot shown. All dimensions are in inches.

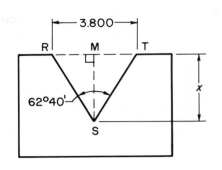

Connect a line between points R and T. Sides RS = TS; therefore, △RST is isosceles.

Project line SM from point S $\perp$ to RT. Side RT and ∠RST are bisected. In right △RMS,

$$\angle RSM = \frac{62°40'}{2} = 31°20'$$

$$RM = \frac{3.800 \text{ in}}{2} = 1.900 \text{ in}$$

Solve for depth of cut MS.

$$\cot \angle RSM = \frac{MS}{RM}$$

$$\cot 31°20' = \frac{MS}{1.900}$$

$$1.6426 = \frac{MS}{1.900}$$

$$MS = 1.6426(1.900)$$

$$MS = 3.121 \text{ in}$$

$x = MS = 3.121$ in   Ans

─TANGENTS TO CIRCLES APPLICATIONS: V-BLOCKS, THREAD WIRE CHECKING DIMENSIONS, DOVETAILS, AND ANGLE CUTS ────────

A tangent is perpendicular to a radius of a circle at its tangent point. Solutions to many applied trigonometry problems are based on this principle.

Example 1: A 75.00-millimeter diameter pin is used to inspect the groove machined in the block shown. Determine dimension x. The sides of the groove are equal. All dimensions are in millimeters.

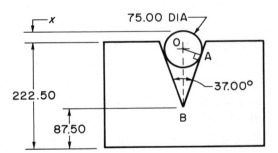

Project a line from center O to point B. Project radius AO from center O to tangent point A. Since a radius is $\perp$ to a tangent line at the point of tangency, △AOB is a right triangle. In right △AOB, OA = $\frac{75.00 \text{ mm}}{2}$ = 37.50 mm

Since the angle formed by two tangents to a circle from an outside point is bisected by a line from the point to the center of the circle, $\angle ABO = \frac{37.00°}{2} = 18.50° = 18°30'$

Solve for side OB.

$$\csc \angle ABO = \frac{OB}{OA}$$

$$\csc 18°30' = \frac{OB}{37.50}$$

$$3.1515 = \frac{OB}{37.50}$$

$$OB = 3.1515(37.50)$$

$$OB = 118.18 \text{ mm}$$

Find the height from the base of the block to the top of the pin.

87.50 mm + OB + radius of pin =
87.50 mm + 118.18 mm + 37.50 mm = 243.18 mm

x = 243.18 mm – 222.50 mm = 20.68 mm   Ans

Example 2: An internal dovetail is shown. Two pins or balls are used to check the dovetail for both location and angular accuracy. Calculate check dimension *x*. All dimensions are in inches.

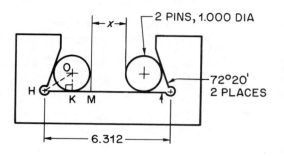

Project line HO from point H to the pin center O; HO bisects the 72°20′ angle. Project a radius from point O to the point of tangency K; ∠HKO is a right angle since a radius is perpendicular to a tangent at the point of tangency.

In right △HOK, $\angle KHO = \frac{72°20′}{2} = 36°10′$

$KO = \frac{1.000 \text{ in}}{2} = 0.500 \text{ in}$

Solve for side HK.

$$\cot \angle KHO = \frac{HK}{KO}$$

$$\cot 36°10′ = \frac{HK}{0.500}$$

$$1.3680 = \frac{HK}{0.500}$$

$$HK = 0.500(1.3680)$$
$$HK = 0.684 \text{ in}$$

KM = pin radius = 0.500 in

HM = HK + KM = 0.684 in + 0.500 in = 1.184 in

x = 6.312 in − 2(HM) = 6.312 in − 2(1.184 in) = 3.944 in

APPLICATION

A Trigonometric Function Table can be found in the Appendix.

Sine Bars and Sine Plates

1. Determine the height of gage blocks required to set the following angles on a 10″ sine plate.

   a. 25° _____   d. 7°20′ _____   g. 0°40′ _____
   b. 13°10′ _____   e. 28°30′ _____   h. 2°20′ _____
   c. 36°50′ _____   f. 44°20′ _____   i. 19°50′ _____

3. Determine the height of gage blocks required to set the following angles with a 5″ sine bar.

   a. 40°40′ _____   d. 0°30′ _____   g. 37°20′ _____
   b. 5° _____   e. 21°50′ _____   h. 44°50′ _____
   c. 12°10′ _____   f. 9° _____   i. 8°10′ _____

Tapers and Bevels

Solve the following problems. For English unit dimensioned problems, calculate angles to the nearer minute and lengths to the nearer thousandths inch. For metric unit dimensioned problems, calculate angles to the nearer hundredth degree and lengths to the nearer hundredth millimeter.

3. Find the included taper ∠*x*. All dimensions are in inches. _____

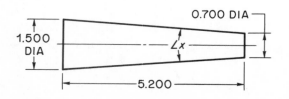

4. Find length *x*. All dimensions are in inches. _____

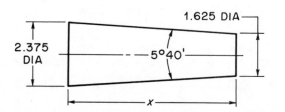

5. Find diameter *y*. All dimensions
    are in millimeters. _____

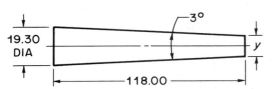

6. Find diameter *x*. All dimensions
    are in inches. _____

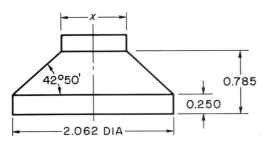

7. Find ∠*x*. All dimensions are in
    inches. _____

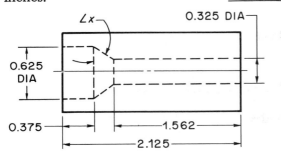

8. Find dimension *y*. All dimensions
    are in inches. _____

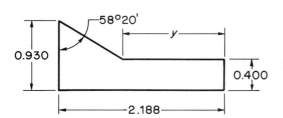

Distance between Holes and V-Slots

9. Find center distance *y*. All dimen-
    sions are in millimeters. _____

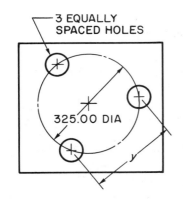

10. Find inside caliper dimension
    x. All dimensions are in inches. _____

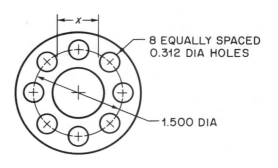

11. Find radius *r*. All dimensions
    are in millimeters. _____

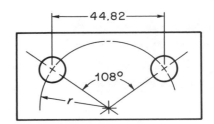

12. Find arc dimension *x*. All
    dimensions are in inches. _____

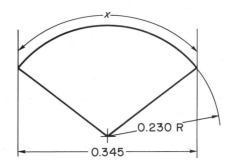

13. Find the depth of cut *x*. All
    dimensions are in inches. _____

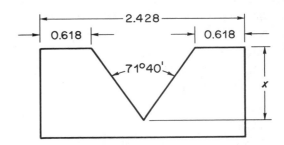

**V-Blocks, Thread Wire Checking
Dimensions, Dovetails, and
Angle Cuts**

14. Find ∠x. All dimensions
     are in inches.   _____

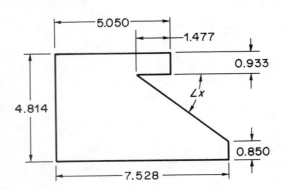

15. Find gage dimension y. All
     dimensions are in millimeters.   _____

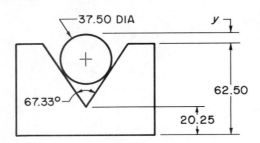

16. Find ∠y. All dimensions
     are in inches.   _____

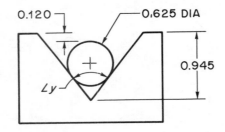

17. Find gage dimension x. All
     dimensions are in inches.   _____

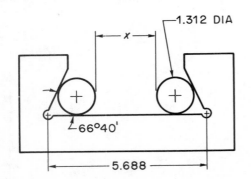

18. Find ∠x. All dimensions
     are in millimeters.   _____

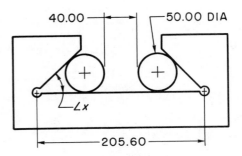

19. Find dimension y. All
     dimensions are in inches.   _____

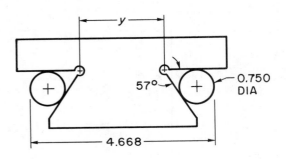

Miscellaneous Applications

20. Find ∠y. All dimensions
     are in inches.   _____

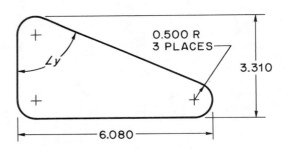

21. Find dimension x. All
     dimensions are in millimeters.   _____

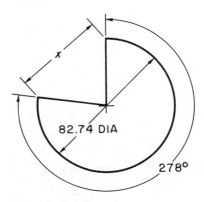

22. Find ∠x. All dimensions are in inches. _____

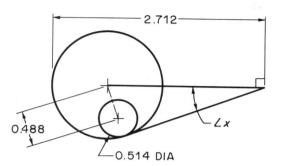

0.488

2.712

∠x

0.514 DIA

23. Find distance y. All dimensions are in millimeters. _____

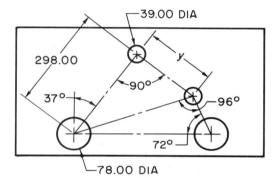

39.00 DIA

298.00

y

90°

37°

96°

72°

78.00 DIA

24. Find dimension y. All dimensions are in inches. _____

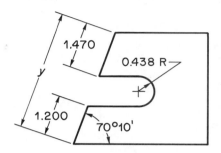

1.470

0.438 R

y

1.200

70°10'

UNIT 52 COMPLEX PRACTICAL MACHINE APPLICATIONS

OBJECTIVE

After studying this unit you should be able to

• Solve complex applied machine shop problems which require forming two or more right triangles by the projection of auxiliary lines.

The problems in this unit are more challenging than those in the last unit and are typical of those found in actual practice when working directly from engineering drawings. The solutions of these problems require the projection of auxiliary lines to form two or more right triangles.

Study the procedures which are given in detail for solving the examples. There is a common tendency to begin writing computations before analyzing the problem. This tendency must be avoided. As problems become more complex, a greater proportion of time and effort is required in the analyses. The written computations must be developed in clear and orderly steps.

METHOD OF SOLUTION

Analyze the problem before writing computations.

• Relate given dimensions to the unknown and determine whether other dimensions in addition to the given dimensions are required in the solution.

• Determine the auxiliary lines which are required to form right triangles which contain dimensions that are needed for the solution.

- Determine whether sufficient dimensions are known to obtain required values within the right triangles. If enough information is not available for solving a triangle, continue the analysis until enough information is obtained.
- Check each step in the analysis to verify that there are no gaps or false assumptions.

Write the computations.

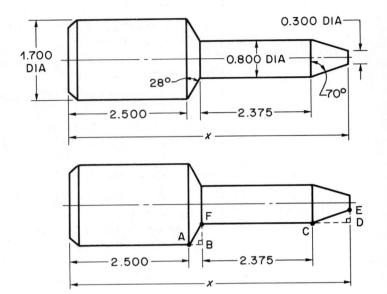

Example 1: Determine length x of the part shown. All dimensions are in inches.

Analyze the problem:

Project auxiliary lines to form right $\triangle ABF$ and right $\triangle CDE$. If distances AB and CD can be determined, length x can be computed.
$x = 2.500\ \text{in} + AB + 2.375\ \text{in} + CD$

Determine whether enough information is given to solve for AB. In right $\triangle ABF$:
$\angle FAB = 90° - 28° = 62°$ (complementary angles)

$$BF = \frac{1.700\ \text{in} - 0.800\ \text{in}}{2} = 0.450\ \text{in}$$

There is enough information to determine AB.

Determine whether enough information is given to solve for CD. In right $\triangle CDE$:
$\angle ECD = 90° - 70° = 20°$ (complementary angles)

$$ED = \frac{0.800\ \text{in} - 0.300\ \text{in}}{2} = 0.250\ \text{in}$$

There is enough information to determine CD.

Computations:

Solve for AB.

$$\cot \angle FAB = \frac{AB}{BF}$$

$$\cot 62° = \frac{AB}{0.450}$$

$$0.53171 = \frac{AB}{0.450}$$

$$AB = 0.450(0.53171)$$
$$AB = 0.2393\ \text{in}$$

Solve for CD.

$$\cot \angle ECD = \frac{CD}{DE}$$

$$\cot 20° = \frac{CD}{0.250}$$

$$2.7475 = \frac{CD}{0.250}$$

$$CD = 0.250(2.7475)$$
$$CD = 0.6869\ \text{in}$$

Solve for x.

$x = 2.500\ \text{in} + 0.2393\ \text{in} + 2.375\ \text{in} + 0.6869\ \text{in} =$
$5.801\ \text{in}$ Ans

Example 2: Determine ∠x of the plate shown. All dimensions are in millimeters.

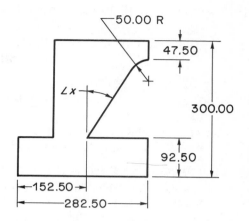

Note: Generally, when solving problems which involve an arc which is tangent to one or more lines, it is necessary to project the radius of the arc to the tangent point and to project a line from the vertex of the unknown angle to the center of the arc.

Analyze the problem:

Project auxiliary lines between the points A and O, from point O to the tangent point B, and from point O to point C. Right △ACO and right △ABO are formed. If ∠1 and ∠2 can be computed, ∠x can be determined. ∠x = 90° − (∠1 + ∠2)

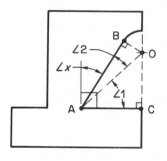

Determine whether enough information is given to solve for ∠1.
In right △ACO:
AC = 282.50 mm − 152.50 mm = 130.00 mm
CO = 300.00 mm − (92.50 mm + 50.00 mm + 47.50 mm) = 110.00 mm
There is enough information to determine ∠1.

Determine whether enough information is given to solve for ∠2.
In right △ABO: BO = 50.00 mm
AO can be determined after solving for ∠1.
There is enough information to determine ∠2.

Computations:

Solve for ∠1.

$$\tan \angle 1 = \frac{CO}{AC} = \frac{110.00}{130.00} = 0.846\ 15$$
$$\angle 1 = 40°14'$$

Solve for AO.

$$\csc \angle 1 = \frac{AO}{CO}$$

$$\csc 40°14' = \frac{AO}{110.00}$$

$$1.5482 = \frac{AO}{110.00}$$
$$AO = 1.5482(110.00)$$
$$AO = 170.30 \text{ mm}$$

Solve for ∠2.

$$\sin \angle 2 = \frac{BO}{AO} = \frac{50.00}{170.30} = 0.293\ 60$$
$$\angle 2 = 17°4'$$

Solve for ∠x.

$$\angle x = 90° - (\angle 1 + \angle 2)$$
$$\angle x = 90° - (40°14' + 17°4') = 32°42' = 32.70° \quad \text{Ans}$$

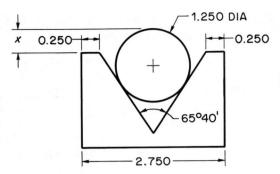

Example 3: The front view of a piece with a V-groove is shown. A 1.250-inch diameter pin is used to check the cut for depth and angular accuracy. Compute check dimension *x*. All dimensions are in inches.

Analyze the problem:

Dimension *x* is determined by the pin size, the points of tangency where the pin touches the groove, the angle of the V-groove, and the depth of the groove. Therefore, these dimensions and locations must be part of the calculations.

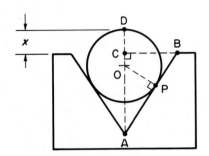

Project auxiliary lines from point A through the center of the pin O, from point O to the tangent point P, and from point B horizontally intersect vertical line AD at point C. Right △APO and right △ACB are formed. If AO and AC can be determined, check dimension *x* can be computed.

DO = radius of pin = 0.625 in

$x = (AO + DO) - AC$

Determine whether enough information is given to solve for AO.

In right △APO: $PO = \frac{1.250 \text{ in}}{2} = 0.625$ in

$$\angle OAP = \frac{65°40'}{2} = 32°50'$$

There is enough information to determine AO.

Determine whether enough information is given to solve for AC.

In right △ACB: $BC = \frac{2.750 \text{ in}}{2} - 0.250$ in $= 1.125$ in

$$\angle BAC = 32°50'$$

There is enough information to determine AC.

Computations:

Solve for AO.

$$\text{csc} \angle OAP = \frac{AO}{PO}$$

$$\text{csc } 32°50' = \frac{AO}{0.625}$$

$$1.8443 = \frac{AO}{0.625}$$

$$AO = 1.8443(0.625)$$
$$AO = 1.1527 \text{ in}$$

Solve for AC.

$$\text{cot} \angle BAC = \frac{AC}{BC}$$

$$\text{cot } 32°50' = \frac{AC}{1.125}$$

$$1.5497 = \frac{AC}{1.125}$$

$$AC = 1.5497(1.125)$$
$$AC = 1.7434 \text{ in}$$

Solve for check dimension *x*.

$x = (AO + DO) - AC$
$x = (1.1527 \text{ in} + 0.625 \text{ in}) - 1.7434 \text{ in} = 0.034 \text{ in}$ Ans

Example 4: Determine ∠x in the series of holes shown in this plate. All dimensions are in inches.

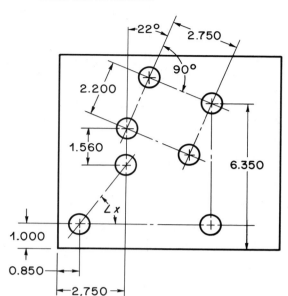 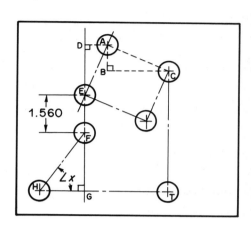

Project auxiliary lines AD, AB, BC. Right △ABC, right △ADE, and right △FGH are formed. If HG and FG can be determined, ∠x can be computed.

HG = 2.750 in – 0.850 in = 1.900 in
FG = (TC + AB) – (DE + 1.560 in)
FG = [(6.350 in – 1.000 in) + AB] – (DE + 1.560 in)

Solve for AB.

In right △ABC: AC = 2.750 in
                 ∠ACB = 22° (Two angles whose corresponding sides are
                            perpendicular and equal.)

$$\text{Sin } 22° = \frac{AB}{2.750}$$

$$0.37461 = \frac{AB}{2.750}$$

AB = 0.37461(2.750)
AB = 1.0302 in

Solve for DE.

In right △ADE: ∠DEA = 22°
                 AE = 2.200 in

$$\text{Cos } 22° = \frac{DE}{2.200}$$

$$0.92718 = \frac{DE}{2.200}$$

DE = 0.92718(2.200)
DE = 2.0398 in

Solve for FG.

FG = [(6.350 in – 1.000 in) + AB] – (DE + 1.560 in)
FG = (5.350 in + 1.0302 in) – (2.0398 in + 1.560 in) = 2.7804 in

Solve for ∠x.
$$\text{Tan } ∠x = \frac{FG}{HG} = \frac{2.7804}{1.900} = 1.4634$$
    ∠x = 55°39′   Ans

Example 5: Determine dimension x of the template shown. All dimensions are in inches.

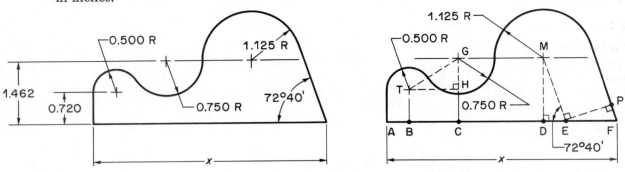

Project auxiliary lines to form right $\triangle GHT$, right $\triangle DEM$, and right $\triangle EFP$.

$x = AB + BC + CD + DE + EF$

$AB = 0.500$ in

$CD = GM = 0.750$ in $+ 1.125$ in $= 1.875$ in

(A line connecting the centers of two externally tangent circles passes through the point of tangency.)

If BC, DE, and EF can be determined, x can be computed.

Solve for BC. (BC = TH)

In right $\triangle GHT$: $\quad$ GH $= 1.462$ in $- 0.720$ in $= 0.742$ in

$\qquad\qquad\quad$ GT $= 0.500$ in $+ 0.750$ in $= 1.250$ in

$\qquad\qquad\qquad$ (GT passes through the point of tangency.)

$\sin \angle GTH = \dfrac{GH}{GT} = \dfrac{0.742}{1.250} = 0.59360$

$\quad\; \angle GTH = 36°25'$

$\cot \angle GTH = \dfrac{TH}{GH}$

$\cot 36°25' = \dfrac{TH}{0.742}$

$\quad\; 1.3555 = \dfrac{TH}{0.742}$

$\qquad TH = 0.742(1.3555)$

$\qquad TH = 1.0058$ in

$\qquad BC = 1.0058$ in

Solve for DE.

In right $\triangle DEM$: $\angle DEM = 72°40'$

$\qquad\qquad\qquad\; DM = 1.462$ in

$\cot \angle DEM = \dfrac{DE}{DM}$

$\cot 72°40' = \dfrac{DE}{1.462}$

$\quad\; 0.31210 = \dfrac{DE}{1.462}$

$\qquad DE = 0.31210(1.462)$

$\qquad DE = 0.4563$ in

Solve for EF.

In right $\triangle EFP$: $\angle F = 72°40'$

$\qquad\qquad\qquad EP = 1.125$ in

$\qquad\qquad\qquad$ (1.125 radius is $\perp$ to tangent line at the point of tangency).

$\csc 72°40' = \dfrac{EF}{EP}$

$\quad\; 1.0476 = \dfrac{EF}{1.125}$

$\qquad EF = 1.0476(1.125)$

$\qquad EF = 1.1786$ in

Solve for x.

$x = AB + BC + CD + EF$

$x = 0.500$ in $+ 1.0058$ in $+ 1.875$ in $+ 0.4563$ in $+ 1.1786$ in $= 5.016$ in $\quad$ Ans

APPLICATION

A Trigonometric Function Table can be found in the Appendix.

Complex Practical Machine Applications

Solve the following problems. For English unit dimensioned problems, calculate angles to the nearer minute and lengths to the nearer thousandth inch. For metric unit dimensioned problems, calculate angles to the nearer hundredth degree and lengths to the nearer hundredth millimeter.

1. Find length *x*. All dimensions are in inches. _____

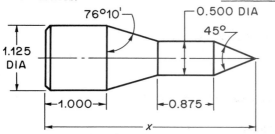

2. Find ∠*x*. All dimensions are in millimeters. _____

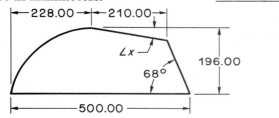

3. Find ∠*x*. All dimensions are in inches. _____

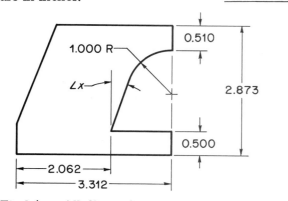

4. Find ∠*y*. All dimensions are in millimeters. _____

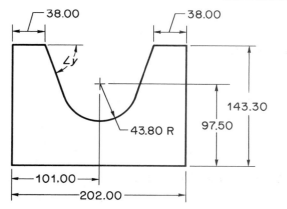

5. Find gage dimension *y*. All dimensions are in inches. _____

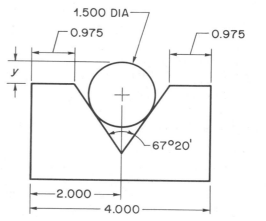

6. Find dimension *x*. All dimensions are in inches. _____

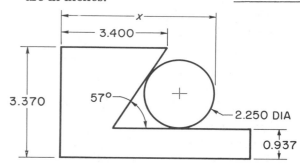

7. Find ∠*x*. All dimensions are in inches. _____

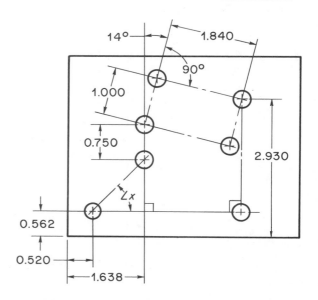

8. Find ∠y. All dimensions are in inches. _____

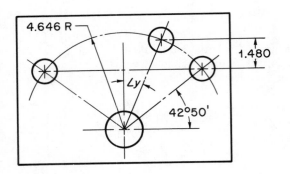

9. Find length x. All dimensions are in millimeters. _____

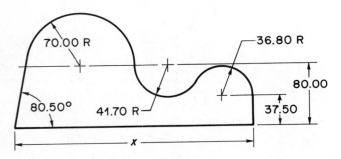

10. Find ∠y. All dimensions are in inches. _____

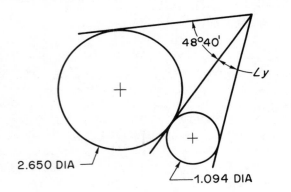

11. Find dimension x. All dimensions are in inches. _____

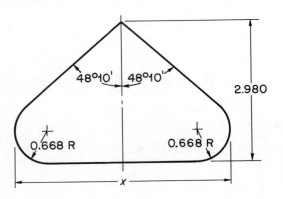

12. Find dimension y. All dimensions are in inches. _____

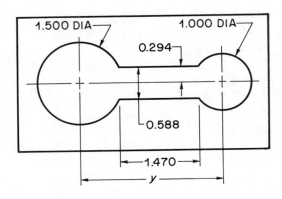

13. Find dimension y. All dimensions are in inches. _____

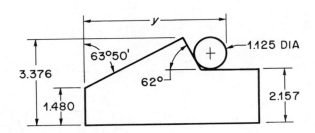

14. Find ∠x. All dimensions are in millimeters. _____

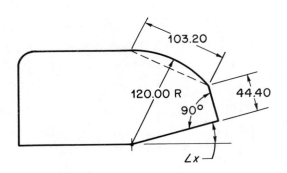

15. Find ∠x. All dimensions are in inches. _____

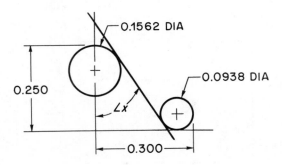

16. Find dimension *y*. All dimensions are in inches. _____

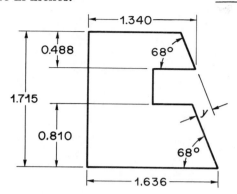

17. Find dimension *y*. All dimensions are in inches. _____

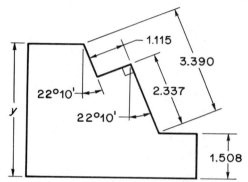

18. Find dimension *x*. All dimensions are in inches. _____

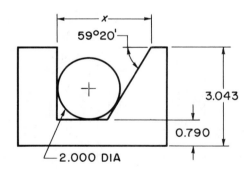

19. Find dimension *x*. All dimensions are in inches. _____

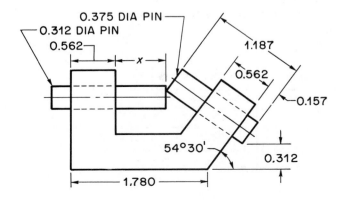

20. Find ∠*x*. All dimensions are in millimeters. _____

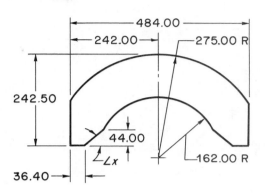

21. Find ∠*x*. All dimensions are in inches. _____

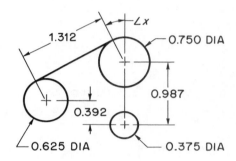

22. Find check dimension *y*. All dimensions are in inches. _____

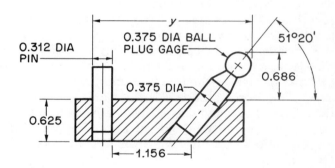

23. Find dimension *y*. All dimensions are in inches. _____

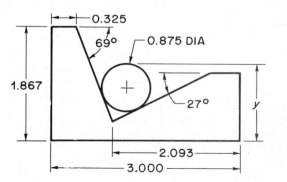

24. Find x. All dimensions are in millimeters. _____

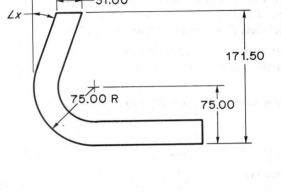

⌐UNIT 53 *THE CARTESIAN COORDINATE SYSTEM*_____

⌐OBJECTIVE_____

After studying this unit you should be able to
  • Compute functions of angles greater than 90°.

⌐CARTESIAN (RECTANGULAR) COORDINATE SYSTEM_____

It is sometimes necessary to determine functions of angles greater than 90°. In a triangle that is not a right triangle, one of the angles can be greater than 90°. Computations using functions of angles greater than 90° are often required in order to solve oblique triangle problems.

Functions of any angles are easily described in reference to the Cartesian Coordinate System. A fixed point (O) called the *origin* is located at the intersection of a vertical and horizontal axes. The horizontal axis is the x-axis and the vertical axis is the y-axis. The x and y axes divide a plane into four parts which are called *quadrants*. Quadrant I is the upper right section. In a counterclockwise direction from Quadrant I are Quadrants II, III, and IV.

All points located to the right of the y-axis have positive (+) x values; all points to the left of the y-axis have negative (−) x

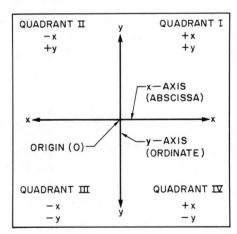

values. All points above the x-axis have positive (+) y values; all points below the x-axis have negative (−) y values. The x value is called the *abscissa* and the y value is called the *ordinate*.

The x and y values for each quadrant are listed in the table.

Quadrant I	Quadrant II	Quadrant III	Quadrant IV
+x	− x	−x	+x
+y	+ y	−y	−y

⌐DETERMINING FUNCTIONS OF ANGLES IN ANY QUADRANT_____

As a ray is rotated through any of the four quadrants, functions of an angle are determined as follows:

- The ray is rotated in a counterclockwise direction with its vertex at the origin (O). Zero degrees is on the x-axis in quadrant I.

- From a point on the rotated ray, a line segment is projected perpendicular to the x-axis. A right triangle is formed of which the rotated side (ray) is the hypotenuse, the projected line segment is the opposite side, and the side on the x-axis is the adjacent side. The *reference angle* is the acute angle of the triangle which has the vertex at the origin (O).

- The sign of the functions of a reference angle is determined by noting the signs (+ or –) of the opposite and adjacent sides of the right triangle. The hypotenuse (r) is always positive in all four quadrants.

These examples illustrate the method of determining functions of angles greater than 90° in the various quadrants.

Example 1: Determine the sine and cosine functions of 115°.

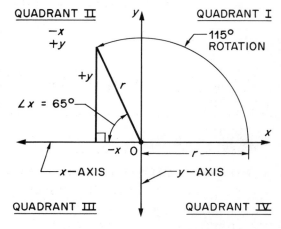

With the endpoint of the ray (r) at the origin (O), the ray is rotated 115° in a counterclockwise direction.

From a point on r, side y is projected perpendicular to the x-axis. In the right triangle formed, in relation to the reference angle ($\angle x$), r is the hypotenuse, y is the opposite side, and x is the adjacent side.
$\angle x = 180° - 115° = 65°$

Sin $\angle x = \dfrac{\text{opposite side}}{\text{hypotenuse}}$. In Quadrant II, y is positive and r is always positive. Therefore, $\sin \angle x = \dfrac{+y}{+r}$. In Quadrant II, the sine is a positive (+) function.

$\sin 115° = \sin (180° - 115°) = \sin 65° = 0.90631$ Ans

Cos $\angle x = \dfrac{\text{adjacent side}}{\text{hypotenuse}}$. Side x is negative (–), therefore, $\cos \angle x = \dfrac{-x}{+r}$. Since the quotient of a negative value divided by a positive value is negative, in Quadrant II, the cosine is a negative (–) function.

$\cos 115° = -\cos(180° - 115°) = -\cos 65°$

Look up the cosine of 65° in the function table and prefix a negative sign.

$\cos 65° = 0.42262$
$-\cos 65° = -0.42262$
$\cos 115° = -\cos 65° = -0.42262$ Ans

Note: A negative function of an angle does <u>not</u> mean that the angle is negative; it is a negative function of a positive angle. For example, $-\cos 65°$ does not mean $\cos(-65°)$.

Example 2: Determine the tangent and secant functions of 218°.

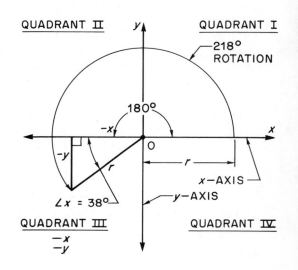

Rotate 218° counterclockwise.

Project side $y \perp$ to the x-axis.

Reference $\angle x = 218° - 180° = 38°$

$\tan \angle x = \dfrac{-y}{-x} = +$ function

$\tan 218° = \tan 38° = 0.78129$ Ans

$\sec \angle x = \dfrac{+r}{-x} = -$ function

$\sec 218° = -\sec 38° = -1.2690$ Ans

Example 3: Determine the cotangent and cosecant functions
of 310°.

Rotate 310° counterclockwise.

Project side $y \perp$ to the x-axis.

Reference $\angle x = 360° - 310° = 50°$

$\cot \angle x = \dfrac{+x}{-y} = -$ function

$\cot 310° = -\cot 50° = -0.83910$ Ans

$\csc \angle x = \dfrac{+r}{-y} = -$ function

$\csc 310° = -\csc 50° = -1.3054$ Ans

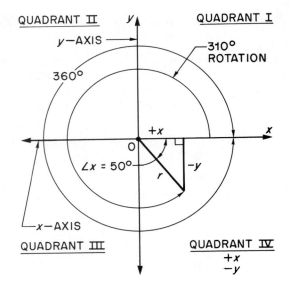

APPLICATION

Determining Functions of Angles in any Quadrant

Determine the sine, cosine, tangent, cotangent, secant, and cosecant functions for
each of these angles. For each angle, sketch a right triangle. Label the sides of
the triangles + or −. Determine the reference angles and look up the functions of
the angles. Keep in mind that a function of an angle greater than 90° may be a
negative value.

1. 125°

2. 207°

3. 260°

4. 168°

5. 300°

6. 350°

7. 216°20′

8. 96°50′

9. 146°10′

10. 202.60°

11. 313.20°

12. 179.90°

UNIT 54 OBLIQUE TRIANGLES: LAW OF SINES AND LAW OF COSINES

OBJECTIVES

After studying this unit you should be able to

- Solve simple oblique triangles using the Law of Sines and the Law of Cosines.
- Solve practical shop problems by applying the Law of Sines and the Law of Cosines.

OBLIQUE TRIANGLES

An *oblique triangle* is one that does not contain a right angle. The machinist must often solve practical machine shop problems which involve oblique triangles. These problems can be reduced to a series of right triangles, but the process can be cumbersome and time consuming. Two formulas, the Law of Sines and the Law of Cosines, can be used to simplify such computations. In order to use either formula, three parts of an oblique triangle must be known; at least one part must be a side.

LAW OF SINES

The Law of Sines states that in any triangle, the sides are proportional to the sines of the opposite angles. In reference to the triangle shown, the formula is stated:

$$\frac{a}{\sin A} = \frac{b}{\sin B} = \frac{c}{\sin C}$$

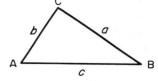

The Law of Sines is used to solve the following two kinds of problems:

- Problems where any two angles and any one side of an oblique triangle are known.
- Problems where any two sides and an angle opposite one of the given sides of an oblique triangle are known.

Solving Oblique Triangle Problems Given Two Angles and a Side

Example 1: Given two angles and a side, determine side x of the oblique triangle shown. All dimensions are in inches.

Set up a proportion and solve for x.

$$\frac{x}{\sin 36°} = \frac{3.500}{\sin 58°}$$
$$\frac{x}{0.58779} = \frac{3.500}{0.84805}$$
$$0.84805x = 3.500(0.58779)$$
$$x = \frac{3.500(0.58779)}{0.84805}$$
$$x = 2.426 \text{ inches} \quad \text{Ans}$$

Example 2: Given two angles and a side of the oblique triangle shown. All dimensions are in millimeters.

    a. Determine $\angle A$.

    b. Determine side a.

    c. Determine side b.

a. Determine $\angle A$.
   $\angle A = 180° - (37.3° + 24.5°) = 118.2°$ Ans

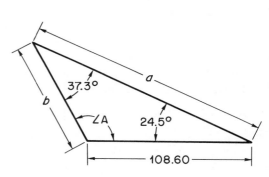

b. Determine side a. Set up a proportion and solve for side a.

$$\frac{a}{\sin 118.2°} = \frac{108.60}{\sin 37.3°}$$

$\sin 118.2° = \sin 118°12' = \sin (180° - 118°12') = \sin 61°48' = 0.881\ 30$

$\sin 37.3° = \sin 37°18' = 0.605\ 99$

$$\frac{a}{0.881\ 30} = \frac{108.60}{0.605\ 99}$$

$$a = 157.94 \text{ mm} \quad \text{Ans}$$

c. Determine side b. Set up a proportion and solve for side b.

$$\frac{b}{\sin 24.5°} = \frac{108.60}{\sin 37.3°}$$

$\sin 24.5° = \sin 24°30' = 0.414\ 69$

$$\frac{b}{0.414\ 69} = \frac{108.60}{0.605\ 99}$$

$$b = 74.32 \text{ mm} \quad \text{Ans}$$

Solving Oblique Triangle Problems Given Two Sides and an Angle Opposite One of the Given Sides

A special condition exists when solving certain problems in which two sides and an angle opposite one of the sides is given. If triangle data are given in word form or if a triangle is inaccurately sketched, there may be two solutions to a problem.

It is possible to have two different triangles with the same two sides and the same angle opposite one of the given sides. A situation of this kind is called an ambiguous case. The following example illustrates the ambiguous case or a problem with two solutions.

Example (the Ambiguous Case or 2 solutions): A triangle has a 1.5-inch side, a 2.5-inch side and an angle of 32° which is opposite the 1.5-inch side.

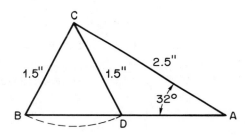

Using the given data, a figure is accurately drawn. Observe that two different triangles are constructed using identical given data. Both △BCA and △DCA have a 1.5-inch side, a 2.5-inch side, and a 32° angle opposite the 1.5-inch side. The two different triangles are shown.

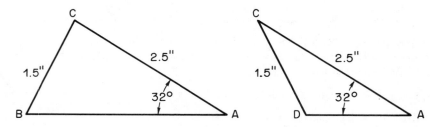

The only conditions under which a problem can have two solutions is when the given angle is acute and the given side opposite the given angle is smaller than the other given side. For example, in the problem illustrated the 32° angle is acute, and the 1.5-inch side opposite the 32° angle is smaller than the 2.5-inch side.

In most problems you do not get involved with two solutions. Even under the conditions in which there can be two solutions, if the problem is shown in picture form as an accurately drawn triangle, it can readily be observed that there is only one solution.

Example 1: Given two sides and an opposite angle of the oblique triangle shown. All dimensions are in inches.

    a. Determine $\angle x$.

    b. Determine side y.

a. Determine $\angle x$.

$$\frac{4.500}{\sin \angle x} = \frac{6.000}{\sin 63°50'}$$

$$\frac{4.500}{\sin \angle x} = \frac{6.000}{0.89752}$$

$$6.000(\sin \angle x) = 4.500(0.89752)$$

$$\sin \angle x = \frac{4.500(0.89752)}{6.000}$$

$$\sin \angle x = 0.67314$$

$$\angle x = 42°19' \quad \text{Ans}$$

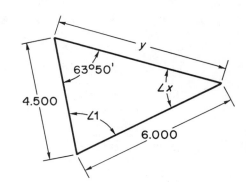

b. Determine side y.

$$\angle 1 = 180° - (63°50' + \angle x) = 180° - 106°9' = 73°51'$$

$$\frac{6.000}{\sin 63°50'} = \frac{y}{\sin 73°51'}$$

$$\frac{6.000}{0.89752} = \frac{y}{0.96054}$$

$$0.89752y = 6.000(0.96054)$$

$$y = \frac{6.000(0.96054)}{0.89752}$$

$$y = 6.421 \text{ inches} \quad \text{Ans}$$

Example 2: Given two sides and an opposite angle, determine $\angle x$ of the oblique triangle shown. All dimensions are in millimeters.

$$28.17° = 28°10'$$

$$\frac{140.00}{\sin 28°10'} = \frac{275.00}{\sin \angle x}$$

$$\frac{140.00}{0.472\,04} = \frac{275.00}{\sin \angle x}$$

$$\sin \angle x = 0.927\,22$$

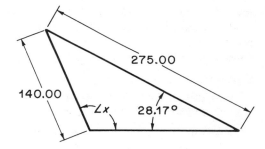

The angle that corresponds to the sine function 0.927 22 is 68°. Because $\angle x$ is greater than 90°, $\angle x$ = the supplement of 68°.

$$\angle x = 180° - 68° = 112° \quad \text{Ans}$$

¬LAW OF COSINES

The Law of Cosines states that in any triangle, the square of any side is equal to the sum of the squares of the other two sides minus twice the product of these two sides multiplied by the cosine of their included angle.

In reference to the triangle shown the formula is stated:

$$a^2 = b^2 + c^2 - 2bc(\cos A)$$
$$b^2 = a^2 + c^2 - 2ac(\cos B)$$
$$c^2 = a^2 + b^2 - 2ab(\cos C)$$

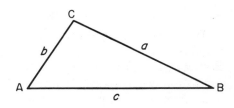

The equations can be rearranged in this form:

$$\cos A = \frac{b^2 + c^2 - a^2}{2bc}$$

$$\cos B = \frac{a^2 + c^2 - b^2}{2ac}$$

$$\cos C = \frac{a^2 + b^2 - c^2}{2ab}$$

The Law of Cosines is used to solve the following two kinds of problems:
• Problems where two sides and the included angle of an oblique triangle are known.
• Problems where three sides of an oblique triangle are known.

Solving Oblique Triangle Problems Given Two Sides and the Included Angle

Example 1: Given two sides and the included angle, determine side x of the oblique triangle shown. All dimensions are in inches.

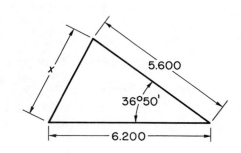

Substitute the values in their appropriate places in the formula and solve for x.

$x^2 = 5.600^2 + 6.200^2 - 2(5.600)(6.200)(\cos 36°50')$
$x^2 = 31.360 + 38.440 - 2(5.600)(6.200)(0.80038)$
$x^2 = 69.8000 - 55.5784$
$x^2 = 14.2216$
$x = 3.771$ inches   Ans

Example 2: Given two sides and the included angle of the oblique triangle shown. All dimensions are in inches.

    a. Determine side x.

    b. Determine $\angle 1$.

a. Substitute values to solve for side x.

$x^2 = 3.800^2 + 4.100^2 - 2(3.800)(4.100)(\cos 123°)$

Since the given angle is greater than $90°$, the cosine of the angle is equal to the negative cosine of its supplement.

Therefore, $\cos 123° = -\cos(180° - 123°) = -\cos 57° = -0.54464$. This negative value must be used in computing side x.

$x^2 = 3.800^2 + 4.100^2 - 2(3.800)(4.100)(-0.54464)$
$x^2 = 31.250 - (-16.971)$

Recall that subtracting a negative value is the same as adding a positive value.
$x^2 = 31.250 + 16.971$
$x^2 = 48.221$
$x = 6.944$ inches   Ans

b. Solve for $\angle 1$ using the Law of Sines.

$$\frac{3.800}{\sin \angle 1} = \frac{\text{side } x}{\sin 123°}$$

$$\sin 123° = \sin(180° - 123°) = \sin 57° = 0.83867$$

$$\frac{3.800}{\sin \angle 1} = \frac{6.944}{0.83867}$$

$$6.944(\sin \angle 1) = (3.800)(0.83867)$$

$$\sin \angle 1 = \frac{3.800(0.83867)}{6.944}$$

$$\sin \angle 1 = 0.45895$$
$$\angle 1 = 27°19'   \text{Ans}$$

Note: Since side x has been determined, $\angle 1$ may be computed by using the Law of Cosines (given three sides), but it is simpler to use the Law of Sines.

Solving Oblique Triangle Problems Given Three Sides

Example 1: Given three sides, determine $\angle x$ of the oblique triangle shown. All dimensions are in millimeters.

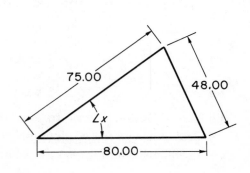

Substitute values in their appropriate places in the formula.

$$\cos \angle x = \frac{75.00^2 + 80.00^2 - 48.00^2}{2(75.00)(80.00)}$$

$$\cos \angle x = \frac{5625 + 6400 - 2304}{12\,000}$$

$$\cos \angle x = 0.810\,08$$
$$\angle x = 35°54' = 35.90°   \text{Ans}$$

Example 2: Given three sides, determine $\angle x$ of the oblique triangle shown. All dimensions are in inches.

Substitute values in the formula.

$$\cos \angle x = \frac{3.000^2 + 3.700^2 - 5.300^2}{2(3.000)(3.700)}$$

$$\cos \angle x = \frac{9.000 + 13.690 - 28.090}{22.200}$$

$$\cos \angle x = \frac{22.690 - 28.090}{22.200}$$

$$\cos \angle x = \frac{-5.400}{22.200}$$

$$\cos \angle x = -0.24324$$

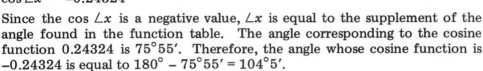

Since the $\cos \angle x$ is a negative value, $\angle x$ is equal to the supplement of the angle found in the function table. The angle corresponding to the cosine function 0.24324 is $75°55'$. Therefore, the angle whose cosine function is -0.24324 is equal to $180° - 75°55' = 104°5'$.

$$\angle x = 104°5' \quad \text{Ans}$$

APPLICATION

A Trigonometric Function Table can be found in the Appendix. For English unit dimensioned problems, calculate angles to the nearer minute and lengths to the nearer thousandth inch. For metric unit dimensioned problems, calculate angles to the nearer hundredth degree and lengths to the nearer hundredth millimeter.

Law of Sines

Solve the following problems using the Law of Sines.

1. Find side x. All dimensions are in inches. _____

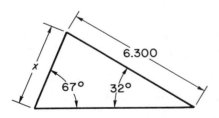

2. Find side x. All dimensions are in inches. _____

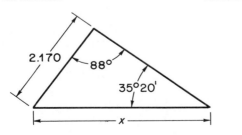

3. Find side x. All dimensions are in inches. _____

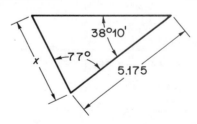

4. Find side x. All dimensions are in millimeters. _____

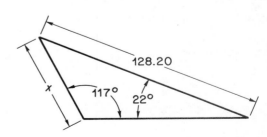

5. Find ∠x. All dimensions
   are in inches.   _____

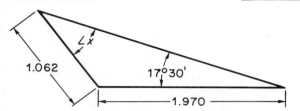

6. Find ∠x. All dimensions
   are in millimeters.   _____

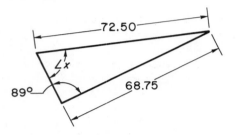

7. Find ∠x. All dimensions
   are in inches.   _____

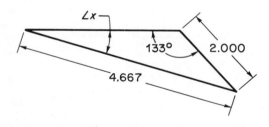

8. Find ∠x. All dimensions
   are in inches.   _____

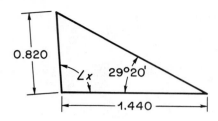

9. Find side x. All dimensions
   are in millimeters.   _____

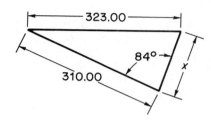

Identifying Problems with One or Two Solutions

Two sides and an angle opposite one of the sides of triangles are given in the following problems. Identify each problem as to whether it has one or two solutions.
Do not solve the problems for angles and sides.

10. A 3″ side, a 5″ side, a 37° angle opposite the 3″ side.   _____

11. A 95.00-mm side, a 98.00-mm side, a 75° angle opposite the 95.00-mm side.   _____

12. A 21-mm side, a 29-mm side, a 41° angle opposite the 29-mm side.   _____

13. A 0.943″ side, a 1.612″ side, and a 82°15′ angle opposite the 1.612″ side.   _____

14. A 2.10-ft side, a 3.05-ft side, a 29°30′ angle opposite the 3.05-ft side.   _____

15. A 16.35-mm side, a 23.86-mm side, a 115° angle opposite the 23.86-mm side. _____

16. An 87.60-mm side, a 124.80-mm side, a 12.90° angle opposite the 87.60-mm side. _____

17. A 33.860″ side, a 34.090″ side, a 46°18′ angle opposite the 33.860″ side.   _____

Law of Cosines

Solve the following problems using the Law of Cosines.

18. Find side x. All dimensions
    are in inches.   _____

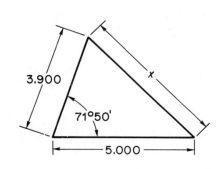

19. Find side x. All dimensions
    are in millimeters.   _____

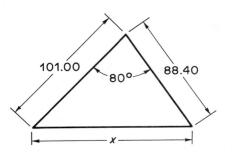

20. Find side x. All dimensions
    are in inches.   _____

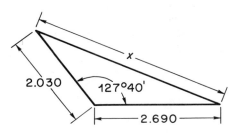

21. Find ∠x. All dimensions
    are in millimeters.   _____

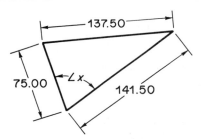

22. Find ∠x. All dimensions
    are in inches.   _____

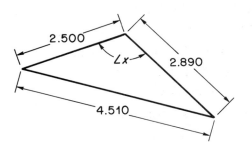

23. Find ∠x. All dimensions
    are in inches   _____

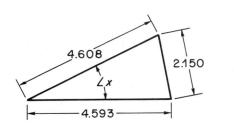

24. Find ∠x. All dimensions
    are in inches.   _____

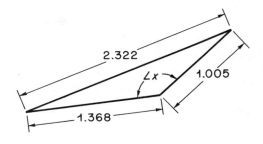

25. Find side x. All dimensions
    are in millimeters.   _____

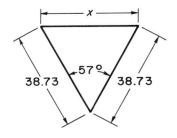

26. Find side x. All dimensions
    are in inches.   _____

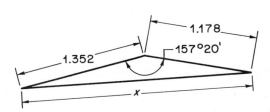

**Combination of the Law of Cosines
and the Law of Sines**

Solve the following problems using a combination
of the Law of Cosines and the Law of Sines.

27. All dimensions are in inches.

    a. Find side x.   _____

    b. Find ∠y.   _____

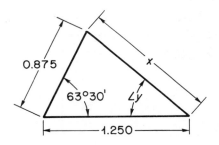

28. All dimensions are in inches.
    a. Find side x.   _____
    b. Find $\angle y$.   _____

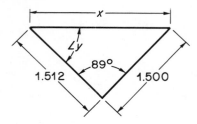

29. All dimensions are in millimeters.
    a. Find side x.   _____
    b. Find $\angle y$.   _____

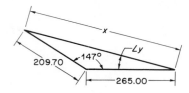

30. All dimensions are in inches.
    a. Find $\angle x$.   _____
    b. Find $\angle y$.   _____

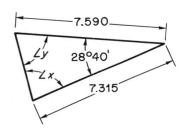

31. All dimensions are in inches.
    a. Find $\angle y$.   _____
    b. Find $\angle y$.   _____

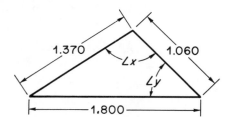

32. All dimensions are in millimeters.
    a. Find $\angle x$.   _____
    b. Find $\angle y$.   _____

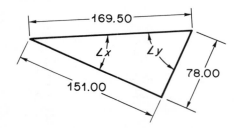

Practical Machine Shop Problems

Solve the following machine shop problems.

33. Find $\angle x$. All dimensions
    are in millimeters   _____

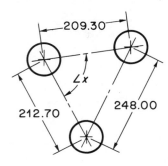

34. Find distance y. All dimen-
    sions are in inches.   _____

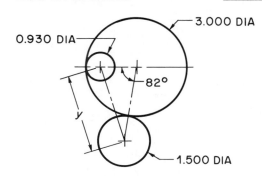

35. Find distance x. All dimen-
    sions are in inches.   _____

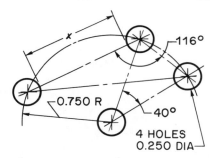

36. Find $\angle x$. All dimensions
    are in inches.   _____

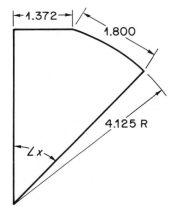

37. Find dimension y. All dimensions are in millimeters. _____

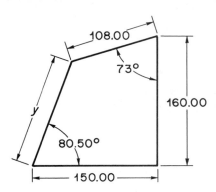

41. Find $\angle y$. All dimensions are in millimeters. _____

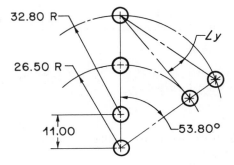

38. Find dimension y. All dimensions are in inches. _____

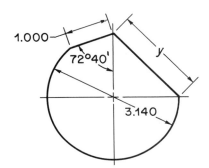

39. Find dimension y. All dimensions are in inches. _____

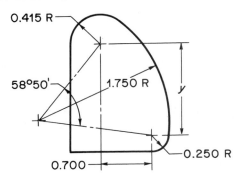

40. Find $\angle x$. All dimensions are in inches. _____

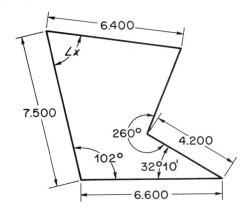

─UNIT 55 ACHIEVEMENT REVIEW —
SECTION 5

A Trigonometric Function Table can be found in the Appendix.

─OBJECTIVE

> You should be able to solve the exercises and problems in this Achievement Review by applying the principles and methods covered in units 48–54.

1. With reference to ∠1, name the sides of each of the following triangles as opposite, adjacent, or hypotenuse.

   a. _____   b. _____   c. _____   d. _____

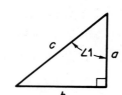
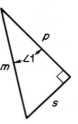
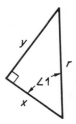

2. Determine the functions of the following angles.

   a. sin 25° _____   e. cos 63°18′ _____

   b. cot 46°20′ _____   f. tan 12°46′ _____

   c. sec 37°50′ _____   g. sin 7.43° _____

   d. tan 0°18′ _____   h. csc 57.82° _____

3. Determine the values of ∠A in degrees and minutes that correspond to the following functions.

   a. cos A = 0.69675 _____   d. cot A = 1.1340 _____

   b. tan A = 0.49858 _____   e. sec A = 1.5753 _____

   c. sin A = 0.98531 _____   f. cos A = 0.16440 _____

4. Determine the values of ∠A in decimal-degrees to 2 decimal places that correspond to the following functions.

   a. sin A = 0.746 64 _____   c. cos A = 0.340 38 _____

   b. tan A = 1.3925 _____

5. For each of the following functions of angles, write the cofunction of the complement of the angle.

   a. sin 36° _____   c. cos 16°53′ _____

   b. tan 51°18′ _____   d. cot 76.43° _____

6. Solve the following problems. Compute angles to the nearer minute in triangles with English unit sides. Compute angles to the nearer hundredth degree in triangles with metric unit sides. Compute sides to 3 decimal places.

   a. Determine ∠A. All dimensions are in inches. _____

   b. Determine side *a*. All dimensions are in inches. _____

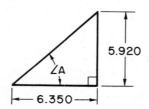
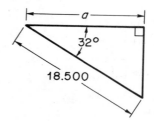

c. Determine ∠D. All dimensions
   are in millimeters. _____

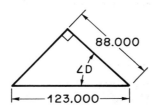

d. Determine ∠1. All dimensions
   are in millimeters. _____

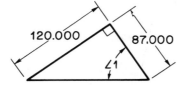

e. All dimensions are in millimeters.
   (1) Determine side g. _____
   (2) Determine side h. _____
   (3) Determine ∠H. _____

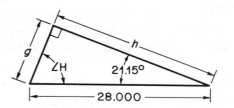

f. All dimensions are in inches.
   (1) Determine ∠A. _____
   (2) Determine ∠B. _____
   (3) Determine side c. _____

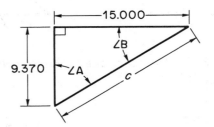

7. Solve the following applied right triangle problems. Compute linear values
   to 3 decimal places, English unit angles to the nearer minute, and metric
   angles to the nearer hundredth degree.

   a. All dimensions are in inches.
      (1) Determine dimension c. _____
      (2) Determine dimension d. _____

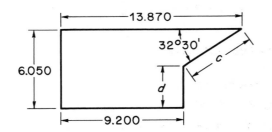

   b. Determine ∠T. All dimensions
      are in millimeters. _____

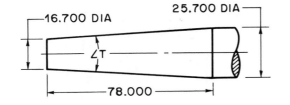

   c. Determine dimension x. All
      dimensions are in millimeters. _____

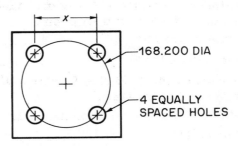

   d. Determine dimension d. All
      dimensions are in inches. _____

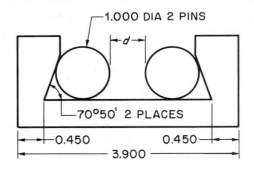

e. All dimensions are in millimeters.

(1) Determine ∠A. _____

(2) Determine distance x. _____

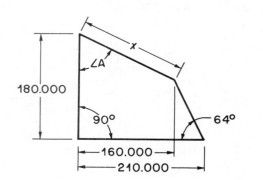

f. Determine check dimension y.
All dimensions are in inches. _____

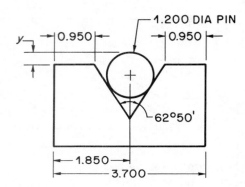

g. Determine ∠x. All dimensions
are in inches. _____

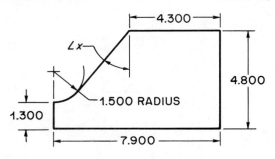

h. Determine ∠y. All dimensions
are in millimeters. _____

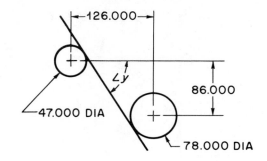

8. Determine the sine, cosine, tangent, cotangent, secant, and cosecant of each
of the following angles.

a. 120° _____

b. 225° _____

c. 312°30′ _____

9. Solve the following problems using the Law of Sines and/or the Law of Cosines. Compute side lengths to 3 decimal places, English unit angles to the
nearer minute, and metric unit angles to the nearer hundredth degree.

a. Determine side a. All dimensions are in millimeters. _____

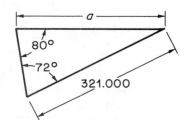

b. Determine ∠D. All dimensions
are in inches. _____

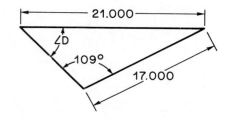

c. All dimensions are in inches.

   (1) Determine ∠A.   _____

   (2) Determine side a.   _____

   (3) Determine side b.   _____

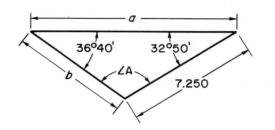

d. Determine side d. All dimensions are in millimeters.   _____

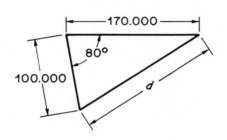

e. Determine ∠E. All dimensions are in millimeters.   _____

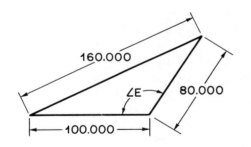

f. All dimenisons are in inches.

   (1) Determine side m.   _____

   (2) Determine ∠N.   _____

   (3) Determine ∠P.   _____

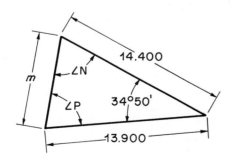

g. Determine dimension d.
All dimensions are in inches.   _____

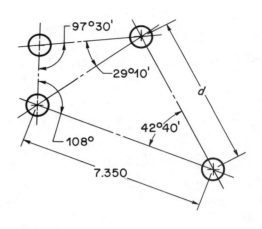

h. Determine ∠x. All dimensions
are in millimeters.   _____

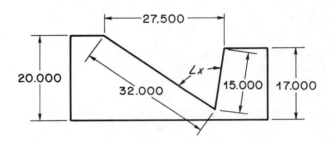

i. Determine ∠x. All dimensions
are in inches.   _____

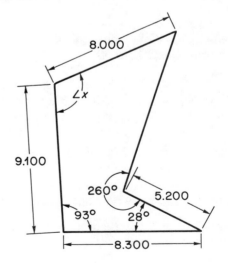

SECTION 6
COMPOUND ANGLES

─UNIT 56 *INTRODUCTION TO COMPOUND ANGLES* _____

─OBJECTIVES_____

After studying this unit you should be able to

- Compute true lengths of diagonals of rectangular solids.

- Compute true angles of diagonals of rectangular solids.

In the machine trades, the application of principles of solid or three-dimensional trigonometry is commonly called *compound angles.* Generally compound angle problems require the computation of an unknown angle in a plane which is the resultant of two or more known angles lying in different planes.

Applications of compound angles are frequently required in machining fixture parts, die sections, and cutting tools. An understanding of compound angle procedures is necessary in setting up parts for drilling or boring compound-angular holes.

Often, compound angle problems are encountered when machining parts as shown on engineering drawings. Usually, the top, front, and right side views of orthographic projections are shown. Wherever applicable, compound angle examples and problems in this text are given in relation to these views.

Formulas for specific compound angle applications can be found in certain trade handbooks. These formulas are useful provided the particular compound angle applications are properly visualized and identified. There are variations in compound angle situations. Merely plugging in values in given formulas without fully visualizing the components of a problem can result in costly errors.

Certain basic compound angle situations are presented in this text. A comprehensive study of compound angles is not intended. An understanding of applications is emphasized. Visualization of a problem with its components is stressed.

Pictorial views of compound angle situations with their components located and identified in rectangular solids or pyramids are shown. The student should make sketches in pictorial form to develop understanding. Formulas should be used in the solution of a problem only after the problem has been clearly visualized.

─DIAGONAL OF A RECTANGULAR SOLID_____

A pictorial view of a rectangular solid with diagonal AB is shown. A rectangular solid has six rectangular faces.

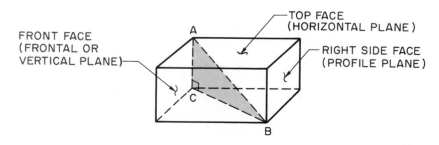

FRONT FACE
(FRONTAL OR
VERTICAL PLANE)

TOP FACE
(HORIZONTAL PLANE)

RIGHT SIDE FACE
(PROFILE PLANE)

The top face (horizontal plane), front face (frontal or vertical plane), and right side face (profile plane) are identified. These faces correspond to the top, front, and right side views as they appear on an engineering drawing as shown.

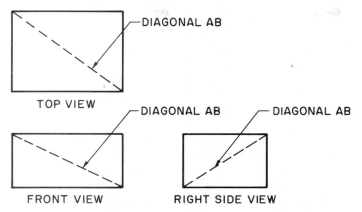

Observe that although AB appears as a diagonal in each of the three views, it does not appear in its actual (true) length in any of the views. Neither does the true angle made by AB with either a vertical or horizontal plane appear in any of the three views. The true length of a line is shown when the line is contained in a plane which is viewed perpendicular to the line of sight.

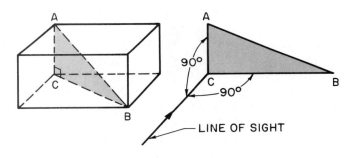

‾COMPUTING TRUE LENGTHS AND TRUE ANGLES

Example: Compute true length AB and true ∠CAB shown. All
          dimensions are in inches.

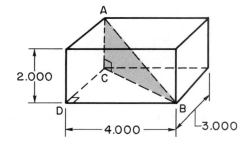

In right △CDB:
Compute CB.
Apply the Pythagorean Theorem:

$$CB^2 = DB^2 + DC^2$$
$$CB^2 = 4.000^2 + 3.000^2$$
$$CB^2 = 16.000 + 9.000$$
$$CB^2 = 25.000$$
$$CB = 5.000 \text{ in}$$

In right △ACB:
Compute AB.

$$AB^2 = AC^2 + CB^2$$
$$AB^2 = 2.000^2 + 5.000^2$$
$$AB^2 = 4.000 + 25.000$$
$$AB^2 = 29.000$$
$$AB = 5.385 \text{ in} \quad \text{Ans}$$

Compute ∠CAB.

$$\tan \angle CAB = \frac{CB}{AC} = \frac{5.000}{2.000} = 2.5000$$

$$\angle CAB = 68°12' \quad \text{Ans}$$

APPLICATION

A Trigonometric Function Table can be found in the Appendix. In each of the following problems a diagonal is shown within a rectangular solid.

a. Compute the true length of diagonal AB

b. Compute ∠CAB

Use this figure for # 1 and # 2.

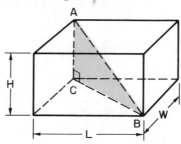

1. Given: H = 1.500 in
            L = 2.700 in
            W = 2.000 in

   a. _____

   b. _____

2. Given: H = 50.00 mm
            L = 100.00 mm
            W = 80.00 mm

   a. _____

   b. _____

Use this figure for # 3 and # 4.

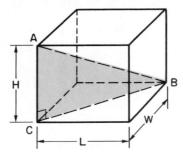

3. Given: H = 3.100 in
            L = 3.500 in
            W = 3.000 in

   a. _____

   b. _____

4. Given: H = 75.00 mm
            L = 90.00 mm
            W = 70.00 mm

   a. _____

   b. _____

Use this figure for # 5 and # 6.

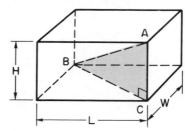

5. Given: H = 0.800 in
            L = 1.400 in
            W = 1.000 in

   a. _____

   b. _____

6. Given: H = 20.00 mm
            L = 36.00 mm
            W = 28.00 mm

UNIT 57 DRILLING AND BORING COMPOUND-ANGULAR HOLES: COMPUTING ANGLES OF ROTATION AND TILT USING GIVEN LENGTHS _____

OBJECTIVE _____

After studying this unit you should be able to
- Compute the angles of rotation and angles of tilt of hole axes in given rectangular solids.
- Sketch, dimension, and label compound-angular components within rectangular solids and compute angles of rotation and angles of tilt.

COMPUTING ANGLES OF ROTATION AND ANGLES OF TILT FOR DRILLING AND BORING COMPOUND-ANGULAR HOLES _____

A part is usually positioned on an angle plate when drilling or boring compound-angular holes. In order to position a part, the angle of rotation and the angle of tilt must be computed.

The *angle of rotation*, $\angle R$, is the angle that the piece is rotated so the hole axis is in a plane perpendicular to the pivot axis of the angle plate to which the piece is mounted.

The *angle of tilt*, $\angle T$, is the angle that the angle plate is raised to put the axis of the hole in a vertical position.

The following example shows the procedure, using given length dimensions, for finding the angle of rotation and the angle of tilt.

Example: Three views of a compound-angular hole are shown. All dimensions are in inches.
   a. Determine the angle of rotation.
   b. Determine the angle of tilt.

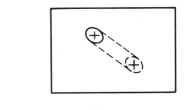

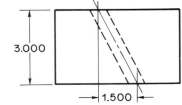
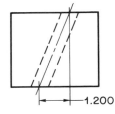

Sketch, dimension, and label a rectangular solid showing a right triangle within the solid which contains the hole axis as a side and the true angle.

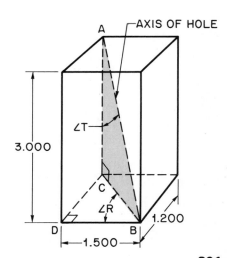

a. Compute the angle of rotation, $\angle R$. In right $\triangle BDC$:

$$\tan \angle R = \frac{DC}{DB} = \frac{1.200}{1.500} = 0.80000$$
$$\angle R = 38°40' \quad \text{Ans}$$

b. Compute angle of tilt, $\angle T$.
   In right $\triangle BDC$:
$$\sin \angle R = \frac{DC}{CB}$$
$$\sin 38°40' = \frac{1.200}{CB}$$
$$0.62479 = \frac{1.200}{CB}$$
$$CB = 1.9206 \text{ in}$$

In right $\triangle ACB$:
$$\tan \angle T = \frac{CB}{AC} = \frac{1.9206}{3.000} = 0.6402$$
$$\angle T = 32°38' \quad \text{Ans}$$

Procedure for Positioning the Part on an Angle Plate for Drilling

- Rotate the part to the angle of rotation, ∠R as shown. Care must be taken as to whether the part is rotated to the computed ∠R or the complement of ∠R. Rotate the part 38°40′. Note the position of right △ACB is shown with hidden lines.

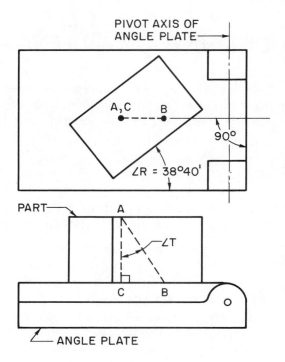

- Raise the angle plate to tilt angle, ∠T. Tilt to 32°38′ as shown. Care must be taken as to whether the part is tilted to the computed ∠T or the complement of ∠T. Observe that the position of hole axis AB is vertical.

With the part set to the angle of rotation and to the angle of tilt it is positioned to drill the hole on vertical axis AB.

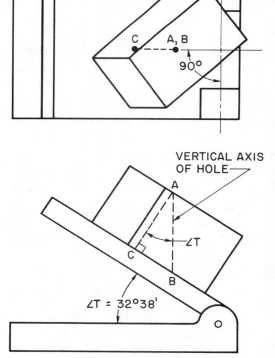

―APPLICATION―――――――――――――――――――

A Trigonometric Function Table can be found in the Appendix. In each of the following problems, the axis of a hole is shown in a rectangular solid. In order to position the hole axis for drilling, the angle of rotation and the angle of tilt must be determined.

a. Compute the angle of rotation, ∠R.

b. Compute the angle of tilt, ∠T.

1. Given: H = 2.600 in
            L = 2.400 in
            W = 1.900 in

    a. _____

    b. _____

2. Given: H = 55.00 mm
            L = 48.00 mm
            W = 30.00 mm

    a. _____

    b. _____

Use this figure for # 1 and # 2.

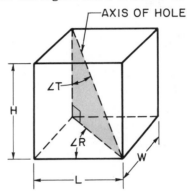

3. Given: H = 1.900 in
            L = 1.600 in
            W = 1.500 in

    a. _____

    b. _____

4. Given: H = 42.00 mm
            L = 37.00 mm
            W = 32.00 mm

    a. _____

    b. _____

Use this figure for # 3 and # 4.

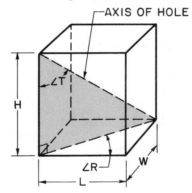

5. Given: H = 0.970 in
            L = 0.860 in
            W = 0.750 in

    a. _____

    b. _____

6. Given: H = 23.00 mm
            L = 19.00 mm
            W = 16.00 mm

    a. _____

    b. _____

Use this figure for # 5 and # 6.

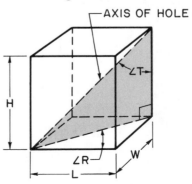

In each of the following problems, the top, front, and right side views of a compound-angular hole are shown. For each problem do the following:

a. Sketch, dimension, and label a rectangular solid. Within the solid, show the right triangle which contains the hole axis as a side and the angle of tilt. Show the position of the angle of rotation.

b. Compute the angle of rotation, $\angle R$.

c. Compute the angle of tilt, $\angle T$.

7. All dimensions are in inches.

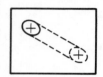

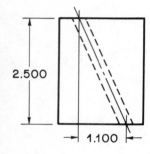

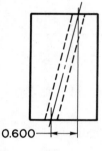

2.500

1.100   0.600

a.

b. _____

c. _____

9. All dimensions are in inches.

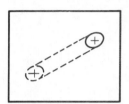

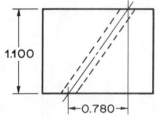

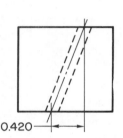

1.100

0.780   0.420

a.

b. _____

c. _____

8. All dimensions are in millimeters.

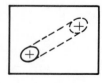

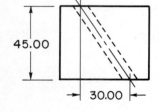

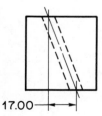

45.00

30.00   17.00

a.

b. _____

c. _____

10. All dimensions are in millimeters.

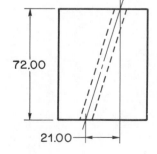

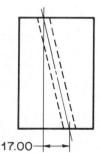

72.00

21.00   17.00

a.

b. _____

c. _____

UNIT 58 *DRILLING AND BORING COMPOUND-ANGULAR HOLES: COMPUTING ANGLES OF ROTATION AND TILT USING GIVEN ANGLES* _____

OBJECTIVES _____

After studying this unit you should be able to

- Compute angles of rotation and angles of tilt of hole axes in rectangular solids. No length dimensions are known.
- Sketch, dimension, and label compound-angular components within rectangular solids and compute angles of rotation and angles of tilt. No length dimensions are known.
- Compute angles of rotation and angles of tilt by use of formulas.
- Compute front view and side view angles by use of formulas.

COMPUTING ANGLES OF ROTATION AND ANGLES OF TILT WHEN NO LENGTH DIMENSIONS ARE KNOWN _____

In certain compound angle problems no length dimensions are known, instead, angles in two different planes are known. In problems of this type where no length dimensions are known, it is necessary to assign a value of 1 (unity) to one of the sides in order to compute with trigonometric functions.

The side which is assigned a value of 1 must be a side which is common to two of the formed right triangles. One of the right triangles must have a known angle. The other right triangle must have either a known angle or an angle which is to be computed, ∠R or ∠T.

Example: Three views of a compound-angular hole are shown. Hole angles are given in the front and right side views. No length dimensions are given.

   a. Determine the angle of rotation.
   b. Determine the angle of tilt.

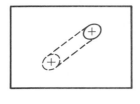

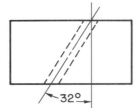 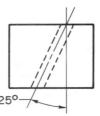

Sketch a rectangular solid. Project auxiliary lines which form right triangles containing the given angles, the axis of the hole, and the angles to be computed, ∠R and ∠T.

BC is a side of right △BCD which contains the given 25° angle. BC is also a side of right △BCE which contains ∠R which is to be computed. Make BC = 1.

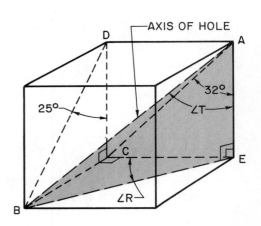

305

a. Compute angle of rotation, $\angle R$.

In right $\triangle BCD$: $BC = 1, \angle D = 25°$

$$\cot 25° = \frac{DC}{BC}$$

$$2.1445 = \frac{DC}{1}$$

$$DC = 2.1445$$

In right $\triangle AEC$: $AE = DC = 2.1445, \angle A = 32°$

$$\tan 32° = \frac{CE}{AE}$$

$$0.62487 = \frac{CE}{2.1445}$$

$$CE = 1.3400$$

In right $\triangle BCE$: $BC = 1, CE = 1.3400$

$$\cot \angle R = \frac{CE}{BC} = \frac{1.3400}{1} = 1.3400$$

$$\angle R = 36°44'\quad \text{Ans}$$

The part is rotated $36°44'$ ($\angle R$) on the angle plate and the angle plate is raised $37°57'$ ($\angle T$).

b. Compute angle of tilt, $\angle T$.

In right $\triangle BCE$: $\angle R = 36°44', BC = 1$

$$\csc 36°44' = \frac{BE}{BC}$$

$$1.6720 = \frac{BE}{1}$$

$$BE = 1.6720$$

In right $\triangle AEB$: $AE = 2.1445, BE = 1.6720$

$$\tan \angle T = \frac{BE}{AE} = \frac{1.6720}{2.1445} = 0.77967$$

$$\angle T = 37°57'\quad \text{Ans}$$

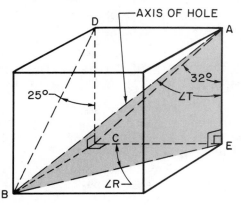

FORMULAS FOR COMPUTING ANGLES OF ROTATION AND ANGLES OF TILT USED IN DRILLING

Formulas for determining angles of rotation and angles of tilt have been computed. These formulas reduce the amount of computation required in solving compound angle problems. The formulas should <u>not</u> be used unless a problem is completely visualized and the method of solution shown in the previous example is fully understood.

To use formulas for the angles of rotation and angles of tilt, $\angle A$ and $\angle B$ shown in the figure must be identified.

$\angle A$ is the given angle in the front view (frontal plane) in relation to the vertical.

$\angle B$ is the given angle in the side view (profile plane) in relation to the vertical.

Formula for the Angle of Rotation in Relation to the Frontal Plane (Plane of $\angle A$) Used in Drilling

$$\tan \angle R = \frac{\tan \angle B}{\tan \angle A}$$

Formula for the Angle of Tilt Used in Drilling

$$\tan \angle T = \sqrt{\tan^2 \angle A + \tan^2 \angle B}$$

In using the formula given for the angle of rotation, $\angle R$ must be determined in relation to the frontal plane (plane of $\angle A$). If $\angle R$ is to be determined in relation to the profile plane (plane of $\angle B$), the complement of the computed formula $\angle R$ must be used.

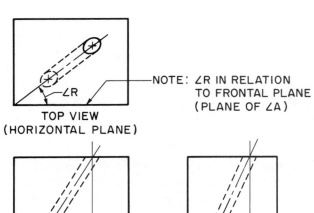

Example 1: Three views of a compound-angular hole are shown. (This is the same compound-angular hole used in the previous example.) The angle in the front view in relation to the vertical is $32°$ ($\angle A = 32°$). The angle in the right side view in relation to the vertical is $25°$ ($\angle B = 25°$).

    a. Compute $\angle R$.

    b. Compute $\angle T$.

a. $\tan \angle R = \dfrac{\tan \angle B}{\tan \angle A}$

    $\tan \angle R = \dfrac{\tan 25°}{\tan 32°}$

    $\tan \angle R = \dfrac{0.46631}{0.62487}$

    $\tan \angle R = 0.74625$

    $\angle R = 36°44'$ Ans

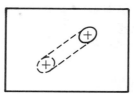

b. $\tan \angle T = \sqrt{\tan^2 \angle A + \tan^2 \angle B}$

    $\tan \angle T = \sqrt{\tan^2 32° + \tan^2 25°}$

    $\tan \angle T = \sqrt{(0.62487)^2 + (0.46631)^2}$

    $\tan \angle T = \sqrt{0.3904625 + 0.217445}$

    $\tan \angle T = \sqrt{0.6079075}$

    $\tan \angle T = 0.77968$

    $\angle T = 37°57'$ Ans

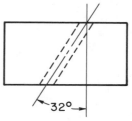
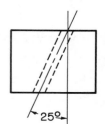

    Observe that the values of $\angle R = 36°44'$ and $\angle T = 37°57'$ are the same as those computed in the previous example.

    Occasionally a problem requires computing a front view angle when the side view angle and the angle of tilt or rotation are known. Also, it may be required to compute a side view angle in a problem when the front view angle and angle of tilt or rotation are known. The formulas for angle of rotation and tilt are used.

Example 2: Given: $\angle B = 20°$, and $\angle R = 24°$
            Compute: $\angle A$ and $\angle T$

$\tan \angle R = \dfrac{\tan \angle B}{\tan \angle A}$
                    $\tan \angle T = \sqrt{\tan^2 \angle A + \tan^2 \angle B}$

$\tan 24° = \dfrac{\tan 20°}{\tan \angle A}$
                    $\tan \angle T = \sqrt{\tan^2 39°16' + \tan^2 20°}$

$0.44523 = \dfrac{0.36397}{\tan \angle A}$
                    $\tan \angle T = \sqrt{(0.81749)^2 + (0.36397)^2}$

$\tan \angle A = \dfrac{0.36397}{0.44523}$
                    $\tan \angle T = \sqrt{0.6682899 + 0.1324741}$

$\tan \angle A = 0.81749$
                      $\tan \angle T = \sqrt{0.8007640}$

$\angle A = 39°16'$ Ans
                       $\tan \angle T = 0.89485$

                                         $\angle T = 41°49'$ Ans

Example 3: Given: $\angle A = 40°$ and $\angle T = 42.50°$
            Compute: $\angle B$ and $\angle R$

$\tan \angle T = \sqrt{\tan^2 \angle A + \tan^2 \angle B}$

$\tan 42.50° = \sqrt{\tan^2 40° + \tan^2 \angle B}$

$0.91633 = \sqrt{(0.83910)^2 + \tan^2 \angle B}$

$0.91633^2 = 0.83910^2 + \tan^2 \angle B$

$\tan^2 \angle B = 0.13557$

$\tan \angle B = 0.36820$

$\angle B = 20.22°$ Ans

$\tan \angle R = \dfrac{\tan \angle B}{\tan \angle A} = \dfrac{0.36820}{0.83910} = 0.43880$

        $\angle R = 23.70°$ Ans

‾APPLICATION ―――――――――――――――――――――――

A Trigonometric Function Table can be found in the Appendix.

Computing Angles of Rotation and Tilt without Using Drilling Formulas

In each of the following problems, 1–6, the axis of a hole is shown in a rectangular solid. In order to position the hole axis for drilling, the angle of rotation and the angle of tilt must be determined. Do <u>not</u> use drilling formulas in solving these problems.

a. Compute the angle of rotation, ∠R.

b. Compute the angle of tilt, ∠T.

Use this figure for # 1 and # 2.

1.   Given: ∠BDC = 35°
              ∠CAE = 42°

     a. ――――――
     b. ――――――

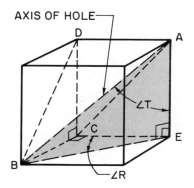

2.   Given: ∠BDC = 29°
              ∠CAE = 36°20′

     a. ――――――
     b. ――――――

Use this figure for # 3 and # 4.

3.   Given: ∠EAC = 18.25°
              ∠CDB = 31°

     a. ――――――
     b. ――――――

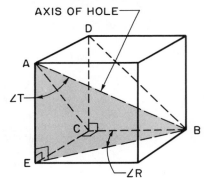

4.   Given: ∠EAC = 21°50′
              ∠CDB = 33°

     a. ――――――
     b. ――――――

Use this figure for # 5 and # 6.

5.   Given: ∠DAE = 30°
              ∠CAE = 42°10′

     a. ――――――
     b. ――――――

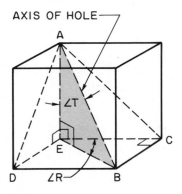

6.   Given: ∠DAE = 26.50°
              ∠CAE = 39°

     a. ――――――
     b. ――――――

In each of the following problems, 7–10, the top, front, and right side views of a compound-angular hole are shown. Do <u>not</u> use drilling formulas in solving these problems. For each problem:

a. Sketch and label a rectangular solid. Within the solid, show the right triangle which contains the hole axis as a side and the angle of tilt. Show the position of the angle of rotation. Show the right triangles which contain the given angles.

b. Compute the angle of rotation, $\angle R$.

c. Compute the angle of tilt, $\angle T$.

7.

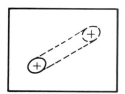

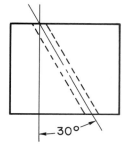

a.

b. _____

c. _____

9.

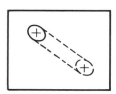

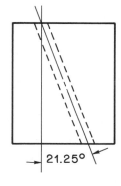

a.

b. _____

c. _____

8.

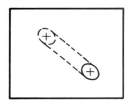

a.

b. _____

c. _____

10.

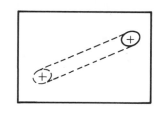

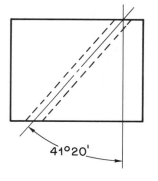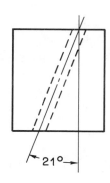

a.

b. _____

c. _____

Computing Angles Using Drilling Formulas

In each of the following problems, the top, front, and right side views of a compound-angular hole are shown. Compute the required angles using these formulas.

$$\tan \angle R = \frac{\tan \angle B}{\tan \angle A}$$

$$\tan \angle T = \sqrt{\tan^2 \angle A + \tan^2 \angle B}$$

Use this figure for # 11, # 12, and # 13.

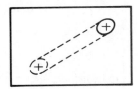

11. Given: $\angle A = 41°$
          $\angle B = 18°$

   a. Compute $\angle R$.   _____

   b. Compute $\angle T$.   _____

12. Given: $\angle B = 22°$
          $\angle R = 31°50'$

   a. Compute $\angle A$.   _____

   b. Compute $\angle T$.   _____

13. Given: $\angle A = 38°$
          $\angle T = 41.30°$

   a. Compute $\angle B$.   _____

   b. Compute $\angle R$.   _____

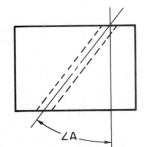
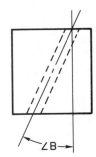

Use this figure for # 14, # 15, and # 16.

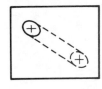

14. Given: $\angle A = 25°$
          $\angle B = 19°10'$

   a. Compute $\angle R$.   _____

   b. Compute $\angle T$.   _____

15. Given: $\angle B = 17°$
          $\angle R = 29.50°$

   a. Compute $\angle A$.   _____

   b. Compute $\angle T$.   _____

16. Given: $\angle A = 23°20'$
          $\angle T = 29°30'$

   a. Compute $\angle B$.   _____

   b. Compute $\angle R$.   _____

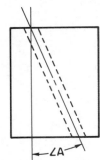
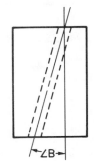

¯UNIT 59 MACHINING COMPOUND-ANGULAR SURFACES: COMPUTING ANGLES OF ROTATION AND TILT_____

¯OBJECTIVES_____

After studying this unit you should be able to

- Compute angles of rotation and angles of tilt in angle plate positioning for machining compound-angular surfaces as given in rectangular solids.

- Sketch, dimension, and label compound-angular surface components within rectangular solids and compute angles of rotation and angles of tilt.

- Compute angles of rotation and angles of tilt by use of formulas.

- Compute front view and side view surface angles by use of formulas.

¯MACHINING COMPOUND-ANGULAR SURFACES_____

When the surface of a part appears as a diagonal in each of two conventional views, such as the front and right side views, setting up the part for machining involves compound angles. When just the surface (plane) must be considered in a compound angle problem, a single rotation and a single tilt are required.

The setting up of a part in a compound angle problem in which a surface (plane) and a line on the surface must both be considered is more complex. Setups of this type require single rotation and double tilt or single tilt and double rotation.

The presentation of compound-angular surfaces in this text is limited to problems which require only single rotation and single tilt. An understanding of the procedures shown will enable you to set up most compound angle surface cutting problems encountered. The procedures are also the basis for the solution of more complex compound angle problems which require double tilt or double rotation.

Example 1: Three views of a rectangular solid block are shown in which a compound-angular surface is to be machined.

a. Determine the angle of rotation, ∠R.

b. Determine the angle of tilt, ∠T.

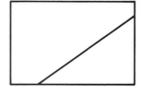

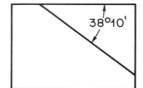 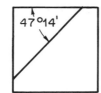

38°10'   47°14'

Sketch the rectangular solid and the pyramid ABCD formed by the surface ABC to be cut and the extended sides of the block. Project auxiliary lines which form right triangles containing the given angles and the angles to be computed, ∠R and ∠T.

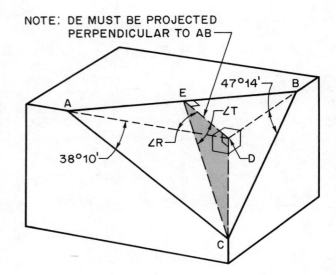

NOTE: DE MUST BE PROJECTED PERPENDICULAR TO AB

In cutout (pyramid) ABCD, *DE must be projected perpendicular to AB*. Right △CDE contains the angle of tilt, ∠T. The angle of rotation, ∠R, is contained in right △AED. Observe that right △AED is contained in the horizontal plane and ∠R is given in reference to line AD which lies in both the horizontal and frontal planes.

Since no length dimensions are given, assign a value of 1 (unity) to a side which is common to two or more sides of the formed *right* triangles. One or more of the triangles must have a known angle. The other right triangle or triangles must have a known angle or an angle to be computed, ∠R or ∠T. Side DC is contained in the following three right triangles.

> Right △ADC which contains the given angle 38°10′.
> Right △BDC which contains the given angle 47°14′.
> Right △CDE which contains ∠T.

Make DC = 1.

a. In right △BDC, DC = 1, ∠B = 47°14′. Compute DB.

$$\cot 47°14' = \frac{DB}{DC}$$

$$0.92493 = \frac{DB}{1}$$

$$DB = 0.92493$$

In right △ADC, DC = 1, ∠A = 38°10′. Compute AD.

$$\cot 38°10' = \frac{AD}{DC}$$

$$1.2723 = \frac{AD}{1}$$

$$AD = 1.2723$$

In right △ADB, DB = 0.92493, AD = 1.2723. Compute ∠A.

$$\tan \angle A = \frac{DB}{AD} = \frac{0.92493}{1.2723} = 0.72697$$

$$\angle A = 36°1'$$

In right △AED compute the angle of rotation, ∠R.

∠R and ∠A are complementary.
∠R = 90° − 36°1′ = 53°59′   Ans

b. In right △AED, ∠A = 36°1′, AD = 1.2723. Compute DE.

$$\sin 36°1' = \frac{DE}{AD}$$

$$0.58778 = \frac{DE}{1.2723}$$

$$DE = 0.74783$$

In right △CDE, CD = 1, DE = 0.74783. Compute the angle of tilt, ∠T.

$$\cot \angle T = \frac{DE}{DC} = \frac{0.74783}{1} = 0.74783$$

$$\angle T = 53°12'   \text{Ans}$$

Procedure for Positioning the Part on an Angle Plate for Machining

- Rotate the part to the angle of rotation, ∠R, as shown. Care must be taken as to whether the part is rotated to the computed ∠R or the complement of ∠R. Rotate the part 53°59′. Note the position of right △CDE is shown with hidden lines.

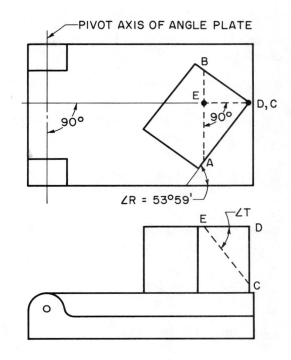

- Raise the angle plate to the tilt angle, ∠T. Tilt to 53°12′ as shown. Care must be taken as to whether the part is tilted to the computed ∠T or the complement of ∠T. Observe that the position of the plane AEBC to be cut is horizontal.

  With the part set to the angle of rotation and to the angle of tilt, it is positioned to machine the surface on the horizontal plane.

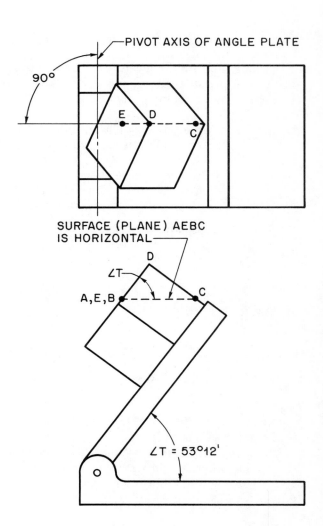

FORMULAS FOR COMPUTING ANGLES OF ROTATION AND ANGLES OF TILT USED IN MACHINING

As with formulas for drilling compound-angular holes, formulas for machining compound-angular surfaces should not be used until the problem is completely visualized and the method of solution shown in the previous example is fully understood.

To use formulas for the angles of rotation and angles of tilt, $\angle A$ and $\angle B$ shown in the figure must be identified.

$\angle A$ is the given angle in the front view (frontal plane) in relation to the horizontal plane.

$\angle B$ is the given angle in the side view (profile plane) in relation to the horizontal plane.

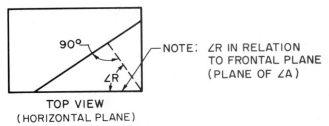

Formula for the Angle of Rotation in Relation to the Frontal Plane (Plane of $\angle A$) Used in Machining

$$\tan \angle R = \frac{\tan \angle B}{\tan \angle A}$$

Formula for the Angle of Tilt Used in Machining

$$\tan \angle T = \frac{\tan \angle A}{\cos \angle R}$$

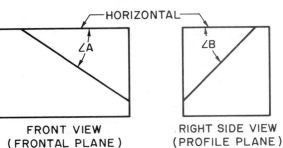

In using the formula given for the angle of rotation, $\angle R$ must be determined in relation to the frontal plane (plane of $\angle A$). If $\angle R$ is to be determined in relation to the profile plane (plane of $\angle B$), the complement of the computed formula $\angle R$ must be used.

Example 1: Three views of a compound-angular surface are shown. (This is the same compound-angular surface used in the previous example.) The angle in the front view in relation to the horizontal is $38°10'$ ($\angle A = 38°10'$). The angle in the right side view in relation to the horizontal is $47°14'$ ($\angle B = 47°14'$).

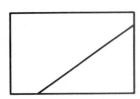

    a. Compute $\angle R$.

    b. Compute $\angle T$.

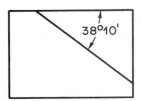 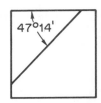

a. $\tan \angle R = \dfrac{\tan \angle B}{\tan \angle A} = \dfrac{\tan 47°14'}{\tan 38°10'} = \dfrac{1.0812}{0.78598} = 1.3756$

    $\angle R = 53°59'$ Ans

b. $\tan \angle T = \dfrac{\tan \angle A}{\cos \angle R} = \dfrac{\tan 38°10'}{\cos 53°59'} = \dfrac{0.78598}{0.58802} = 1.3367$

    $\angle T = 53°12'$ Ans

Observe that the values of $\angle R = 53°59'$ and $\angle T = 53°12'$ are the same as those computed in the previous example.

The same formulas for angles of rotation and tilt are used to compute an unknown front view angle when a side view angle and an angle of rotation or tilt are known. An unknown side view angle may be computed if the front view angle and angle of rotation or tilt are known.

Example 2: Given: $\angle B = 18.15°$, and $\angle R = 27.45°$.
Compute: $\angle A$ and $\angle T$.

$$\tan \angle R = \frac{\tan \angle B}{\tan \angle A}$$

$$\tan 27.45° = \frac{\tan 18.15°}{\tan \angle A}$$

$$0.51946 = \frac{0.32782}{\tan \angle A}$$

$$\tan \angle A = \frac{0.32782}{0.51946}$$

$$\tan \angle A = 0.63108$$
$$\angle A = 32.25° \quad \text{Ans}$$

$$\tan \angle T = \frac{\tan \angle A}{\cos \angle R} = \frac{0.63108}{0.88741} = 0.71115$$
$$\angle T = 35.42° \quad \text{Ans}$$

APPLICATION

A Trigonometric Function Table can be found in the Appendix.

Computing Angles of Rotation and Tilt without Using Machining Formulas

Three views of a rectangular solid block are shown in which a compound-angular surface is to be machined. A pictorial view of the block with auxiliary lines required for computations is also shown. Do **not** use machining formulas in solving these problems. For each of the following problems, 1–4:

a. Determine the angle of rotation, $\angle R$.

b. Determine the angle of tilt, $\angle T$.

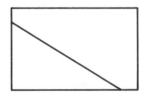

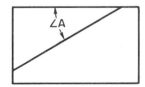

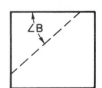

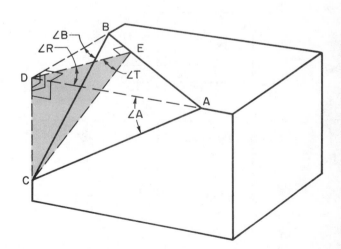

1. Given: $\angle A = 32°$, $\angle B = 44°$.   a. _____   b. _____

2. Given: $\angle A = 27°$, $\angle B = 39°$.   a. _____   b. _____

3. Given: $\angle A = 18°10'$, $\angle B = 27°50'$.   a. _____   b. _____

4. Given: $\angle A = 22.50°$, $\angle B = 36.20°$.   a. _____   b. _____

In each of the following problems, 5–8, the top, front, and right side views of a compound-angular surface are shown. Do <u>not</u> use machining formulas in solving these problems. For each problem:

a. Sketch and label a rectangular solid and the pyramid formed by the surface to be cut and the extended sides of the block. Show the right triangles which contain ∠T and the right triangles which contain the given angles. Identify ∠T, ∠R, and the given angles.

b. Compute the angle of rotation, ∠R.

c. Compute the angle of tilt, ∠T.

5.

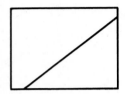

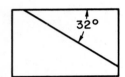 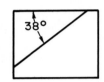

a.

b. _____

c. _____

6.

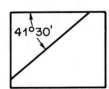

a.

b. _____

c. _____

7.

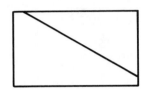

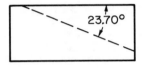

a.

b. _____

c. _____

8.

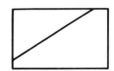

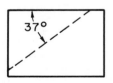

a.

b. _____

c. _____

Computing Angles Using Machining Formulas

In each of the following problems, the top, front, and right side views of a compound-angular surface are shown. Compute the required angles using these formulas.

$$\tan \angle R = \frac{\tan \angle B}{\tan \angle A}$$

$$\tan \angle T = \frac{\tan \angle A}{\cos \angle R}$$

9. Given: $\angle A = 41°$
       $\angle B = 46°$

   a. Compute $\angle R$.

   b. Compute $\angle T$.

10. Given: $\angle B = 43°20'$
        $\angle R = 46°$

   a. Compute $\angle A$.

   b. Compute $\angle T$.

11. Given: $\angle A = 39.50°$
        $\angle T = 50°$

   a. Compute $\angle R$.

   b. Compute $\angle B$.

Use this figure for # 9, # 10, and # 11.

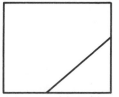

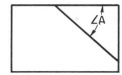 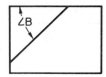

12. Given: $\angle A = 19°$
        $\angle B = 23°10'$

   a. Compute $\angle R$.

   b. Compute $\angle T$.

13. Given: $\angle B = 17°$
        $\angle R = 21°$

   a. Compute $\angle A$.

   b. Compute $\angle T$.

14. Given: $\angle A = 15.60°$
        $\angle T = 26.50°$

   a. Compute $\angle R$.

   b. Compute $\angle B$.

Use this figure for # 12, # 13, and # 14.

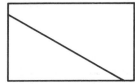

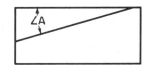

UNIT 60 COMPUTING ANGLES MADE BY THE INTERSECTION OF TWO ANGULAR SURFACES

OBJECTIVES

After studying this unit you should be able to

- Compute the true angles of compound-angular edges made by the intersection of two angular surfaces as given in rectangular solids.
- Sketch and label compound-angular surface edge components within rectangular solids and compute true angles.
- Compute true angles, front view angles and side view angles by the use of formulas.

COMPUTING ANGLES MADE BY THE INTERSECTION OF TWO ANGULAR SURFACES

For design or inspection purposes it may be required to compute angles which are made by the intersection of two cut surfaces in reference to the horizontal plane.

Example: Three views of a part are shown. A pictorial view of the angular portion of the part with auxiliary lines required for computations is also shown. The surfaces are to be machined in reference to the horizontal plane at angles of 32° and 40° as shown in the front and right side views.

a. Compute ∠R.

b. Compute ∠C.

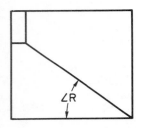

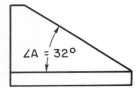

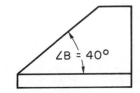

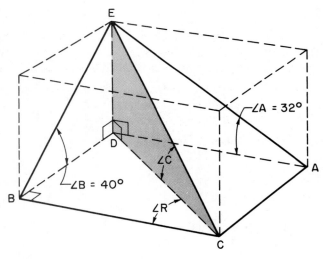

Since DE is a side of right △BDE, right △ADE, and right △CDE, make DE = 1.

a. In right △BDE, DE = 1, ∠B = 40°.
   Compute BD:

$$\cot 40° = \frac{BD}{DE}$$

$$1.1918 = \frac{BD}{1}$$

$$BD = 1.1918$$

In right △ADE, DE = 1, ∠A = 32°.
   Compute DA:

$$\cot 32° = \frac{DA}{DE}$$

$$1.6003 = \frac{DA}{1}$$

$$DA = 1.6003$$

In right △CBD, BD = 1.1918, BC = DA = 1.6003.
   Compute ∠R:

$$\tan \angle R = \frac{BD}{BC}$$

$$\tan \angle R = \frac{1.1918}{1.6003}$$

$$\tan \angle R = 0.74474$$

$$\angle R = 36°41' \text{Ans}$$

b. In right △CBD, ∠R = 36°41′, BD = 1.1918.   In right △CDE, DE = 1, DC = 1.9950.
Compute DC: Compute ∠C:

$$\sin 36°41' = \frac{BD}{DC}$$
$$0.59739 = \frac{1.1918}{DC}$$
$$DC = \frac{1.1918}{0.59739}$$
$$DC = 1.9950$$

$$\cot \angle C = \frac{DC}{DE}$$
$$\cot \angle C = \frac{1.9950}{1}$$
$$\angle C = 26°37' \quad \text{Ans}$$

FORMULAS FOR COMPUTING ANGLES OF INTERSECTING ANGULAR SURFACES

Apply the formulas for intersecting angular surfaces only after a problem has been completely visualized and the previous method of solution is fully understood.

To use formulas for intersecting angular surfaces, ∠A and ∠B must be identified.

∠A is the given angle in the front view (frontal plane) in relation to the horizontal plane.

∠B is the given angle in the side view (profile plane) in relation to the horizontal plane.

Formula for ∠R in Relation to the Frontal Plane (Plane of ∠A) Used for Intersecting Angular Surfaces

$$\tan \angle R = \frac{\cot \angle B}{\cot \angle A}$$

Formula for ∠C Used for Intersecting Angular Surfaces

$$\cot \angle C = \sqrt{\cot^2 \angle A + \cot^2 \angle B}$$

In using the formula given for the angle of rotation, ∠R must be determined in relation to the frontal plane (plane of ∠A). If ∠R is to be determined in relation to the profile plane (plane of ∠B), the complement of the computed formula ∠R must be used.

Example 1: Three conventional views and a pictorial view of the intersection of two angular surfaces are shown. (These are the same intersecting angular surfaces used in the previous example.)

a. Compute ∠R.

b. Compute ∠C.

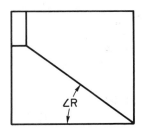

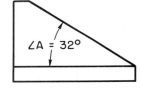
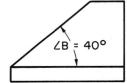

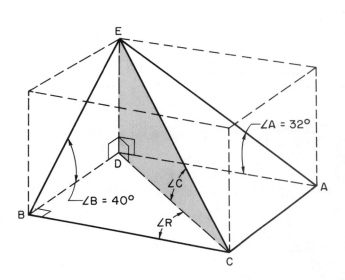

a. $\tan \angle R = \dfrac{\cot \angle B}{\cot \angle A} = \dfrac{\cot 40°}{\cot 32°} = \dfrac{1.1918}{1.6003} = 0.74474$

$\angle R = 36°41' \quad \text{Ans}$

b. $\cot \angle C = \sqrt{\cot^2 \angle A + \cot^2 \angle B}$

$\cot \angle C = \sqrt{\cot^2 32° + \cot^2 40°}$

$\cot \angle C = \sqrt{1.6003^2 + 1.1918^2}$

$\cot \angle C = \sqrt{3.9813}$

$\cot \angle C = 1.9953$

$\angle C = 26°37'$ Ans

Observe that the values of $\angle R = 36°41'$ and $\angle C = 26°37'$ are the same as those computed in the previous example.

The same formulas for $\angle R$ and $\angle C$ may be used to compute an unknown front view angle, $\angle A$, when a side view angle, $\angle B$, and $\angle R$ or $\angle C$ are known. An unknown side view angle, $\angle B$, may be computed if the front view angle, $\angle A$, and $\angle R$ or $\angle C$ are known.

Example 2: Given: $\angle B = 35.50°$ and $\angle R = 28.30°$.
              Compute: $\angle A$.

$\tan \angle R = \dfrac{\cot \angle B}{\cot \angle A}$

$\tan 28.30° = \dfrac{\cot 35.50°}{\cot \angle A}$

$0.53844 = \dfrac{1.4020}{\cot \angle A}$

$\cot \angle A = 2.6038$

$\angle A = 21°1' = 21.01°$ Ans

Example 3: Given: $\angle A = 23°10'$ and $\angle C = 17°40'$.
              Compute: $\angle B$.

$\cot \angle C = \sqrt{\cot^2 \angle A + \cot^2 \angle B}$

$\cot 17°40' = \sqrt{\cot^2 23°10' + \cot^2 \angle B}$

$3.1397 = \sqrt{2.3369^2 + \cot^2 \angle B}$

$3.1397^2 = 2.3369^2 + \cot^2 \angle B$

$\cot \angle B = \sqrt{4.3966}$

$\cot \angle B = 2.0968$

$\angle B = 25°30'$ Ans

APPLICATION

A Trigonometric Function Table can be found in the Appendix.

Computing Angles without Using Formulas for Intersecting Angular Surfaces

Three views of a part are shown in which surfaces are to be machined in reference to the horizontal plane at $\angle A$ and $\angle B$ as shown in the front and right side views. A pictorial view of the angular portion of the part with auxiliary lines required for computations is also shown. Do <u>not</u> use intersecting angular surface formulas in solving these problems. For each of the following problems, 1–4:

a. Compute $\angle R$.

b. Compute $\angle C$.

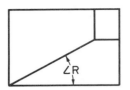

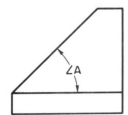 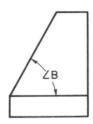

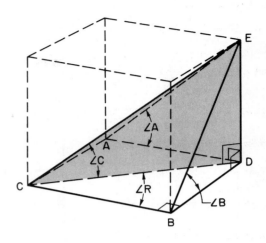

1. Given: $\angle A = 42°$, $\angle B = 55°$.

2. Given: $\angle A = 39°$, $\angle B = 47°$.

3. Given: $\angle A = 50°10'$, $\angle B = 61°40'$.

4. Given: $\angle A = 43.35°$, $\angle B = 52.70°$.

a. _____ b. _____

a. _____ b. _____

a. _____ b. _____

a. _____ b. _____

In each of the following problems, 5–8, three views of a part are shown. Two surfaces are to be machined in reference to the horizontal plane at the angles shown in the front and right side views. Do <u>not</u> use intersecting angular surface formulas in solving these problems. For each problem:

a. Sketch and label a rectangular solid and the pyramid formed by the angular surface edges. Show the right triangle which contains $\angle C$ and the right triangles which contain the given angles and $\angle R$. Identify $\angle C$, $\angle R$, and the given angles.

b. Compute $\angle R$.

c. Compute $\angle C$.

5.

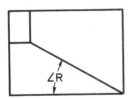

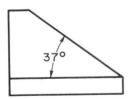 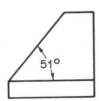

a.

6.

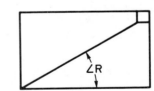

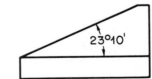 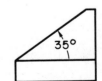

a.

b. _____

c. _____

b. _____

c. _____

7.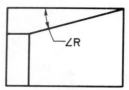

8.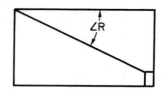

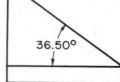 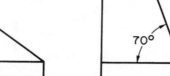

a.

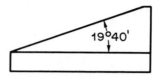

a.

b. _____

c. _____

b. _____

c. _____

Computing Angles Using Formulas for Intersecting Angular Surfaces

In each of the following problems, three views of a part are shown. The angular surfaces are to be machined in reference to the horizontal plane at $\angle A$ and $\angle B$ as shown in the front and right side views. Compute the required angles using these formulas.

$$\tan \angle R = \frac{\cot \angle B}{\cot \angle A}$$

$$\cot \angle C = \sqrt{\cot^2 \angle A + \cot^2 \angle B}$$

9. Given: $\angle A = 36°$
   $\angle B = 43°50'$

   a. Compute $\angle R$.   _____

   b. Compute $\angle C$.   _____

Use this figure for # 9, # 10, and # 11.

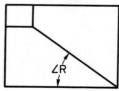

10. Given: $\angle B = 48°10'$
    $\angle R = 40°$

    a. Compute $\angle A$.   _____

    b. Compute $\angle C$.   _____

11. Given: $\angle A = 29.40°$
    $\angle C = 26°$

    a. Compute $\angle B$.   _____

    b. Compute $\angle R$.   _____

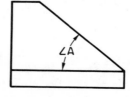

12. Given: ∠A = 17°40′
     ∠B = 25°

a. Compute ∠R. _____

b. Compute ∠C. _____

Use this figure for # 12, # 13, and # 14.

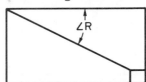

13. Given: ∠B = 31°
     ∠R = 27°50′

a. Compute ∠A. _____

b. Compute ∠C. _____

14. Given: ∠A = 14.10°
     ∠C = 12°

a. Compute ∠B. _____

b. Compute ∠R. _____

UNIT 61 *COMPUTING COMPOUND ANGLES ON CUTTING AND FORMING TOOLS*

OBJECTIVES

After studying this unit you should be able to

- Compute compound angles required for cutting and forming tools as given in rectangular solids.
- Sketch and label tool angular-surface-edge components within rectangular solids and compute true angles.
- Compute true angles by use of formulas.

COMPUTING TRUE ANGLES FOR CUTTING AND FORMING TOOLS

The following examples show methods of computing compound angles that are often required in making die sections, cutting tools and forming tools.

Example: Three conventional views and a pictorial view of the angular portion of a tool are shown. Compute ∠C.

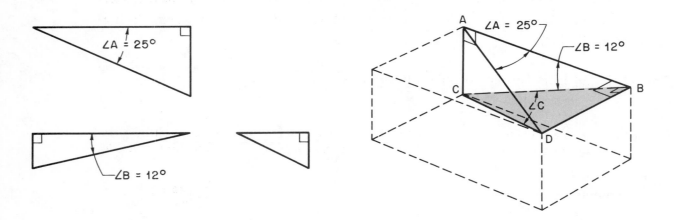

Since AB is a side of both right △ABD and right △CAB which contain given angles of 25° and 12°, make AB = 1.

In right △ABD, AB = 1, ∠A = 25°. Compute DB:

$$\tan 25° = \frac{DB}{AB}$$

$$0.46631 = \frac{DB}{1}$$

$$DB = 0.46631$$

In right △CAB, AB = 1, ∠B = 12°. Compute CB:

$$\cos 12° = \frac{AB}{CB}$$

$$0.97815 = \frac{1}{CB}$$

$$CB = 1.0223$$

In right △CBD, DB = 0.46631, CB = 1.0223. Compute ∠C:

$$\tan \angle C = \frac{DB}{CB} = \frac{0.46631}{1.0223} = 0.45614$$

$$\angle C = 24°31' \quad \text{Ans}$$

FORMULA FOR COMPUTING ∠C USED FOR CUTTING AND FORMING TOOLS

Apply the formula for finding ∠C used for cutting and forming tools only after a problem has been completely visualized and the previous solution is fully understood.

To use the formula for cutting and forming tools, $\angle A$ and $\angle B$ must be identified.

$\angle A$ is the given angle in the top view (horizontal plane) in relation to the frontal plane.

$\angle B$ is the given angle in the front view (frontal plane) in relation to the horizontal plane.

In the front view, a right angle is made with either the left or right edge and the horizontal plane.

Formula for $\angle C$ Used for Cutting and Forming Tools

$$\tan \angle C = (\tan \angle A)(\cos \angle B)$$

Example 1: Three conventional views and a pictorial view of an angular portion of a tool are shown. Compute $\angle C$. (This is the same tool used in the previous example.)

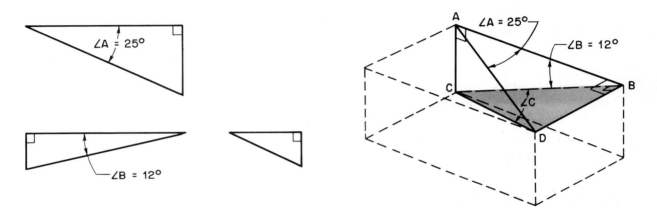

$$\tan \angle C = (\tan \angle A)(\cos \angle B)$$
$$\tan \angle C = (\tan 25°)(\cos 12°)$$
$$\tan \angle C = (0.46631)(0.97815)$$
$$\tan \angle C = 0.45612$$
$$\angle C = 24°31' \quad \text{Ans}$$

Observe that the value of $\angle C = 24°31'$ is the same value as computed in the previous example.

The same formula may be used to compute an unknown top view angle, $\angle A$, when the front view angle, $\angle B$, and $\angle C$ are known. The front view angle, $\angle B$, may be computed when the top view angle, $\angle A$, and $\angle C$ are known.

Example 2: Given: $\angle B = 14°$, $\angle C = 28.75°$.
         Compute: $\angle A$.

$$\tan \angle C = (\tan \angle A)(\cos \angle B)$$
$$\tan 28.75° = (\tan \angle A)(\cos 14°)$$
$$0.54862 = (\tan \angle A)(0.97030)$$
$$\tan \angle A = 0.56541$$
$$\angle A = 29°29' \quad \text{Ans}$$

⎯COMPUTING TRUE ANGLES IN FRONT-CLEARANCE-ANGLE APPLICATIONS⎯⎯⎯⎯⎯⎯⎯⎯⎯⎯⎯

The following type of compound angle problem is often found in cutting tool situations where a front clearance angle is required, such as in thread cutting.

Example: Three conventional views and a pictorial view of the angular portion of a cutting tool with a front clearance angle of 10° are shown. Compute ∠C.

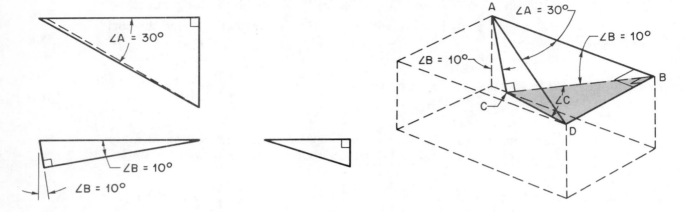

Since AB is a side of both right △ACB and right △ABD which contain given angles of 10° and 30°, make AB = 1.

In right △ABD, AB = 1, ∠A = 30°. Compute DB:

$$\tan 30° = \frac{DB}{AB}$$

$$0.57735 = \frac{DB}{1}$$

$$DB = 0.57735$$

In right △ACB, AB = 1, ∠B = 10°. Compute CB:

$$\cos 10° = \frac{CB}{AB}$$

$$0.98481 = \frac{CB}{1}$$

$$CB = 0.98481$$

In right △CBD, DB = 0.57735, CB = 0.98481. Compute ∠C:

$$\tan ∠C = \frac{DB}{CB} = \frac{0.57735}{0.98481} = 0.58626$$

$$∠C = 30°23'   \text{Ans}$$

⎯FORMULA FOR COMPUTING ∠C IN FRONT-CLEARANCE-ANGLE APPLICATIONS⎯

Apply the formula for finding ∠C in front-clearance-angle applications only after a problem has been completely visualized and the previous method of solution is fully understood.

To use the formula, ∠A and ∠B must be identified.

∠A is the given angle in the top view (horizontal plane) in relation to the frontal plane.

∠B is the given angle in the front view (frontal plane) in relation to the horizontal plane. ∠B is also the front clearance angle made with the vertical.

Formula for ∠C Used for Front-Clearance-Angle Applications

$$\tan ∠C = \frac{\tan ∠A}{\cos ∠B}$$

Example: Three conventional views and a pictorial view of the angular portion of a cutting tool with a front clearance angle of 10° are shown. Compute ∠C. (This is the same front-clearance-angle application used on the previous example.)

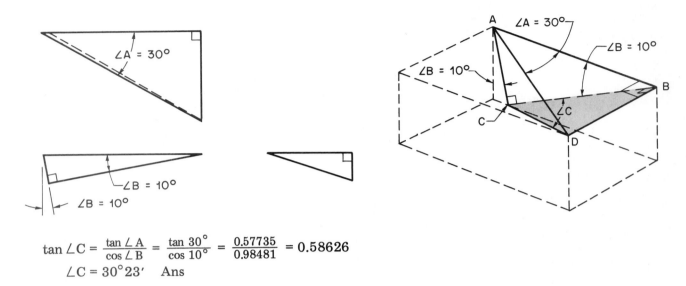

$$\tan \angle C = \frac{\tan \angle A}{\cos \angle B} = \frac{\tan 30°}{\cos 10°} = \frac{0.57735}{0.98481} = 0.58626$$
$$\angle C = 30°23'   \text{Ans}$$

Observe that the value of ∠C = 30°23′ is the same value as computed in the previous example.

APPLICATION

A Trigonometric Function Table can be found in the Appendix.

Computing Angles without Using Formulas for Cutting and Forming Tools

Three views of the angular portion of a tool are shown. A pictorial view with auxiliary lines forming the right triangles which are required for computations is also shown. Do <u>not</u> use cutting and forming tool formulas in solving these problems. For each of the following problems, 1–4, compute ∠C.

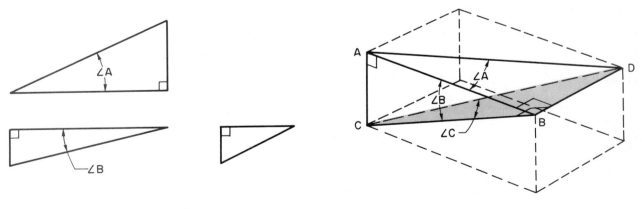

1. Given: ∠A = 30°, ∠B = 15°.   _____
2. Given: ∠A = 26°, ∠B = 12°.   _____
3. Given: ∠A = 28°, ∠B = 10°.   _____
4. Given: ∠A = 30°, ∠B = 24°.   _____

In each of the following problems, 5–8, three views of the angular portion of a tool are shown. Do <u>not</u> use cutting and forming tool formulas in solving these problems. For each problem:

a. Sketch and label a rectangular solid and the pyramid formed by the angular surface edges. Show the right triangles which contain ∠A, ∠B and ∠C. Identify the angles.

b. Compute ∠C.

Use this figure for # 5 and # 6.

5. Given: ∠A = 26°
              ∠B = 15°

   a.

   b. _____

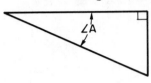

6. Given: ∠A = 20°
              ∠B = 8°

   a.

   b. _____

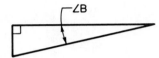

Use this figure for # 7 and # 8.

7. Given: ∠A = 30°
              ∠B = 12°

   a.

   b. _____

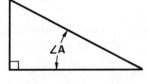

8. Given: ∠A = 25°
              ∠B = 10°

   a.

   b. _____

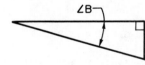

Computing Angles Using Formulas for Cutting and Forming Tools

For each of the following problems, 9–12, compute the required angle using this formula.

$$\tan \angle C = (\tan \angle A)(\cos \angle B)$$

9. Given: ∠A = 35°, ∠B = 15°.
    Compute: ∠C.   _____

Use this figure for # 9–# 12.

10. Given: ∠B = 10°, ∠C = 28°30′.
     Compute: ∠A.   _____

11. Given: ∠A = 26°, ∠B = 14°.
     Compute: ∠C.   _____

12. Given: ∠A = 30°, ∠C = 29.50°.
     Compute: ∠B.   _____

Computing Angles without Using Front-Clearance-Application Formulas

Three views of the angular portion of a tool with front clearance are shown. A pictorial view with auxiliary lines forming the right triangles which are required for computations is also shown. Do <u>not</u> use front-clearance-application formulas for solving these problems. Compute ∠C for each of the following problems, 13–16.

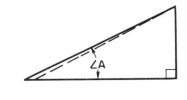
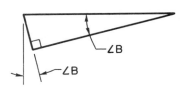

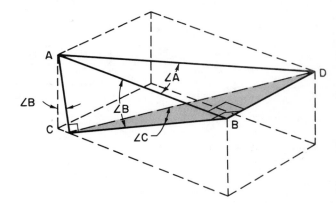

13. Given: ∠A = 30°, ∠B = 12°.   _____
14. Given: ∠A = 28°, ∠B = 15°.   _____
15. Given: ∠A = 32°, ∠B = 10°.   _____
16. Given: ∠A = 25°, ∠B = 14°.   _____

In each of the following problems, 17–20, three views of the angular portion of a tool with front clearance are shown. Do <u>not</u> use front-clearance-application formulas for solving these problems. For each problem:

a. Sketch and label a rectangular solid and the pyramid formed by the angular surface edges. Show the right triangles which contain ∠A, ∠B, and ∠C. Identify the angles.

b. Compute ∠C.

17. Given: ∠A = 30°
          ∠B = 15°

    a.

Use this figure for # 17 and # 18.

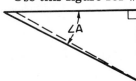

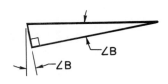

    b. _____

18. Given: ∠A = 40°
          ∠B = 14°

    a.

    b. _____

19. Given: ∠A = 32°
            ∠B = 15°

    a.

    Use this figure for # 19 and # 20.

    b. _____

20. Given: ∠A = 25°
            ∠B = 12°

    a.

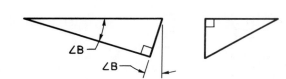

    b. _____

Computing Angles Using Front-Clearance-Application Formulas

For each of the following problems, 21–24, compute the required angle using this formula.

$$\tan \angle C = \frac{\tan \angle A}{\cos \angle B}$$

21. Given: ∠A = 30°, ∠B = 14°.
    Compute: ∠C.   _____

    Use this figure for # 21–# 24.

22. Given: ∠B = 15°, ∠C = 30°40′.
    Compute: ∠A.   _____

23. Given: ∠A = 36°, ∠B = 10°.
    Compute: ∠C.   _____

24. Given: ∠A = 28°, ∠C = 28.60°
    Compute: ∠B.   _____

UNIT 62 ACHIEVEMENT REVIEW — SECTION 6

A Trigonometric Function Table can be found in the Appendix.

OBJECTIVE

You should be able to solve the problems in this Achievement Review by applying the principles and methods covered in units 56–61.

1. Three views of a compound-angular hole are shown. All dimensions are in inches.

   a. Compute the angle of rotation, ∠R. _____

   b. Compute the angle of tilt, ∠T. _____

   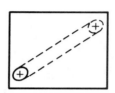
   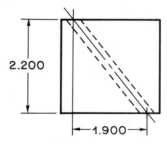

2. Three views of a compound-angular hole are shown.

   a. Compute the angle of rotation, ∠R. _____

   b. Compute the angle of tilt, ∠T. _____

   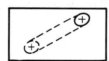
   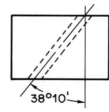
   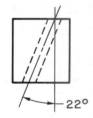

3. Three views of a rectangular solid block are shown in which a compound-angular surface is to be machined.

   a. Compute the angle of rotation, ∠R. _____

   b. Compute the angle of tilt, ∠T. _____

   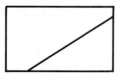
   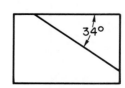
   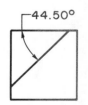

4. Three views of a part are shown. Two surfaces are to be machined in reference to the horizontal plane at the angles shown in the front and right side views.

    a. Compute ∠R.   _____

    b. Compute ∠C.   _____

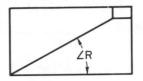

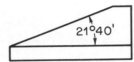

5. Three views of the angular portion of a tool are shown. Compute ∠C.   _____

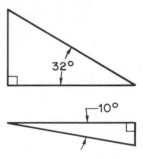

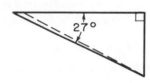

6. Three views of the angular portion of a tool with front clearance are shown. Compute ∠C.   _____

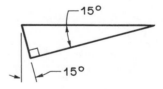

SECTION 7
NUMERICAL CONTROL

‾UNIT 63 *INTRODUCTION TO NUMERICAL CONTROL* _____

‾OBJECTIVES _____

After studying this unit you should be able to

- Locate points in a Cartesian coordinate system.
- Plot points in a Cartesian coordinate system.

Numerical control machines are widely used in the manufacture of machined parts with the trend toward greater application in the future. Therefore, at least a fundamental knowledge of numerical control should be acquired. Although the machinist does not usually write a program of operations, the basics of numerical control programming should be understood.

Numerical control machines are designed for a wide range of machining operations. Numerical control milling machines, engine lathes, turret lathes, and punch presses are widely used in industry, as are more specialized machines such as riveting, drafting, flame cutting, and inspection machines.

Although numerical control machines may vary as to the type and complexity of operations, the basic principles of operations are, in general, common to all machines. The operations to be performed by the machines are first programmed. Programming consists of writing a program manuscript (order of instruction) from a conventional process sheet or blueprint. The data on the manuscript is then transferred to either punched or magnetic tape codes. The tape is fed through a tape reader which converts the tape codes into electrical signals. These signals are sent to a central control unit. The control unit then converts the signals into commands which are sent to the machine. By means of electrical circuits, switches, motors, and cams, the commands from central control determine the operation of the machine. As the numerical control machine operates, feedback signals are sent back to the central control unit. These signals are compared to the original tape signals and must agree before the machine proceeds to the next operation.

To write a program for numerical control machines, the programmer should have an understanding of proper machining practices such as feeds, speeds, properties of metals, and cutter characteristics. The programmer must be trained in numerical control programming methods. A knowledge of how the numerical control machine functions and how to write a program using proper programming coding is essential. Coding determines where the signals are ultimately set in the machine. Some programs require only simple arithmetic, while others require a knowledge of advanced mathematics.

Because programming for numerical control machinery is a study within itself, no attempt is made in this text to discuss even a simple program thoroughly or to teach coding. However, two important concepts which are fundamental to numerical control machines can be discussed. They are numerical control programming with reference to the Cartesian coordinate system and the binary system of numeration.

‾LOCATION OF POINTS _____

Programming is based on locating points within the Cartesian coordinate system, which is discussed in unit 53. In a plane, a point can be located from a fixed point by two dimensions. For example, a point can be located by stating that it is three units up and five units to the right of a fixed point. The Cartesian coordinate system gives point location by using positive and negative values rather than location stated as being up or down and left or right from a fixed point.

A point is located in reference to the origin by giving the point an *x* and *y* value. The *x* value is always given first. The *x* and *y* values are called the *coordinates* of the point. The following examples locate points in the Cartesian coordinate system shown.

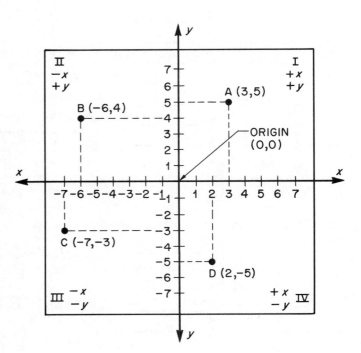

Example 1: Locate point A which has coordinates of (3, 5).

The *x* value is +3 units and the *y* value is +5 units. Therefore, point A is located in Quadrant I.

Example 2: Locate point B which has coordinates of (−6, 4).

The *x* value is −6 units and the *y* value is +4 units. Therefore, point B is located in Quadrant II.

Example 3: Locate point C which has coordinates of (−7, −3).

The *x* value is −7 units and the *y* value is −3 units. Therefore, point C is located in Quadrant III.

Example 4: Locate point D which has coordinates of (2, −5).

The *x* value is +2 units and the *y* value is −5 units. Therefore, point D is located in Quadrant IV.

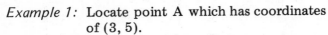

APPLICATION

Plotting Points

1. Using graph paper, plot the following coordinates.

   A = (−2, 5)   D = (0, 3)   G = (5, −7)   J = (0, 0)
   B = (3, 9)   E = (−5, 0)   H = (−7, 5)   K = (−3, −4)
   C = (−7, −2)   F = (−2, −2)   I = (−1, 0)   L = (8, −6)

2. Graph the following points: (−5, −5), (−3, −3), (0, 0), (2, 2), (4, 4), (7, 7). Connect these points.

   a. What kind of geometric figure is formed?   _____

   b. What is the value of the angle formed in reference to the *x*-axis?   _____

3. Graph the following points. Connect these points in the order that they are given. What kind of a geometric figure is formed?   _____

   Point 1:   (−9, −7)   Point 6:   (7, 2.5)   Point 11: (−2, 7)
   Point 2:   (−6, −5.3)   Point 7:   (6, 3)   Point 12: (−3, 5)
   Point 3:   (−3, −3.5)   Point 8:   (4, 4)   Point 13: (−5, 1)
   Point 4:   (1, −1)   Point 9:   (2, 5)   Point 14: (−6.5, −2)
   Point 5:   (4.5, 1)   Point 10: (0, 6)   Point 15: (−8, −5)

Coordinates of Points

4. Refer to the points plotted on the illustrated Cartesian Coordinate plane. Give coordinates of the following points.

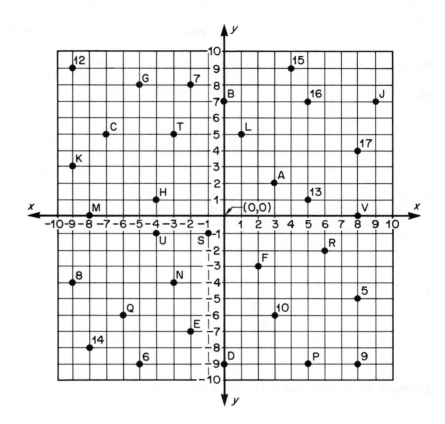

A _____	S _____
B _____	T _____
C _____	U _____
D _____	V _____
E _____	5 _____
F _____	6 _____
G _____	7 _____
H _____	8 _____
J _____	9 _____
K _____	10 _____
L _____	12 _____
M _____	13 _____
N _____	14 _____
P _____	15 _____
Q _____	16 _____
R _____	17 _____

⌐UNIT 64 *POINT-TO-POINT PROGRAMMING ON TWO-AXIS MACHINES⌐*

⌐OBJECTIVE⌐

After studying this unit you should be able to

*   Program dimension from engineering drawings using a point-to-point control system with two-axis machines. Both absolute and incremental dimensioning are applied.

⌐TYPES OF SYSTEMS⌐

Some numerical control machines are designed so that they can be programmed for as many as six axes. Table movement left and right is the *x*-axis and table movement forward and back is the *y*-axis. Other axes are movement up and down by either the table or the spindle, the tilting of the table, the swiveling of the head, and the rotation of the table during the machining process.

A thorough study of the subject of numerical control programming is required in order to program for all axes. The discussion in this text is limited to the principles of programming for two-axis machines, the *x*- and *y*-axis. All other motions are controlled by the operator and not by tape commands.

Control systems are generally either point-to-point or continuous path (contouring) systems. The continuous path system is the more complex of the two systems, and programming for this system requires an analysis of the system which is beyond the scope of this text. This discussion of the principles of programming is limited to the fundamentals of programming for machining holes using the point-to-point system, although most point-to-point machines can also be used for straight cut milling (cuts parallel to the *x*- and *y*-axes).

POINT-TO-POINT PROGRAMMING ON TWO-AXIS MACHINES

The movement of the machine as it performs the operations of machining holes in a part is similar to that of conventional machinery. The part to be machined is located for its various operations by either a movable table with a fixed spindle or a movable spindle with a fixed table.

The part to be machined must be positioned with respect to a zero reference point on the table. This zero reference point is either fixed or movable (floating zero) depending upon the design of the machine. Only the fixed zero is considered here.

The methods of dimensioning for the hole locations are either absolute or incremental.

ABSOLUTE DIMENSIONING

All dimensions are given from the fixed zero of the machine table. The workpiece must be located in reference to this fixed zero. For example, the *x*- and *y*-dimensions are established from two finished edges (setup point) of the workpiece to the fixed zero of the table. All other machining locations are then given from the fixed zero.

Example 1: The figure shows a part as it is dimensioned on a blueprint before programming for numerical control. All dimensions are in inches. Program dimension the hole locations in the part.

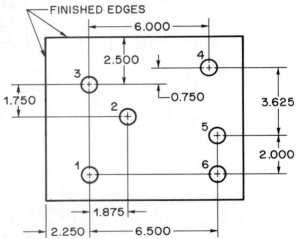

The workpiece is positioned at a convenient distance from the fixed zero of the machine table to the setup point (top left corner) of the workpiece. In this example, the setup point is located at $x = 5''$ and $y = 4''$.

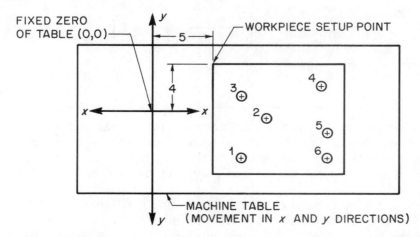

All the hole locations are programmed from the fixed zero of the table. Generally, the order in which the holes are machined is that which requires the least amount of machine movement. Note positive and negative *x*- and *y*-directions from (0, 0). The coordinates of the hole locations from (0, 0) are listed in the table shown.

Hole 1: $x = 5'' + 2.250'' = 7.250''$
$y = 4'' - (2.500'' - 0.750'') - 3.625'' - 2.000'' = -3.375''$

Hole 2: $x = 5'' + 2.250'' + 1.875'' = 9.125''$
$y = 4'' - (2.500'' + 1.750'') = -0.250''$

Hole 3: $x = 5'' + 2.250'' = 7.250''$
$y = 4'' - 2.500'' = 1.500''$

Hole 4: $x = 5'' + 2.250'' + 6.000'' = 13.250''$
$y = 4'' - (2.500'' - 0.750'') = 2.250''$

Hole 5: $x = 5'' + 2.250'' + 6.500'' = 13.750''$
$y = 4'' - (2.500'' - 0.750'') - 3.625'' = -1.375''$

Hole 6: $x = 5'' + 2.250'' + 6.500'' = 13.750''$
$y = 4'' - (2.500'' - 0.750'') - 3.625'' - 2.000'' = -3.375''$

Hole	x	y
1	7.250"	−3.375"
2	9.125"	−0.250"
3	7.250"	1.500"
4	13.250"	2.250"
5	13.750"	−1.375"
6	13.750"	−3.375"

Example 2: The figure shows a part as it is dimensioned before programming for numerical control. All dimensions are in millimeters. Program dimension the hole locations.

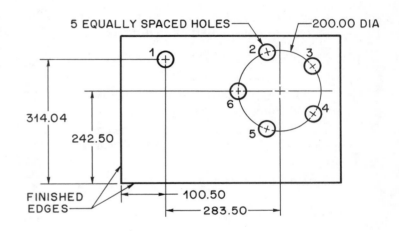

The workpiece is positioned at a convenient distance from the fixed zero of the machine table to the setup point (bottom left corner) of the workpiece. In this example, the setup point is located at $x = -200.00$ mm and $y = -160.00$ mm.

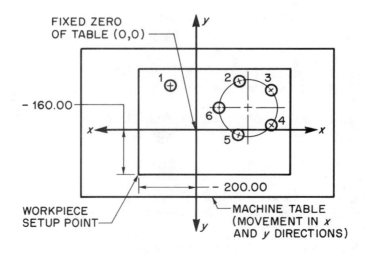

All hole locations are programmed from the fixed zero of the table.

Hole 1: $x = -200.00$ mm $+ 100.50$ mm $= -99.50$ mm
$y = 314.04$ mm $- 160.00$ mm $= 154.04$ mm

Hole 2: x = x distance to the center of the 200.00-mm diameter circle – b

   y = y distance to the center of the 200.00-mm diameter circle + a

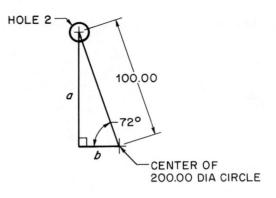

Calculate the number of degrees between two consecutive holes on the 200.00-mm diameter circle.

$$\frac{360°}{5} = 72°$$

From the center of the 200.00-mm diameter circle calculate a and b dimensions.

$$\sin 72° = \frac{a}{100.00 \text{ mm}} \qquad \cos 72° = \frac{b}{100.00 \text{ mm}}$$

$$0.95106 = \frac{a}{100.00 \text{ mm}} \qquad 0.30902 = \frac{b}{100.00 \text{ mm}}$$

$$a = 95.11 \text{ mm} \qquad b = 30.90 \text{ mm}$$

x = –200.00 mm + 100.50 mm + 283.50 mm – 30.90 mm = 153.10 mm
y = 242.50 mm – 160.00 mm + 95.11 mm = 177.61 mm

Hole 3: x = x distance to the center of the 200.00-mm diameter circle + b

   y = y distance to the center of the 200.00-mm diameter circle + a

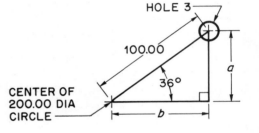

From the center of the 200.00-mm diameter circle calculate the angle formed by the horizontal centerline and Hole 3.
180° – 2(72°) = 36°

Calculate a and b dimensions.

$$\sin 36° = \frac{a}{100.00 \text{ mm}} \qquad \cos 36° = \frac{b}{100.00 \text{ mm}}$$

$$0.58779 = \frac{a}{100.00 \text{ mm}} \qquad 0.80902 = \frac{b}{100.00 \text{ mm}}$$

$$a = 58.78 \text{ mm} \qquad b = 80.90 \text{ mm}$$

x = –200.00 mm + 100.50 mm + 283.50 mm + 80.90 mm = 264.90 mm
y = 242.50 mm – 160.00 mm + 58.78 mm = 141.28 mm

Hole 4: x = 264.90 mm (the same as x of Hole 3)

   y = y distance to the center of the 200.00-mm diameter circle – a

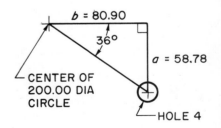

From the center of the 200.00-mm diameter circle calculate the angle formed by the horizontal centerline and Hole 4.
3(72°) – 180° = 36°
Since both Hole 4 and Hole 3 are projected 36° from the horizontal, the a and b dimensions of Hole 4 are the same as Hole 3.
x = 264.90 mm
y = 242.50 mm – 160.00 mm – 58.78 mm = 23.72 mm

Hole 5: x = 153.10 mm (the same as x of Hole 2)

   y = y distance to the center of the 200.00-mm diameter circle – a

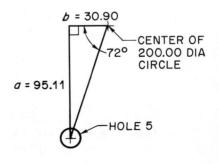

Since both Hole 5 and Hole 2 are projected 72° from the horizontal, the a and b dimensions of Hole 5 are the same as Hole 2.

x = 153.10 mm
y = 242.50 mm – 160.00 mm – 95.11 mm = –12.61 mm

Hole 6: x = x distance to the center of the 200.00-mm diameter
        circle – 100.00 mm

      x = –200.00 mm + 100.50 mm + 283.50 mm – 100.00 mm = 84.00 mm

      y = 242.50 mm – 160.00 mm = 82.50 mm

This table lists the coordinates of the hole locations from (0, 0).

Hole	x	y
1	–99.50 mm	154.04 mm
2	153.10 mm	177.61 mm
3	264.90 mm	141.28 mm
4	264.90 mm	23.72 mm
5	153.10 mm	–12.61 mm
6	84.00 mm	82.50 mm

INCREMENTAL DIMENSIONING

In incremental dimensioning, the dimensions to each location are given from the immediate previous location. The location of a hole is considered the origin (0, 0) of the x and y axes. From this origin, x and y dimensions are given to the next hole. Each new location in turn becomes the origin for the x and y dimensions to the next hole. The direction of travel, positive and negative, must be noted and is based upon the Cartesian coordinate system just as it was with absolute dimensioning. The first hole location is dimensioned with reference to the fixed zero of the table, while each subsequent hole is dimensioned from the hole directly preceding it. Each hole becomes the origin for the next hole to be machined.

Example: Dimension the part shown using incremental dimensioning. All dimensions are in inches. (This same part was used to illustrate absolute dimensioning.)

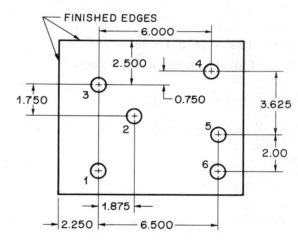

The workpiece is positioned on the table with the distance from the fixed zero of the machine table to the setup point of the workpiece as x = 5″ and y = 4″. Be aware of direction of travel as + or – from one hole to the next hole.

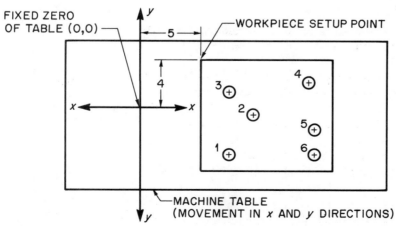

Each hole is the origin for the next hole to be machined. The x and y dimensions for Hole 1 are identical to those using absolute dimensioning. This is true for the first hole only. The table shown lists the coordinates using incremental dimensioning.

Hole 1: $x = 5'' + 2.250'' = 7.250''$
           $y = 4'' - (2.500'' - 0.750'') - 3.625'' - 2.000'' = -3.375''$

Hole 2:   The center of Hole 1 is the origin for locating Hole 2.
           $x = 1.875''$
           $y = (2.500'' - 0.750'') + 3.625'' + 2.000'' - (2.500'' + 1.750'') = 3.125''$

Hole 3:   The center of Hole 2 is the origin for locating Hole 3.
           $x = -1.875''$
           $y = 1.750''$

Hole 4:   The center of Hole 3 is the origin for locating Hole 4.
           $x = 6.000''$
           $y = 0.750''$

Hole 5:   The center of Hole 4 is the origin for locating Hole 5.
           $x = 6.500'' - 6.000'' = 0.500''$
           $y = -3.625''$

Hole 6:   The center of Hole 5 is the origin for locating Hole 6.
           $x = 0$
           $y = -2.000''$

Hole	x	y
1	7.250"	-3.375"
2	1.875"	3.125"
3	-1.875"	1.750"
4	6.000"	0.750"
5	0.500"	-3.625"
6	0	-2.000"

APPLICATION

Program Dimensioning

Program dimension the following parts. The dimensions given in the tables are taken from blueprints before programming for numerical control. Dimensions X', and Y', are the positioning dimensions from the fixed zero (0, 0) of the table to the setup point of the workpiece. Use the hole location dimensions given in the tables. Write the program in table form listing the holes in sequence. Program dimension the part shown using

a. absolute dimensioning.

b. incremental dimensioning.

Use this figure for # 1, # 2, and # 3.

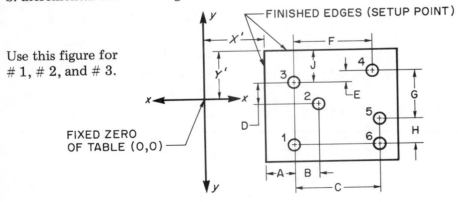

			Hole Location Dimensions								
	X'	Y'	A	B	C	D	E	F	G	H	J
1.*	5	4	2.500	2.100	6.750	1.875	0.825	6.125	3.875	2.125	2.375
2.*	6	5	2.710	2.615	7.010	2.070	0.920	6.475	4.307	2.416	2.300
3.**	100	100	60.00	56.24	148.06	45.06	20.16	132.32	92.24	52.30	51.00

  *All dimensions are in inches.
  **All dimensions are in millimeters.

Use this figure for
4, # 5, and # 6.

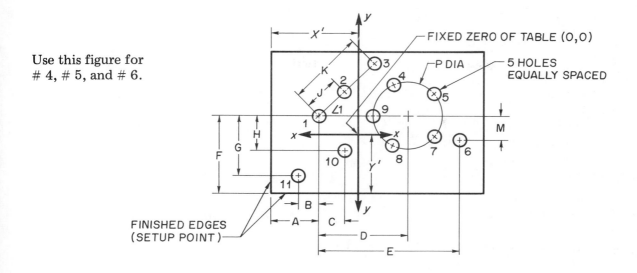

	X'	Y'	A	B	C	D	E	F	G	H	J	K	M	P DIA	∠1
4.*	−10	−8	5.175	1.300	3.250	14.250	22.100	12.500	9.150	5.150	5.500	11.250	4.625	10.000	42°0′
5.*	−10	−8	5.250	1.412	3.562	14.400	22.250	12.750	9.375	5.270	5.600	11.300	4.850	10.200	43°0′
6.**	−180	−140	122.40	30.00	72.00	302.50	446.00	257.40	190.24	112.40	104.00	228.00	98.20	196.00	41.75°

Hole Location Dimensions

*All dimensions are in inches.
**All dimensions are in millimeters.

Use this figure for
7, # 8, and # 9.

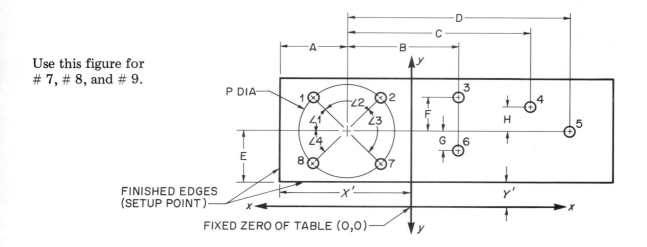

	X'	Y'	A	B	C	D	E	F	G	H	P DIA	∠1	∠2	∠3	∠4
7.*	−18	5	10.185	13.700	19.215	26.750	7.500	5.750	3.170	4.250	12.200	75°45′	55°30′	95°15′	20°10′
8.*	−20	5	10.520	13.975	19.504	27.315	7.615	5.942	2.623	4.514	12.400	77°10′	57°15′	93°25′	15°0′
9.**	−400	80	196.30	255.00	378.34	521.40	142.50	106.80	59.50	79.68	224.00	72.67°	61.50°	98.83°	18.33°

Hole Location Dimensions

*All dimensions are in inches.
**All dimensions are in millimeters.

UNIT 65 *BINARY NUMERATION SYSTEM AND BINARY CODED TAPE*

OBJECTIVES

After studying this unit you should be able to

- Express binary numbers as decimal numbers.
- Express decimal numbers as binary numbers.
- Code numerical control tape using a binary-decimal system.

After the numerical control programming manuscript is made, the data on the program manuscript is transferred to tape. The signals which control the operation of the machine are received from the tape. Numerical control tape is either punched or magnetic; both operate basically the same. Punched tapes are supplied in rolls of varying diameters which are usually one inch wide. Holes are punched in the tape in rows and columns (tracks). Normally the tape has eight tracks with one additional column for the tape-feeding sprockets.

A hole punched in the tape signals an open circuit; the absence of a hole signals a closed circuit. Because a circuit is either open or closed (ON or OFF), it can be represented by either of two digits: 0 to represent OFF and 1 to represent ON. By the use of a sufficient number of circuits, combinations of these two digits can be used to represent any number.

The mathematical system which uses only the digits 1 and 0 is called the *binary numeration system*. An understanding of the structure of the decimal system is helpful in discussing the binary system.

STRUCTURE OF THE DECIMAL SYSTEM

The elements of a mathematical system are the base of the system, the particular digits used, and the locations of the digits with respect to the decimal point (place value). In the decimal system, all numbers are combinations of the digits 0–9. The decimal system is built on powers of the base 10. Each place value is ten times greater than the place value directly to its right. Since any number with an exponent of 0 equals 1, 10^0 equals 1.

An analysis of the number 64,216 shows this structure.

6	4	2	1	6	Number
$10^4 = 10,000$	$10^3 = 1000$	$10^2 = 100$	$10^1 = 10$	$10^0 = 1$	Place Value
$6 \times 10^4 =$ $6 \times 10,000 =$ $60,000$	$4 \times 10^3 =$ $4 \times 1000 =$ 4000	$2 \times 10^2 =$ $2 \times 100 =$ 200	$1 \times 10^1 =$ $1 \times 10 =$ 10	$6 \times 10^0 =$ $6 \times 1 =$ 6	Value
60,000 +	4000 +	200 +	10 +	6 =	64,216

Examples: Analyze the following numbers.
1. $16 = 1(10^1) + 6(10^0) = 10 + 6$ Ans
2. $216 = 2(10^2) + 1(10^1) + 6(10^0) = 200 + 10 + 6$ Ans
3. $4216 = 4(10^3) + 2(10^2) + 1(10^1) + 6(10^0) = 4000 + 200 + 10 + 6$ Ans
4. $64,216 = 6(10^4) + 4(10^3) + 2(10^2) + 1(10^1) + 6(10^0) = 60,000 + 4000 + 200 + 10 + 6$ Ans

The same principles of structure hold true for numbers that are less than one. A number less than one can be expressed by using negative exponents. A number with a negative exponent is equal to its positive reciprocal. When the number is inverted and the negative exponent changed to a positive exponent, the result is as follows.

342

$$10^{-1} = \frac{1}{10^1} = 0.1$$

$$10^{-2} = \frac{1}{10^2} = \frac{1}{100} = 0.01$$

$$10^{-3} = \frac{1}{10^3} = \frac{1}{1000} = 0.001$$

$$10^{-4} = \frac{1}{10^4} = \frac{1}{10,000} = 0.0001$$

An analysis of the number 0.8502 shows this structure.

• 8	5	0	2	Number
$10^{-1} = 0.1$	$10^{-2} = 0.01$	$10^{-3} = 0.001$	$10^{-4} = 0.0001$	Place Value
$8 \times 10^{-1} =$ $8 \times 0.1 =$ 0.8	$5 \times 10^{-2} =$ $5 \times 0.01 =$ 0.05	$0 \times 10^{-3} =$ $0 \times 0.001 =$ 0	$2 \times 10^{-4} =$ $2 \times 0.0001 =$ 0.0002	Value
0.8 +	0.05 +	0 +	0.0002 =	0.8502

STRUCTURE OF THE BINARY SYSTEM

The same principles of structure apply to the binary system as to the decimal system. The binary system is built upon the base 2 and uses only the digits 0 and 1. Numbers are shown as binary numbers by putting a 2 to the right and below the number (subscript) as shown; 11_2, 100_2, 1_2, and 10001_2 are binary numbers. As with the decimal system, the elements which must be considered are the base, the particular digits used, and the place value of the digits. The binary system is built on the powers of the base 2, each place value is twice as large as the place value directly to its right.

Place Values of Binary Numbers

2^6	2^5	2^4	2^3	2^2	2^1	2^0		2^{-1}	2^{-2}	2^{-3}	2^{-4}
64	32	16	8	4	2	1	•	0.5	0.25	0.125	0.0625

EXPRESSING BINARY NUMBERS AS DECIMAL NUMBERS

Numbers in the decimal system are usually shown without a subscript. It is understood the number is in the decimal system. In certain instances, for clarity, decimal numbers are shown with the subscript 10. The following examples show the method of expressing binary numbers as equivalent decimal numbers. Remember that 0 and 1 are the only digits in the binary system.

Examples: Express each binary number as an equivalent decimal number.

1. $11_2 = 1(2^1) + 1(2^0) = 2 + 1 = 3_{10}$ Ans
2. $111_2 = 1(2^2) + 1(2^1) + 1(2^0) = 4 + 2 + 1 = 7_{10}$ Ans
3. $11101_2 = 1(2^4) + 1(2^3) + 1(2^2) + 0(2^1) + 1(2^0)$
    $= 16 + 8 + 4 + 0 + 1 = 29_{10}$ Ans
4. $101.11_2 = 1(2^2) + 0(2^1) + 1(2^0) + 1(2^{-1}) + 1(2^{-2})$
    $= 4 + 0 + 1 + 0.5 + 0.25 = 5.75_{10}$ Ans

EXPRESSING DECIMAL NUMBERS AS BINARY NUMBERS

The following examples show the method of expressing decimal numbers as equivalent binary numbers.

Example 1: Express 25_{10} as an equivalent binary number.

Determine the largest power of 2 in 25; $2^4 = 16$. There is one 2^4. Subtract 16 from 25; $25 - 16 = 9$.

Determine the largest power of 2 in 9; $2^3 = 8$. There is one 2^3. Subtract 8 from 9; $9 - 8 = 1$.

Determine the largest power of 2 in 1; $2^0 = 1$. There is one 2^0. Subtract 1 from 1; $1 - 1 = 0$.

There are no 2^2 and 2^1. The place positions for these values must be shown as zeros.

$$25_{10} = 1(2^4) + 1(2^3) + 0(2^2) + 0(2^1) + 1(2^0)$$
$$25_{10} = \quad 1 \qquad 1 \qquad 0 \qquad 0 \qquad 1$$
$$25_{10} = 11001_2 \quad \text{Ans}$$

Example 2: Express 11.625_{10} as an equivalent binary number.

$2^3 = 8; 11.625 - 8 = 3.625$
$2^1 = 2; 3.625 - 2 = 1.625$
$2^0 = 1; 1.625 - 1 = 0.625$
$2^{-1} = 0.5; 0.625 - 0.5 = 0.125$
$2^{-3} = 0.125; 0.125 - 0.125 = 0$

There are no 2^2 and 2^{-2}.

$$11.625_{10} = 1(2^3) + 0(2^2) + 1(2^1) + 1(2^0) + 1(2^{-1}) + 0(2^{-2}) + 1(2^{-3})$$
$$11.625_{10} = \quad 1 \qquad 0 \qquad 1 \qquad 1 \qquad 1 \qquad 0 \qquad 1$$
$$11.625_{10} = 1011.101_2 \quad \text{Ans}$$

BINARY CODED TAPE

A hole in the numerical control tape indicates an open circuit and represents a 1 in the binary system. The absence of a hole indicates a closed circuit and represents a 0 in the binary system. The tracks of the tape determine place value.

A simplified binary coded tape is shown with its relationship to equivalent decimal and binary numbers. In actual practice the coding on the tape is more involved with codings such as the sequence of operations, direction of machine table movement, spindle movement, and the feeds and speeds. This tape shows the coding for the decimal numbers 12, 7, 17, 23, 5, 31, 16, 15, and 1.

Decimal Number	Binary Equivalent of Decimal Number
12	1100
7	111
17	10001
23	10111
5	101
31	11111
16	10000
15	1111
1	1

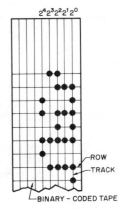

BINARY-DECIMAL SYSTEM

Most numerical control machines use a binary-decimal system rather than the pure binary system that has been discussed. In a pure binary system a large number of places are required to represent numbers.

For example, the decimal number 3173 when converted to the binary system consists of a 2^{11}, 2^{10}, 2^6, 2^5, 2^2, and a 2^0; in binary form it appears as 110001100101. To code this number on tape in pure binary form, 12 tracks are required.

	2^{11}	2^{10}	2^9	2^8	2^7	2^6	2^5	2^4	2^3	2^2	2^1	2^0
Binary Form	1	1	0	0	0	1	1	0	0	1	0	1
Coding on Punched Tape	•	•				•	•			•		•

The binary-decimal system eliminates the need for a large number of tracks. It uses the decimal system for place locations, but converts each digit of the decimal number into a binary number.

Example 1: Express 243_{10} in the binary-decimal system in horizontal form.

	10^2	10^1	10^0
Decimal System	2	4	3
Binary-Decimal System	10	100	11

In the actual coding of the numerical control tape, the number should be positioned vertically rather than horizontally.

Example 2: Express 243_{10} in the binary-decimal system in vertical form.

Position the number vertically.

The vertical positioning of 243_{10} is 2
4
3

The coding is shown on the punched tape.

	Decimal Number	Vertical Binary-Decimal Number
10^2	2	10
10^1	4	100
10^0	3	11

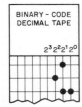

Example 3: Express 3109_{10} in the binary-decimal system in vertical form.

Position the number vertically.

The vertical positioning of 3109_{10} is 3
1
0
9

The coding is shown on the punched tape.

	Decimal Number	Vertical Binary-Decimal Number
10^3	3	11
10^2	1	1
10^1	0	0
10^0	9	1001

APPLICATION

Structure of the Decimal System

Analyze the following numbers.

1. 285
2. 2855
3. 90,500

4. 0.704
5. 23.023
6. 105.009

7. 4751.107
8. 3006.0204
9. 175.0753

Expressing Binary Numbers as Decimal Numbers

Express the following binary numbers as decimal numbers.

10. 10_2 _____
11. 1_2 _____
12. 100_2 _____
13. 1011_2 _____
14. 1101_2 _____
15. 1111_2 _____
16. 10100_2 _____

17. 1001_2 _____
18. 11000_2 _____
19. 10101_2 _____
20. 101010_2 _____
21. 110011_2 _____
22. 111010_2 _____
23. 0.1_2 _____

24. 0.101_2 _____
25. 11.11_2 _____
26. 11.01_2 _____
27. 10.001_2 _____
28. 1111.11_2 _____
29. 1001.0101_2 _____
30. 10011.1001_2 _____

Expressing Decimal Numbers as Binary Numbers

Express the following decimal numbers as binary numbers.

31.	12	_____	38.	98	45.	0.375
32.	100	_____	39.	1	46.	10.5
33.	87	_____	40.	7	47.	81.75
34.	26	_____	41.	51	48.	19.0625
35.	43	_____	42.	270	49.	111.25
36.	4	_____	43.	0.5	50.	1.125
37.	102	_____	44.	0.125	51.	163.875

Binary-Decimal System

Express the given decimal numbers as binary-decimal numbers in vertical form.
Code the tape for each number. (The solution to the first problem is shown).

	DECIMAL NUMBER	VERTICAL BINARY-DECIMAL NUMBER	BINARY-CODE-DECIMAL TAPE
52. 2403 10^3	2	10	
10^2	4	100	
10^1	0	0	
10^0	3	11	
53. 3173 10^3-10^0			
54. 9157 10^3-10^0			
55. 803 10^2-10^0			
56. 736 10^2-10^0			
57. 74,932 10^4-10^0			
58. 87 10^1-10^0			
59. 1029 10^3-10^0			
60. 5005 10^3-10^0			
61. 8321 10^3-10^0			
62. 1000 10^3-10^0			
63. 429 10^2-10^0			
64. 202 10^2-10^0			
65. 60,070 10^4-10^0			
66. 24 10^1-10^0			
67. 2097 10^3-10^0			

⁻OBJECTIVE_____

You should be able to solve the exercises and problems in this Achievement Review by applying the principles and methods covered in units 63–65.

1. Using graph paper, draw an x- and a y-axis and plot the following coordinates.

   A = (6, –8)   C = (–2, 0)   E = (–5, –5)
   B = (–4, 8)   D = (0, –9)   F = (3, 3)

2. Refer to the points plotted on the illustrated Cartesian Coordinate plane. Write the x and y coordinates of the following points, A–M.

   A = _____   H = _____
   B = _____   I = _____
   C = _____   J = _____
   D = _____   K = _____
   E = _____   L = _____
   F = _____   M = _____
   G = _____

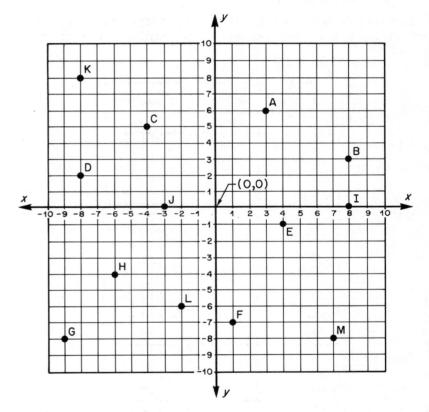

3. Program dimension the hole locations on the part shown. All dimensions are in inches. Use

   a. absolute dimensioning
   b. incremental dimensioning

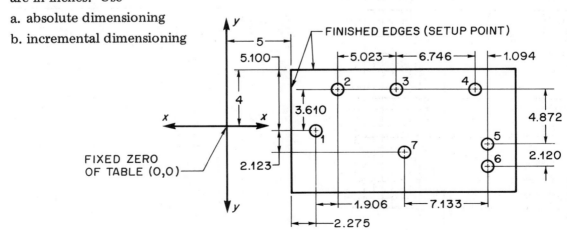

4. Program dimension the hole locations on the part shown. All dimensions are
    in millimeters. Use

    a. absolute dimensioning

    b. incremental dimensioning

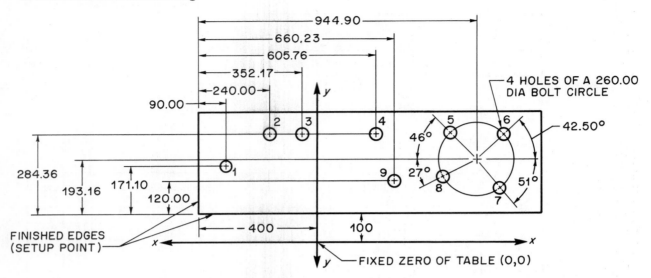

5. Express the following binary numbers as decimal numbers.

    a. 1_2 _____   c. 10101_2 _____   e. 1001.1001_2 _____

    b. 101_2 _____   d. 10.01_2 _____

6. Express the following decimal numbers as binary numbers.

    a. 5 _____   c. 157 _____   e. 74.25 _____

    b. 32 _____   d. 0.125 _____

7. Express the given decimal numbers as binary-decimal numbers in the vertical
    form. Code the tape for each number.

	DECIMAL NUMBER	VERTICAL BINARY-DECIMAL NUMBER	BINARY-CODE-DECIMAL TAPE $2^3 2^2 2^1 2^0$
a. 5624	10^3		
	10^2		
	10^1		
	10^0		
b. 8078	10^3		
	10^2		
	10^1		
	10^0		
c. 1005	10^3		
	10^2		
	10^1		
	10^0		
d. 723	10^2		
	10^1		
	10^0		

	DECIMAL NUMBER	VERTICAL BINARY-DECIMAL NUMBER	BINARY-CODE-DECIMAL TAPE $2^3 2^2 2^1 2^0$
e. 906	10^2		
	10^1		
	10^0		
f. 36,840	10^4		
	10^3		
	10^2		
	10^1		
	10^0		
g. 62	10^1		
	10^0		
h. 2057	10^3		
	10^2		
	10^1		
	10^0		

DECIMAL EQUIVALENT TABLE

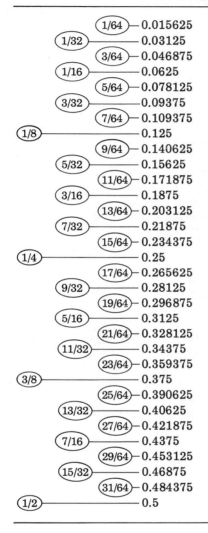

Fraction	Decimal		Fraction	Decimal
1/64	0.015625		33/64	0.515625
1/32	0.03125		17/32	0.53125
3/64	0.046875		35/64	0.546875
1/16	0.0625		9/16	0.5625
5/64	0.078125		37/64	0.578125
3/32	0.09375		19/32	0.59375
7/64	0.109375		39/64	0.609375
1/8	0.125		5/8	0.625
9/64	0.140625		41/64	0.640625
5/32	0.15625		21/32	0.65625
11/64	0.171875		43/64	0.671875
3/16	0.1875		11/16	0.6875
13/64	0.203125		45/64	0.703125
7/32	0.21875		23/32	0.71875
15/64	0.234375		47/64	0.734375
1/4	0.25		3/4	0.75
17/64	0.265625		49/64	0.765625
9/32	0.28125		25/32	0.78125
19/64	0.296875		51/64	0.796875
5/16	0.3125		13/16	0.8125
21/64	0.328125		53/64	0.828125
11/32	0.34375		27/32	0.84375
23/64	0.359375		55/64	0.859375
3/8	0.375		7/8	0.875
25/64	0.390625		57/64	0.890625
13/32	0.40625		29/32	0.90625
27/64	0.421875		59/64	0.921875
7/16	0.4375		15/16	0.9375
29/64	0.453125		61/64	0.953125
15/32	0.46875		31/32	0.96875
31/64	0.484375		63/64	0.984375
1/2	0.5		1	1.

ENGLISH UNITS OF LINEAR MEASURE

1 yard (yd) = 3 feet (ft)

1 yard (yd) = 36 inches (in)

1 foot (ft) = 12 inches (in)

1 mile (mi) = 1760 yards (yd)

1 mile (mi) = 5280 feet (ft)

METRIC UNITS OF LINEAR MEASURE

1 millimeter (mm) = 0.001 meter (m)	1000 millimeters (mm) = 1 meter (m)
1 centimeter (cm) = 0.01 meter (m)	100 centimeters (cm) = 1 meter (m)
1 decimeter (dm) = 0.1 meter (m)	10 decimeters (dm) = 1 meter (m)
1 meter (m) = 1 meter (m)	1 meter (m) = 1 meter (m)
1 dekameter (dam) = 10 meters (m)	0.1 dekameter (dam) = 1 meter (m)
1 hectometer (hm) = 100 meters (m)	0.01 hectometer (hm) = 1 meter (m)
1 kilometer (km) = 1000 meters (m)	0.001 kilometer (km) = 1 meter (m)

METRIC-ENGLISH LINEAR EQUIVALENTS (CONVERSION FACTORS)

Metric to English Units	English to Metric Units
1 millimeter (mm) = 0.03937 inch (in)	1 inch (in) = 25.4 millimeters (mm)
1 centimeter (cm) = 0.3937 inch (in)	1 inch (in) = 2.54 centimeters (cm)
1 meter (m) = 39.37 inches (in)	1 foot (ft) = 0.3048 meter (m)
1 meter (m) = 3.2808 feet (ft)	1 yard (yd) = 0.9144 meter (m)
1 kilometer (km) = 0.6214 mile (mi)	1 mile (mi) = 1.609 kilometers (km)

TRIGONOMETRIC FUNCTIONS FOR DEGREES AND MINUTES

Deg	Sin	Cos	Tan	Cot	Sec	Csc	Deg
0° 00′	0.000 00	1.000 0	0.000 00	Infin	1.000 0	Infin	90° 00′
10′	0.002 91	1.000 0	0.002 91	343.77	1.000 0	343.77	50′
20′	0.005 82	0.999 98	0.005 82	171.88	1.000 0	171.89	40′
30′	0.008 73	0.999 96	0.008 73	114.59	1.000 0	114.59	30′
40′	0.011 64	0.999 93	0.011 64	85.940	1.000 1	85.946	20′
50′	0.014 54	0.999 89	0.014 55	68.750	1.000 1	68.757	10′
1° 00′	0.017 45	0.999 85	0.017 46	57.290	1.000 1	57.299	89° 00′
10′	0.020 36	0.999 79	0.020 36	49.104	1.000 2	49.114	50′
20′	0.023 27	0.999 73	0.023 28	42.964	1.000 3	42.976	40′
30′	0.026 18	0.999 66	0.026 19	38.189	1.000 3	38.201	30′
40′	0.029 08	0.999 58	0.029 10	34.368	1.000 4	34.382	20′
50′	0.031 99	0.999 49	0.032 01	31.242	1.000 5	31.257	10′
2° 00′	0.034 90	0.999 39	0.034 92	28.636	1.000 6	28.654	88° 00′
10′	0.037 81	0.999 29	0.037 83	26.432	1.000 7	26.450	50′
20′	0.040 71	0.999 17	0.040 75	24.542	1.000 8	24.562	40′
30′	0.043 62	0.999 05	0.043 66	22.904	1.000 9	22.925	30′
40′	0.046 53	0.998 92	0.046 58	21.470	1.001 1	21.494	20′
50′	0.049 43	0.998 78	0.049 49	20.206	1.001 2	20.230	10′
3° 00′	0.052 34	0.998 63	0.052 41	19.081	1.001 4	19.107	87° 00′
10′	0.055 24	0.998 47	0.055 33	18.075	1.001 5	18.103	50′
20′	0.058 14	0.998 31	0.058 24	17.169	1.001 7	17.198	40′
30′	0.061 05	0.998 13	0.061 16	16.350	1.001 9	16.380	30′
40′	0.063 95	0.997 95	0.064 08	15.605	1.002 0	15.637	20′
50′	0.066 85	0.997 76	0.067 00	14.924	1.002 2	14.958	10′
4° 00′	0.069 76	0.997 56	0.069 93	14.301	1.002 4	14.335	86° 00′
10′	0.072 66	0.997 36	0.072 85	13.727	1.002 6	13.763	50′
20′	0.075 56	0.997 14	0.075 78	13.197	1.002 9	13.235	40′
30′	0.078 46	0.996 92	0.078 70	12.706	1.003 1	12.745	30′
40′	0.081 36	0.996 68	0.081 63	12.251	1.003 3	12.291	20′
50′	0.084 26	0.996 44	0.084 56	11.826	1.003 6	11.868	10′
5° 00′	0.087 16	0.996 19	0.087 49	11.430	1.003 8	11.474	85° 00′
10′	0.090 05	0.995 94	0.090 42	11.059	1.004 1	11.104	50′
20′	0.092 95	0.995 67	0.093 35	10.712	1.004 3	10.758	40′
30′	0.095 85	0.995 40	0.096 29	10.385	1.004 6	10.433	30′
40′	0.098 74	0.995 11	0.099 23	10.078	1.004 9	10.127	20′
50′	0.101 64	0.994 82	0.102 16	9.788 2	1.005 2	9.839 1	10′
6° 00′	0.104 53	0.994 52	0.105 10	9.514 4	1.005 5	9.566 8	84° 00′
10′	0.107 42	0.994 21	0.108 05	9.255 3	1.005 8	9.309 2	50′
20′	0.110 31	0.993 90	0.110 99	9.009 8	1.006 1	9.065 1	40′
30′	0.113 20	0.993 57	0.113 94	8.776 9	1.006 5	8.833 7	30′
40′	0.116 09	0.993 24	0.116 88	8.555 6	1.006 8	8.613 8	20′
50′	0.118 98	0.992 90	0.119 83	8.345 0	1.007 1	8.404 6	10′
7° 00′	0.121 87	0.992 55	0.122 78	8.144 4	1.007 5	8.205 5	83° 00′
10′	0.124 76	0.992 19	0.125 74	7.953 0	1.007 9	8.015 6	50′
20′	0.127 64	0.991 82	0.128 69	7.770 4	1.008 2	7.834 4	40′
30′	0.130 53	0.991 44	0.131 65	7.595 8	1.008 6	7.661 3	30′
40′	0.133 41	0.991 06	0.134 61	7.428 7	1.009 0	7.495 7	20′
50′	0.136 29	0.990 67	0.137 58	7.268 7	1.009 4	7.337 2	10′
8° 00′	0.139 17	0.990 27	0.140 54	7.115 4	1.009 8	7.185 3	82° 00′
10′	0.142 05	0.989 86	0.143 51	6.968 2	1.010 2	7.039 6	50′
20′	0.144 93	0.989 44	0.146 48	6.826 9	1.010 7	6.899 8	40′
30′	0.147 81	0.989 02	0.149 45	6.691 2	1.011 1	6.765 5	30′
40′	0.150 69	0.988 58	0.152 43	6.560 6	1.011 5	6.636 3	20′
50′	0.153 56	0.988 14	0.155 40	6.434 8	1.012 0	6.512 1	10′
9° 00′	0.156 43	0.987 69	0.158 38	6.313 8	1.012 5	6.392 4	81° 00′
10′	0.159 31	0.987 23	0.161 37	6.197 0	1.012 9	6.277 2	50′
20′	0.162 18	0.986 76	0.164 35	6.084 4	1.013 4	6.166 1	40′
30′	0.165 05	0.986 29	0.167 34	5.975 8	1.013 9	6.058 8	30′
40′	0.167 92	0.985 80	0.170 33	5.870 8	1.014 4	5.955 4	20′
50′	0.170 78	0.985 31	0.173 33	5.769 4	1.014 9	5.855 4	10′
10° 00′	0.173 65	0.984 81	0.176 33	5.671 3	1.015 4	5.758 8	80° 00′
10′	0.176 51	0.984 30	0.179 33	5.576 4	1.015 9	5.665 3	50′
20′	0.179 37	0.983 78	0.182 33	5.484 5	1.016 5	5.574 9	40′
30′	0.182 24	0.983 25	0.185 34	5.395 5	1.017 0	5.487 4	30′
40′	0.185 09	0.982 72	0.188 35	5.309 3	1.017 6	5.402 6	20′
50′	0.187 95	0.982 18	0.191 36	5.225 7	1.018 1	5.320 5	10′
11° 00′	0.190 81	0.981 63	0.194 38	5.144 6	1.018 7	5.240 8	79° 00′
10′	0.193 66	0.981 07	0.197 40	5.065 8	1.019 3	5.163 6	50′
20′	0.196 52	0.980 50	0.200 42	4.989 4	1.019 9	5.088 6	40′
30′	0.199 37	0.979 92	0.203 45	4.915 2	1.020 5	5.015 8	30′
40′	0.202 22	0.979 34	0.206 48	4.843 0	1.021 1	4.945 2	20′
50′	0.205 07	0.978 75	0.209 52	4.772 9	1.021 7	4.876 5	10′
12° 00′	0.207 91	0.978 15	0.212 56	4.704 6	1.022 3	4.809 7	78° 00′
10′	0.210 76	0.977 54	0.215 60	4.638 3	1.023 0	4.744 8	50′
20′	0.213 60	0.976 92	0.218 64	4.573 6	1.023 6	4.681 7	40′
30′	0.216 44	0.976 30	0.221 69	4.510 7	1.024 3	4.620 2	30′
40′	0.219 28	0.975 66	0.224 75	4.449 4	1.024 9	4.560 4	20′
50′	0.222 12	0.975 02	0.227 81	4.389 7	1.025 6	4.502 1	10′
13° 00′	0.224 95	0.974 37	0.230 87	4.331 5	1.026 3	4.445 4	77° 00′
10′	0.227 78	0.973 71	0.233 93	4.274 7	1.027 0	4.391 0	50′
20′	0.230 62	0.973 04	0.237 00	4.219 3	1.027 7	4.336 2	40′
30′	0.233 45	0.972 37	0.240 08	4.165 3	1.028 4	4.283 6	30′
40′	0.236 27	0.971 69	0.243 16	4.112 6	1.029 1	4.232 4	20′
50′	0.239 10	0.971 00	0.246 24	4.061 1	1.029 9	4.182 4	10′
14° 00′	0.241 92	0.970 30	0.249 33	4.010 8	1.030 6	4.133 6	76° 00′
10′	0.244 74	0.969 59	0.252 42	3.961 7	1.031 4	4.085 9	50′
20′	0.247 56	0.968 87	0.255 52	3.913 6	1.032 1	4.039 4	40′
30′	0.250 38	0.968 15	0.258 62	3.866 7	1.032 9	3.993 9	30′
40′	0.253 20	0.967 42	0.261 72	3.820 8	1.033 7	3.949 5	20′
50′	0.256 01	0.966 67	0.264 83	3.776 0	1.034 5	3.906 1	10′
Deg	Cos	Sin	Cot	Tan	Csc	Sec	Deg

Deg	Sin	Cos	Tan	Cot	Sec	Csc	Deg
15° 00′	0.258 82	0.965 93	0.267 95	3.732 1	1.035 3	3.863 7	75° 00′
10′	0.261 63	0.965 17	0.271 07	3.689 1	1.036 1	3.822 2	50′
20′	0.264 43	0.964 40	0.274 19	3.647 1	1.036 9	3.781 6	40′
30′	0.267 24	0.963 63	0.277 32	3.605 9	1.037 7	3.742 0	30′
40′	0.270 04	0.962 85	0.280 46	3.565 6	1.038 6	3.703 1	20′
50′	0.272 84	0.962 06	0.283 60	3.526 1	1.039 4	3.665 1	10′
16° 00′	0.275 64	0.961 26	0.286 75	3.487 4	1.040 3	3.627 9	74° 00′
10′	0.278 43	0.960 46	0.289 90	3.449 5	1.041 2	3.591 5	50′
20′	0.281 23	0.959 64	0.293 05	3.412 4	1.042 0	3.555 9	40′
30′	0.284 02	0.958 82	0.296 21	3.375 9	1.042 9	3.520 9	30′
40′	0.286 80	0.957 99	0.299 38	3.340 2	1.043 8	3.486 7	20′
50′	0.289 59	0.957 15	0.302 55	3.305 2	1.044 8	3.453 2	10′
17° 00′	0.292 37	0.956 30	0.305 73	3.270 9	1.045 7	3.420 3	73° 00′
10′	0.295 15	0.955 45	0.308 91	3.237 1	1.046 6	3.388 1	50′
20′	0.297 93	0.954 59	0.312 10	3.204 1	1.047 6	3.356 5	40′
30′	0.300 71	0.953 72	0.315 30	3.171 6	1.048 5	3.325 5	30′
40′	0.303 48	0.952 84	0.318 50	3.139 7	1.049 5	3.295 1	20′
50′	0.306 25	0.951 95	0.321 71	3.108 4	1.050 5	3.265 3	10′
18° 00′	0.309 02	0.951 06	0.324 92	3.077 7	1.051 5	3.236 1	72° 00′
10′	0.311 78	0.950 15	0.328 14	3.047 5	1.052 5	3.207 4	50′
20′	0.314 54	0.949 24	0.331 36	3.017 8	1.053 5	2.179 2	40′
30′	0.317 30	0.948 32	0.334 60	2.988 7	1.054 5	3.151 5	30′
40′	0.320 06	0.947 40	0.337 83	2.960 0	1.055 5	3.124 4	20′
50′	0.322 82	0.946 46	0.341 08	2.931 9	1.056 6	3.097 7	10′
19° 00′	0.325 57	0.945 52	0.344 33	2.904 2	1.057 6	3.071 5	71° 00′
10′	0.328 32	0.944 57	0.347 58	2.877 0	1.058 7	3.045 8	50′
20′	0.331 06	0.943 61	0.350 85	2.850 2	1.059 8	3.020 6	40′
30′	0.333 81	0.942 64	0.354 12	2.823 9	1.060 8	2.995 7	30′
40′	0.336 55	0.941 67	0.357 40	2.798 0	1.061 9	2.971 3	20′
50′	0.339 29	0.940 68	0.360 68	2.772 5	1.063 0	2.947 4	10′
20° 00′	0.342 02	0.939 69	0.363 97	2.747 5	1.064 2	2.923 8	70° 00′
10′	0.344 75	0.938 69	0.367 27	2.722 8	1.065 3	2.900 6	50′
20′	0.347 48	0.937 69	0.370 57	2.698 5	1.066 4	2.877 8	40′
30′	0.350 21	0.936 67	0.373 88	2.674 6	1.067 6	2.855 4	30′
40′	0.352 93	0.935 65	0.377 20	2.651 1	1.068 8	2.833 4	20′
50′	0.355 65	0.934 62	0.380 53	2.627 9	1.069 9	2.811 7	10′
21° 00′	0.358 37	0.933 58	0.383 86	2.605 1	1.071 1	2.790 4	69° 00′
10′	0.361 08	0.932 53	0.387 21	2.582 6	1.072 3	2.769 4	50′
20′	0.363 79	0.931 48	0.390 55	2.560 5	1.073 6	2.748 8	40′
30′	0.366 50	0.930 42	0.393 91	2.538 7	1.074 8	2.728 5	30′
40′	0.369 21	0.929 35	0.397 27	2.517 2	1.076 0	2.708 5	20′
50′	0.371 91	0.928 27	0.400 65	2.496 0	1.077 3	2.688 8	10′
22° 00′	0.374 61	0.927 18	0.404 03	2.475 1	1.078 5	2.669 5	68° 00′
10′	0.377 30	0.926 09	0.407 41	2.454 5	1.079 8	2.650 4	50′
20′	0.379 99	0.924 99	0.410 81	2.434 2	1.081 1	2.631 6	40′
30′	0.382 68	0.923 88	0.414 21	2.414 2	1.082 4	2.613 1	30′
40′	0.385 37	0.922 76	0.417 63	2.394 5	1.083 7	2.594 9	20′
50′	0.388 05	0.921 64	0.421 05	2.375 0	1.085 0	2.577 0	10′
23° 00′	0.390 73	0.920 50	0.424 47	2.355 9	1.086 4	2.559 3	67° 00′
10′	0.393 41	0.919 36	0.427 91	2.336 9	1.087 7	2.541 9	50′
20′	0.396 08	0.918 22	0.431 36	2.318 3	1.089 1	2.524 7	40′
30′	0.398 75	0.917 06	0.434 81	2.299 8	1.090 4	2.507 8	30′
40′	0.401 41	0.915 90	0.438 28	2.281 7	1.091 8	2.491 2	20′
50′	0.404 08	0.914 72	0.441 75	2.263 7	1.093 2	2.474 8	10′
24° 00′	0.406 74	0.913 55	0.445 23	2.246 0	1.094 6	2.458 6	66° 00′
10′	0.409 39	0.912 36	0.448 72	2.228 6	1.096 1	2.442 6	50′
20′	0.412 04	0.911 16	0.452 22	2.211 3	1.097 5	2.426 9	40′
30′	0.414 69	0.909 96	0.455 73	2.194 3	1.098 9	2.411 4	30′
40′	0.417 34	0.908 75	0.459 24	2.177 5	1.100 4	2.396 1	20′
50′	0.419 98	0.907 53	0.462 77	2.160 9	1.101 9	2.381 1	10′
25° 00′	0.422 62	0.906 31	0.466 31	2.144 5	1.103 4	2.366 2	65° 00′
10′	0.425 25	0.905 07	0.469 85	2.128 3	1.104 9	2.351 5	50′
20′	0.427 88	0.903 83	0.473 41	2.112 3	1.106 4	2.337 1	40′
30′	0.430 51	0.902 59	0.476 98	2.096 5	1.107 9	2.322 8	30′
40′	0.433 13	0.901 33	0.480 55	2.080 9	1.109 5	2.308 7	20′
50′	0.435 75	0.900 07	0.484 14	2.065 5	1.111 0	2.294 9	10′
26° 00′	0.438 37	0.898 79	0.487 73	2.050 3	1.112 6	2.281 2	64° 00′
10′	0.440 98	0.897 52	0.491 34	2.035 3	1.114 2	2.267 6	50′
20′	0.443 59	0.896 23	0.494 95	2.020 4	1.115 8	2.254 3	40′
30′	0.446 20	0.894 93	0.498 58	2.005 7	1.117 4	2.241 1	30′
40′	0.448 80	0.893 63	0.502 22	1.991 2	1.119 0	2.228 2	20′
50′	0.451 40	0.892 32	0.505 87	1.976 8	1.120 7	2.215 3	10′
27° 00′	0.453 99	0.891 01	0.509 53	1.962 6	1.122 3	2.202 7	63° 00′
10′	0.456 58	0.889 68	0.513 19	1.948 6	1.124 0	2.190 2	50′
20′	0.459 17	0.888 35	0.516 88	1.934 7	1.125 7	2.177 8	40′
30′	0.461 75	0.887 01	0.520 57	1.921 0	1.127 4	2.165 7	30′
40′	0.464 33	0.885 66	0.524 27	1.907 4	1.129 1	2.153 6	20′
50′	0.466 90	0.884 31	0.527 98	1.894 0	1.130 8	2.141 8	10′
28° 00′	0.469 47	0.882 95	0.531 71	1.880 7	1.132 6	2.130 0	62° 00′
10′	0.472 04	0.881 58	0.535 45	1.867 6	1.134 3	2.118 5	50′
20′	0.474 60	0.880 20	0.539 20	1.854 6	1.136 1	2.107 0	40′
30′	0.477 16	0.878 82	0.542 96	1.841 8	1.137 9	2.095 7	30′
40′	0.479 71	0.877 43	0.546 73	1.829 1	1.139 7	2.084 6	20′
50′	0.482 26	0.876 03	0.550 51	1.816 5	1.141 5	2.073 5	10′
29° 00′	0.484 81	0.874 62	0.554 31	1.804 1	1.143 3	2.062 7	61° 00′
10′	0.487 35	0.873 21	0.558 12	1.791 7	1.145 2	2.051 9	50′
20′	0.489 89	0.871 78	0.561 94	1.779 6	1.147 1	2.041 3	40′
30′	0.492 42	0.870 36	0.565 77	1.767 5	1.148 9	2.030 8	30′
40′	0.494 95	0.868 92	0.569 62	1.755 6	1.150 8	2.020 4	20′
50′	0.497 48	0.867 48	0.573 48	1.743 8	1.152 8	2.010 1	10′
Deg	Cos	Sin	Cot	Tan	Csc	Sec	Deg

TRIGONOMETRIC FUNCTIONS FOR DEGREES AND MINUTES (Cont'd)

Deg	Sin	Cos	Tan	Cot	Sec	Csc	Deg
30° 00'	0.500 00	0.866 03	0.577 35	1.732 1	1.154 7	2.000 0	60° 00'
10'	0.502 52	0.864 57	0.581 24	1.720 5	1.156 6	1.990 0	50'
20'	0.505 03	0.863 10	0.585 13	1.709 0	1.158 6	1.980 1	40'
30'	0.507 54	0.861 63	0.589 04	1.697 7	1.160 6	1.970 3	30'
40'	0.510 04	0.860 15	0.592 97	1.686 4	1.162 6	1.960 6	20'
50'	0.512 54	0.858 66	0.596 91	1.675 3	1.164 6	1.951 0	10'
31° 00'	0.515 04	0.857 17	0.600 86	1.664 3	1.166 6	1.941 6	59° 00'
10'	0.517 53	0.855 67	0.604 83	1.653 4	1.168 7	1.932 2	50'
20'	0.520 02	0.854 16	0.608 81	1.642 6	1.170 7	1.923 0	40'
30'	0.522 50	0.852 64	0.612 80	1.631 9	1.172 8	1.913 9	30'
40'	0.524 98	0.851 12	0.616 81	1.621 3	1.174 9	1.904 8	20'
50'	0.527 45	0.849 59	0.620 83	1.610 7	1.177 0	1.895 9	10'
32° 00'	0.529 92	0.848 05	0.624 87	1.600 3	1.179 2	1.887 1	58° 00'
10'	0.532 38	0.846 50	0.628 92	1.590 0	1.181 3	1.878 3	50'
20'	0.534 84	0.844 95	0.632 99	1.579 8	1.183 5	1.869 7	40'
30'	0.537 30	0.843 39	0.637 07	1.569 7	1.185 7	1.861 1	30'
40'	0.539 75	0.841 82	0.641 17	1.559 7	1.187 9	1.852 7	20'
50'	0.542 20	0.840 25	0.645 28	1.549 7	1.190 1	1.844 3	10'
33° 00'	0.544 64	0.838 67	0.649 41	1.539 9	1.192 4	1.836 1	57° 00'
10'	0.547 08	0.837 08	0.653 55	1.530 1	1.194 6	1.827 9	50'
20'	0.549 51	0.835 49	0.657 71	1.520 4	1.196 9	1.819 8	40'
30'	0.551 94	0.833 89	0.661 89	1.510 8	1.199 2	1.811 8	30'
40'	0.554 36	0.832 28	0.666 08	1.501 3	1.201 5	1.803 9	20'
50'	0.556 78	0.830 66	0.670 28	1.491 9	1.203 9	1.796 0	10'
34° 00'	0.559 19	0.829 04	0.674 51	1.482 6	1.206 2	1.788 3	56° 00'
10'	0.561 60	0.827 41	0.678 75	1.473 3	1.208 6	1.780 6	50'
20'	0.564 01	0.825 77	0.683 01	1.464 1	1.211 0	1.773 0	40'
30'	0.566 41	0.824 13	0.687 28	1.455 0	1.213 4	1.765 5	30'
40'	0.568 80	0.822 48	0.691 57	1.446 0	1.215 8	1.758 1	20'
50'	0.571 19	0.820 82	0.695 88	1.437 0	1.218 3	1.750 7	10'
35° 00'	0.573 58	0.819 15	0.700 21	1.428 2	1.220 8	1.743 4	55° 00'
10'	0.575 96	0.817 48	0.704 55	1.419 3	1.223 3	1.736 2	50'
20'	0.578 33	0.815 80	0.708 91	1.410 6	1.225 8	1.729 1	40'
30'	0.580 70	0.814 12	0.713 29	1.402 0	1.228 3	1.722 0	30'
40'	0.583 07	0.812 42	0.717 69	1.393 4	1.230 9	1.715 1	20'
50'	0.585 43	0.810 72	0.722 11	1.384 8	1.233 5	1.708 1	10'
36° 00'	0.587 79	0.809 02	0.726 54	1.376 4	1.236 1	1.701 3	54° 00'
10'	0.590 14	0.807 30	0.731 00	1.368 0	1.238 7	1.694 5	50'
20'	0.592 48	0.805 58	0.735 47	1.359 7	1.241 3	1.687 8	40'
30'	0.594 82	0.803 86	0.739 96	1.351 4	1.244 0	1.681 2	30'
40'	0.597 16	0.802 12	0.744 47	1.343 2	1.246 7	1.674 6	20'
50'	0.599 49	0.800 38	0.749 00	1.335 1	1.249 4	1.668 1	10'
37° 00'	0.601 82	0.798 64	0.753 55	1.327 0	1.252 1	1.661 6	53° 00'
10'	0.604 14	0.796 88	0.758 12	1.319 0	1.254 9	1.655 2	50'
20'	0.606 45	0.795 12	0.762 72	1.311 1	1.257 7	1.648 9	40'
30'	0.608 76	0.793 35	0.767 33	1.303 2	1.260 5	1.642 7	30'
40'	0.611 07	0.791 58	0.771 96	1.295 4	1.263 3	1.636 5	20'
50'	0.613 37	0.789 80	0.776 61	1.287 6	1.266 1	1.630 3	10'
Deg	Cos	Sin	Cot	Tan	Csc	Sec	Deg

Deg	Sin	Cos	Tan	Cot	Sec	Csc	Deg
38° 00'	0.615 66	0.788 01	0.781 29	1.279 9	1.269 0	1.624 3	52° 00'
10'	0.617 95	0.786 22	0.785 98	1.272 3	1.271 9	1.618 2	50'
20'	0.620 24	0.784 42	0.790 70	1.264 7	1.274 8	1.612 3	40'
30'	0.622 51	0.782 61	0.795 44	1.257 2	1.277 8	1.606 4	30'
40'	0.624 79	0.780 79	0.800 20	1.249 7	1.280 7	1.600 5	20'
50'	0.627 06	0.778 97	0.804 98	1.242 3	1.283 7	1.594 7	10'
39° 00'	0.629 32	0.777 15	0.809 78	1.234 9	1.286 7	1.589 0	51° 00'
10'	0.631 58	0.775 31	0.814 61	1.227 6	1.289 8	1.583 3	50'
20'	0.633 83	0.773 47	0.819 46	1.220 3	1.292 9	1.577 7	40'
30'	0.636 08	0.771 62	0.824 34	1.213 1	1.296 0	1.572 1	30'
40'	0.638 32	0.769 77	0.829 23	1.205 9	1.299 1	1.566 6	20'
50'	0.640 56	0.767 91	0.834 15	1.198 8	1.302 2	1.561 1	10'
40° 00'	0.642 79	0.766 04	0.839 10	1.191 8	1.305 4	1.555 7	50° 00'
10'	0.645 01	0.764 17	0.844 07	1.184 7	1.308 6	1.550 3	50'
20'	0.647 23	0.762 29	0.849 06	1.177 8	1.311 8	1.545 0	40'
30'	0.649 45	0.760 41	0.854 08	1.170 9	1.315 1	1.539 8	30'
40'	0.651 66	0.758 51	0.859 12	1.164 0	1.318 4	1.534 5	20'
50'	0.653 86	0.756 61	0.864 19	1.157 2	1.321 7	1.529 4	10'
41° 00'	0.656 06	0.754 71	0.869 29	1.150 4	1.325 0	1.524 2	49° 00'
10'	0.658 25	0.752 80	0.874 41	1.143 6	1.328 4	1.519 2	50'
20'	0.660 44	0.750 88	0.879 55	1.136 9	1.331 8	1.514 1	40'
30'	0.662 62	0.748 96	0.884 73	1.130 3	1.335 2	1.509 2	30'
40'	0.664 80	0.747 03	0.889 92	1.123 7	1.338 6	1.504 2	20'
50'	0.666 97	0.745 09	0.895 15	1.117 1	1.342 1	1.499 3	10'
42° 00'	0.669 13	0.743 14	0.900 40	1.110 6	1.345 6	1.494 5	48° 00'
10'	0.671 29	0.741 20	0.905 69	1.104 1	1.349 2	1.489 7	50'
20'	0.673 44	0.739 24	0.910 99	1.097 7	1.352 7	1.484 9	40'
30'	0.675 59	0.737 28	0.916 33	1.091 3	1.356 3	1.480 2	30'
40'	0.677 73	0.735 31	0.921 70	1.085 0	1.360 0	1.475 5	20'
50'	0.679 87	0.733 33	0.927 09	1.078 6	1.363 6	1.470 9	10'
43° 00'	0.682 00	0.731 35	0.932 52	1.072 4	1.367 3	1.466 3	47° 00'
10'	0.684 12	0.729 37	0.937 97	1.066 1	1.371 0	1.461 7	50'
20'	0.686 24	0.727 37	0.943 45	1.059 9	1.374 8	1.457 2	40'
30'	0.688 35	0.725 37	0.948 96	1.053 8	1.378 6	1.452 7	30'
40'	0.690 46	0.723 37	0.954 51	1.047 7	1.382 4	1.448 3	20'
50'	0.692 56	0.721 36	0.960 08	1.041 6	1.386 3	1.443 9	10'
44° 00'	0.694 66	0.719 34	0.965 69	1.035 5	1.390 2	1.439 5	46° 00'
10'	0.696 75	0.717 32	0.971 32	1.029 5	1.394 1	1.435 2	50'
20'	0.698 83	0.715 29	0.977 00	1.023 6	1.398 0	1.431 0	40'
30'	0.700 91	0.713 25	0.982 70	1.017 6	1.402 0	1.426 7	30'
40'	0.702 98	0.711 21	0.988 43	1.011 7	1.406 0	1.422 5	20'
50'	0.705 05	0.709 16	0.994 20	1.005 8	1.410 1	1.418 3	10'
45° 00'	0.707 11	0.707 11	1.000 0	1.000 0	1.414 2	1.414 2	45° 00'
Deg	Cos	Sin	Cot	Tan	Csc	Sec	Deg

ANSWERS TO ODD-NUMBERED APPLICATIONS

SECTION 1 COMMON FRACTIONS AND DECIMAL FRACTIONS

UNIT 1 INTRODUCTION TO COMMON FRACTIONS AND MIXED NUMBERS

1. A = 3/32
   B = 7/32
   C = 3/8
   D = 19/32
   E = 27/32
   F = 1
3. a. 1/2
   b. 1/2

3. c. 1/2
   d. 1/2
   e. 3/2
   f. 3/2
   g. 3/2
   h. 5/2
5. a. 8/32
   b. 24/32

5. c. 36/32
   d. 14/32
   e. 42/32
   f. 272/32
   g. 394/32
   h. 132/32
7. a. 8/3
   b. 15/8

7. c. 34/5
   d. 27/8
   e. 169/32
   f. 53/7
   g. 31/3
   h. 42/5
   i. 201/2
   j. 319/64

7. k. 207/4
   l. 6541/16
9. a. 20/8
   b. 28/16
   c. 102/15
   d. 228/18
   e. 632/64
   f. 1984/128

UNIT 2 ADDITION OF COMMON FRACTIONS AND MIXED NUMBERS

1. 12
3. 48
5. 6/12, 8/12, 5/12
7. 18/20, 5/20, 12/20, 10/20

9. A = 1 1/16″
   B = 59/64″
   C = 1 17/32″
   D = 11/16″
   E = 3 9/64″

9. F = 13/16″
11. A = 2 9/32″
    B = 3 55/64″
    C = 4 15/64″
    D = 2 19/32″

11. E = 3 3/8″
    F = 1 21/64″
    G = 4 45/64″
13. 5 17/60 h

UNIT 3 SUBTRACTION OF COMMON FRACTIONS AND MIXED NUMBERS

1. a. 9/32
   b. 1/4
   c. 12/25
   d. 31/64
   e. 23/64

1. f. 35/48
3. A = 7/32″
   B = 5/8″
   C = 25/64″

3. D = 3/16″
   E = 27/64″
   F = 13/32″
5. A = 15/32″

5. B = 21/32″
   C = 7/16″
   D = 15/32″
   E = 9/32″

5. F = 5/16″
   G = 1 7/32″
   H = 31/64″
   I = 39/64″

UNIT 4 MULTIPLICATION OF COMMON FRACTIONS AND MIXED NUMBERS

1. a. 1/9
   b. 1/16
   c. 65/512

1. d. 3/10
   e. 13 1/2
   f. 3/35

3. a. 1595/2048″
   b. 7/256″
5. a. 10 1/2

5. b. 25 43/64
   c. 9 11/16
   d. 6 13/32

5. e. 39/160
   f. 37 1/3

UNIT 5 DIVISION OF COMMON FRACTIONS AND MIXED NUMBERS

1. 8/7
3. 8/97
5. A = 6 threads

5. B = 8 7/16 threads
   C = 3 1/2 threads
   D = 7 threads

5. E = 6 7/8 threads
   F = 6 1/2 threads
   G = 2 13/16 threads

7. 5 cuts
9. 288 revolutions

11. 4/5 foot
13. 4 267/364 lb

UNIT 6 COMBINED OPERATIONS OF COMMON FRACTIONS AND MIXED NUMBERS

1. a. 9/16
   b. 2 1/16
   c. 5 33/50
   d. 19 1/2
   e. 33 49/64
   f. 3 1/2

1. g. 9 1/8
   h. 25 1/8
   i. 4 8/21
   j. 20 77/87
3. a. B = 4 7/8″
      E = 5 13/16″

3. b. A = 9/64″
      G = 6 3/4″
   c. C = 1 1/16″
      D = 5 23/32″
   d. B = 3 29/32″
      E = 5 5/8″

3. e. A = 19/32″
      G = 7 27/64″
   f. C = 63/64″
      D = 5 51/64″

5. 8 31/32″
7. 6 2/5 min
9. 57/64″
11. 1 7/8 lb

UNIT 7 INTRODUCTION TO DECIMAL FRACTIONS

1. A = 0.2
   B = 0.5
   C = 0.8
   D = 0.98
   E = 0.04

3. A = 0.0025
   B = 0.004
   C = 0.007
   D = 0.0077
   E = 0.0004

5. 0.01
7. 10
9. 0.1
11. 100
13. 0.001

15. seven thousandths
17. seventy-five ten-thousandths
19. one and five tenths
21. seventeen and nine ten-thousandths
23. thirteen and one hundred three thousandths

25. 0.7
27. 2.00007
29. 10.2
31. 20.71

33. 0.0003
35. 0.43
37. 0.0999
39. 0.01973

UNIT 8 ROUNDING DECIMAL FRACTIONS AND EQUIVALENT DECIMAL AND COMMON FRACTIONS

1. 0.783
3. 0.240
5. 0.02
7. 0.7201
9. 0.001

11. 0.4063
13. 0.6250
15. 0.6667
17. 0.0400
19. 0.2188

21. 0.5714
23. 0.3333
25. a. 0.125
    b. 0.6154
27. 1/8

29. 3/4
31. 11/16
33. 3/1000
35. 251/500
37. 7/16

39. 8717/10,000
41. 3/50
43. 753/1000
45. 9/200
47. 7/8

49. a. 1/4
    b. 49/80
    c. 1/10
    d. 3/80
    e. 3/8

UNIT 9 ADDITION AND SUBTRACTION OF DECIMAL FRACTIONS

1. a. 18.032
    b. 0.14095
    c. 1.295
    d. 5.129
    e. 892.577

1. f. 4.444
    g. 94.2539
    h. 0.1101
    i. 5.7787
    j. 575.757

3. 3.1758"
5. A = 29.83 mm
    B = 59.56 mm
    C = 91.33 mm

5. D = 16.26 mm
    E = 14.93 mm
7. A = 12.80 mm
    B = 27.02 mm

7. C = 6.58 mm
    D = 20.00 mm
    E = 10.49 mm
    F = 7.33 mm

UNIT 10 MULTIPLICATION OF DECIMAL FRACTIONS

1. a. 0.0465
    b. 3.3

1. c. 6
    d. 2.8576

3. Dia A = 31.751 mm
    Dia B = 19.187 mm
    Dia C = 12.835 mm

3. Dia D = 22.581 mm
    Dia E = 6.751 mm

UNIT 11 DIVISION OF DECIMAL FRACTIONS

1. a. 1.667
    b. 2.56
    c. 0.0100
    d. 10,000.000

1. e. 14.529
    f. 4.29
    g. 135.53
    h. 0.0038

3. A = 11.74 mm
    B = 5.91 mm
    C = 12.46 mm
    D = 10.93 mm

5. 24 complete bushings
7. 0.063 mm
9. 0.125"
11. 38.44 mm

UNIT 12 POWERS

1. 39.304
3. 100,000,000
5. 6 3/4
7. 4.41
9. 64
11. 532.23 mm²

13. 114.49 mm²
15. 1/16 sq in
17. 12 1/4 sq in
19. 189 1/16 sq in
21. 8741.82 mm³
23. 1728 mm³

25. 1/27 cu in
27. 3 3/8 cu in
29. 27/64 cu in
31. 1051.55 mm²
33. 270.41 sq in

35. 0.05 cu in
37. 0.22 cu in
39. 276.06 mm³
41. 16.25 cu in
43. 10.31 cu in

45. 10.72 cu in
47. 469.48 mm³
49. 0.01 cu in
51. 286.11 mm²
53. 1.57 cu in

UNIT 13 ROOTS

1. 6
3. 2/9
5. 3/4
7. 12
9. 4

11. a. 6 mm
    b. 4 in
    c. 7 in
    d. 10 mm
    e. 1 in

13. a. D = 3 in
    b. D = 6 mm
    c. D = 2 in
    d. D = 1 in
    e. D = 10 mm

15. 19.77
17. 1.871
19. 4.42
21. 0.0794

23. a. D = 8.60 mm
    b. D = 6.08 in
    c. D = 21.91 mm
    d. D = 1.02 in
25. D = 1.5 in

UNIT 14 TABLE OF DECIMAL EQUIVALENTS AND COMBINED OPERATIONS OF DECIMAL FRACTIONS

1. 0.78125
3. 0.34375
5. 0.078125
7. 5/16
9. 13/64

11. 49/64
13. 1/2
15. 13/16
17. 14.1
19. 25.12

21. 7.24
23. 9.07
25. 17.38
27. 0.084 mm

29. a. C = 8.62 mm
    b. C = 8.74 mm
    c. C = 6.12 mm
31. H = 0.077"

UNIT 15 ACHIEVEMENT REVIEW – SECTION 1

1. a. 12/32
    b. 30/100
    c. 16/64
    d. 72/128
3. a. 2 1/2
    b. 4 3/5
    c. 18 3/4
    d. 3 19/32
    e. 5 9/64
5. a. 8/32, 6/32, 9/32
    b. 28/64, 2/64, 9/64
    c. 70/100, 75/100,
       36/100, 65/100
7. a. 5/16
    b. 1/5

7. c. 1 245/256
    d. 25 23/40
    e. 13 1/8
    f. 3/4
    g. 3 1/3
    h. 24
    i. 6 3/4
    j. 93/280
9. 51 complete pieces
11. 8 min
13. A = 3 5/16"
    B = 2 15/16"
    C = 3 15/64"
    D = 3 15/16"
    E = 3 15/32"

15. a. 0.3
    b. 0.026
    c. 9.034
    d. 5.0081
17. a. 0.75
    b. 0.875
    c. 0.333
    d. 0.08
    e. 0.85
19. a. 1.203
    b. 6.4274
    c. 12.3069
    d. 9.1053
    e. 23.9077
    f. 0.266

19. g. 0.1444
    h. 0.001
    i. 0.0022
    j. 0.002
21. a. 3.24
    b. 0.125
    c. 0.000036
    d. 4/25
    e. 32.768
23. a. 16.67
    b. 0.935
    c. 0.775
    d. 6.780
25. a. 15/32
    b. 25/32

25. c. 1/32
    d. 31/32
27. A = 1.299 mm
    B = 0.812 mm
    C = 0.325 mm
    D = 0.162 mm
    E = 0.188 mm
    F = 0.375 mm
29. 0.12 mm
31. 18 min

SECTION 2 LINEAR MEASUREMENT: ENGLISH AND METRIC
UNIT 16 ENGLISH AND METRIC UNITS OF MEASURE

1. a. 7 ft
    b. 10.25 ft
    c. 42 in
    d. 10.8 in
    e. 45 in
    f. 4 yd
    g. 6.25 ft
    h. 24 ft
    i. 10.8 ft
    j. 9 yd
    k. 14 yd
    l. 12 in
    m. 21.5 ft
    n. 92 in

1. o. 9 in
    p. 46.75 yd
    q. 9.25 yd
    r. 14.75 ft
    s. 62 ft
    t. 111 in
3. 6 complete lengths
5. a. 29 mm
    b. 146.8 mm
    c. 21.975 cm
    d. 9.783 cm
    e. 93 cm
    f. 170 mm
    g. 0.153 m

5. h. 7.84 m
    i. 0.093 cm
    j. 0.8 mm
    k. 9.8 mm
    l. 104.6 cm
    m. 300.3 mm
    n. 97.976 cm
    o. 2.039 m
    p. 0.0347 m
    q. 56 mm
    r. 732.1 cm
    s. 63.77 mm
    t. 898 mm

7. 46 mm
9. a. 1.339 in
    b. 4.992 in
    c. 6.811 in
    d. 0.382 in
    e. 94.488 in
    f. 3.543 in
    g. 19.685 ft
    h. 33.464 ft
    i. 28.976 in
    j. 1.185 in
    k. 22.165 in
    l. 2.187 yd

9. m. 1.640 ft
    n. 2.559 ft
11. 59.0625 in
13. A = 15.75 mm
     B = 28.58 mm
     C = 327.03 mm
     D = 25.10 mm
     E = 3.30 mm
     F = 12.70 mm
     G = 25.00 mm
     H = 2.38 mm
     I = 17.46 mm
     J = 9.53 mm

UNIT 17 DEGREE OF PRECISION AND GREATEST POSSIBLE ERROR

1. a. 0.1"
    b. 3.55"
    c. 3.65"
3. a. 0.1"
    b. 4.25"
    c. 4.35"
5. a. 0.001"
    b. 15.8845"
    c. 15.8855"

7. a. 0.001"
    b. 12.0015"
    c. 12.0025"
9. a. 0.01"
    b. 7.005"
    c. 7.015"
11. a. 0.1"
     b. 9.05"
     c. 9.15"

13. a. 0.01 mm
     b. 26.865 mm
     c. 26.875 mm
15. a. 0.01 mm
     b. 123.075 mm
     c. 123.085 mm
17. a. 0.01 mm
     b. 48.005 mm
     c. 48.015 mm

19. a. 0.01 mm
     b. 6.995 mm
     c. 7.005 mm
21. a. 0.001 mm
     b. 9.0005 mm
     c. 9.0015 mm

	Greatest Possible Error (inches)	ACTUAL LENGTH	
		Smallest Possible (inches)	Largest Possible (inches)
23.	0.025	8.275	8.325
25.	0.0005	0.7525	0.7535
27.	0.00005	0.93685	0.93695

	Greatest Possible Error (millimeters)	ACTUAL LENGTH	
		Smallest Possible (millimeters)	Largest Possible (millimeters)
29.	0.5	63.5	64.5
31.	0.25	98.25	98.75
33.	0.005	13.365	13.375

UNIT 18 TOLERANCE, CLEARANCE, AND INTERFERENCE

1. a. 1/32"
    b. 1/8"
    c. 16.74"
    d. 0.911"
    e. 0.0003"
    f. 11.004"
3. a. Max. Limit = 3.753"
       Min. Limit = 3.750"
    b. Max. Limit = 5.932"
       Min. Limit = 5.927"
    c. Max. Limit = 2.004"
       Min. Limit = 2.000"
    d. Max. Limit = 4.8739"
       Min. Limit = 4.8727"

3. e. Max. Limit = 1.0884"
       Min. Limit = 1.0875"
    f. Max. Limit = 28.16 mm
       Min. Limit = 28.10 mm
    g. Max. Limit = 43.98 mm
       Min. Limit = 43.94 mm
    h. Max. Limit = 120.95 mm
       Min. Limit = 120.88 mm
    i. Max. Limit = 73.398 mm
       Min. Limit = 73.386 mm
    j. Max. Limit = 45.115 mm
       Min. Limit = 45.106 mm
5. a. 0.943" ± 0.005"
    b. 1.735" ± 0.001"

5. c. 2.998" ± 0.002"
    d. 0.069" ± 0.004"
    e. 4.1880" ± 0.0007"
    f. 0.9975" ± 0.0037"
    g. 0.0006" ± 0.0004"
    h. 8.4660" ± 0.0011"
    i. 38.61 mm ± 0.01 mm
    j. 10.02 mm ± 0.04 mm
    k. 64.92 mm ± 0.03 mm
    l. 38.016 mm ± 0.028 mm
    m. 249.9915 mm ± 0.0085 mm
    n. 43.078 mm ± 0.013 mm
    o. 79.9835 mm ± 0.0045 mm

7. All dimensions are in millimeters.

		Basic Dimension	Maximum Diameter (Max. Limit)	Minimum Diameter (Min. Limit)	Maximum Interference (Allowance)	Minimum Interference
a.	DIA A	20.73	20.75	20.71	0.09	0.01
	DIA B	20.68	20.70	20.66		
b.	DIA A	32.07	32.09	32.05	0.10	0.02
	DIA B	32.01	32.03	31.99		
c.	DIA A	10.82	10.84	10.80	0.11	0.03
	DIA B	10.75	10.77	10.73		

9. All dimensions are in millimeters.

		Basic Dimension	Maximum Diameter (Max. Limit)	Minimum Diameter (Min. Limit)	Maximum Interference (Allowance)	Minimum Interference
a.	DIA A	78.78	78.81	78.75	0.14	0.02
	DIA B	78.70	78.73	78.67		
b.	DIA A	9.94	9.97	9.91	0.15	0.03
	DIA B	9.85	9.88	9.82		
c.	DIA A	130.03	130.06	130.00	0.13	0.01
	DIA B	129.96	129.99	129.93		

11. 18.20 mm
13. Max. thickness = 2.82 mm
    Min. thickness = 2.76 mm

UNIT 19 ENGLISH AND METRIC STEEL RULES

1. a. 3/32"
   b. 5/16"
   c. 1/2"
   d. 19/32"
   e. 3/4"
   f. 29/32"
   g. 1 1/32"
   h. 1 5/16"
   i. 5/64"
   j. 7/32"
   k. 3/8"

1. l. 35/64"
   m. 49/64"
   n. 31/32"
   o. 1 13/64"
   p. 1 29/64"
3. a. 1/4"
   b. 9/16"
   c. 1/2"
   d. 15/32"
   e. 11/32"
   f. 2 3/16"

3. g. 7/32"
   h. 3/4"
   i. 23/32"
   j. 1/2"
   k. 15/32"
   l. 3/8"
   m. 1/8"
   n. 4 13/16"
5. a. 0.08"
   b. 0.22"
   c. 0.40"

5. d. 0.58"
   e. 0.80"
   f. 1.04"
   g. 1.28"
   h. 1.42"
   i. 0.09"
   j. 0.23"
   k. 0.38"
   l. 0.55"
   m. 0.84"
   n. 1.07"

5. o. 1.27"
   p. 1.45"
7. A = 0.54"
   B = 0.42"
   C = 1.40"
   D = 1.20"
   E = 0.34"
   F = 0.28"
   G = 1.02"
   H = 0.22"
   I = 0.12"

9. a. 46 mm
   b. 70 mm
   c. 20 mm
   d. 82 mm
   e. 10 mm
   f. 23 mm
   g. 25 mm

9. h. 121 mm
   i. 17 mm
   j. 22 mm
   k. 36 mm
   l. 10 mm
   m. 52 mm
   n. 6 mm

UNIT 20 ENGLISH VERNIER CALIPERS AND HEIGHT GAGES

1. a. 2.641"
   b. 3.376"
   c. 2.021"
   d. 0.508"
   e. 4.788"
   f. 2.991"
   g. 1.581"
   h. 1.098"

3.

	A (inches)	B (inches)	C
a.	3.225	3.250	17
b.	2.875	2.900	2
c.	5.925	5.950	14
d.	0.600	0.625	11
e.	4.350	4.375	19
f.	0.075	0.100	19
g.	7.850	7.875	7
h.	1.625	1.650	21
i.	6.000	6.025	24
j.	0.000	0.025	22
k.	3.325	3.350	8
l.	5.975	6.000	24
m.	0.375	0.400	13
n.	0.950	0.975	15

5. a. 1.909"
   b. 4.620"
   c. 7.969"
   d. 0.439"
   e. 2.779"
   f. 6.459"
   g. 3.612"
   h. 8.391"

UNIT 21 METRIC VERNIER CALIPERS AND HEIGHT GAGES

1. a. 30.82 mm
   b. 60.52 mm
   c. 11.76 mm
   d. 78.82 mm
   e. 52.42 mm
   f. 18.16 mm

3.

	A (millimeters)	B (millimeters)	C
a.	37.5	38.0	9
b.	19.5	20	13
c.	42.0	42.5	2
d.	88.5	89.0	16
e.	63.5	64.0	12
f.	20.0	20.5	14
g.	43.0	43.5	3
h.	77.0	77.5	20
i.	81.0	81.5	11
j.	96.5	97.0	24

5. a. 30.22 mm
   b. 48.62 mm
   c. 78.60 mm
   d. 65.82 mm
   e. 52.18 mm
   f. 14.76 mm
   g. 8.82 mm
   h. 34.42 mm

UNIT 22 ENGLISH MICROMETERS

1. 0.589″
3. 0.736″
5. 0.808″
7. 0.738″
9. 0.157″
11. 0.949″
13. 0.441
15. 0.153″
17. 0.324″
19. 0.038″
21. 0.981″

	Barrel Scale Setting (inches)	Thimble Scale Setting (inches)
23.	0.375–0.400	0.012
25.	0.950–0.975	0.023
27.	0.050–0.075	0.009
29.	0.025–0.050	0.011
31.	0.500–0.525	0.017

33. 0.3637″
35. 0.0982″
37. 0.3105″
39. 0.1448″
41. 0.5157″
43. 0.2749″
45. 0.3928″
47. 0.9719″
49. 0.2004″
51. 0.0009″
53. 0.8594″

	Barrel Scale Setting (inches)	Thimble Scale Setting (inches)	Vernier Scale Setting (inches)
55.	0.775–0.800	0.009–0.010	0.0006
57.	0.000–0.025	0.007–0.008	0.0009
59.	0.300–0.325	0.000–0.001	0.0001
61.	0.800–0.825	0.000–0.001	0.0008
63.	0.975–1.000	0.014–0.015	0.0004

UNIT 23 METRIC MICROMETERS

1. 7.09 mm
3. 5.69 mm
5. 9.78 mm
7. 0.34 mm
9. 3.12 mm
11. 24.93 mm

	Barrel Scale Setting is Between: (millimeters)	Thimble Scale Setting (millimeters)
13.	12.5–13.0	0.36
15.	15.0–15.5	0.08
17.	0.5–1.0	0.38
19.	18.0–18.5	0.12
21.	8.0–8.5	0.44
23.	23.0–23.5	0.08
25.	21.5–22.0	0.32

27. 4.268 mm
29. 7.218 mm
31. 2.132 mm
33. 8.308 mm
35. 9.484 mm
37. 11.114 mm

	Barrel Scale Setting is Between: (millimeters)	Thimble Scale Setting is Between: (millimeters)	Vernier Scale Setting (millimeters)
39.	14.5–15.0	0.37–0.38	0.004
41.	9.0–9.5	0.23–0.24	0.008
43.	3.0–3.5	0.04–0.05	0.006
45.	7.0–7.5	0.00–0.01	0.004
47.	5.5–6.0	0.20–0.21	0.008
49.	8.0–8.5	0.32–0.33	0.004
51.	14.5–15.0	0.08–0.09	0.002

UNIT 24 ENGLISH AND METRIC GAGE BLOCKS

One combination for each dimension is given. A number of different combinations will produce the given dimensions.

1. 0.1008″, 0.113″, 0.650″, 4.000″
3. 0.1002″, 0.122″, 0.900″, 2.000″
5. 0.1009″, 0.125″, 0.150″
7. 0.123″, 0.850″, 3.000″, 4.000″
9. 0.125″, 0.250″, 1.000″, 2.000″, 3.000″, 4.000″
11. 0.1007″, 0.125″, 0.650″, 4.000″
13. 0.1001″, 0.123″, 0.050″
15. 0.140″, 0.950″, 4.000″
17. 0.1009″, 0.128″, 0.750″, 2.000″
19. 0.1007″, 0.127″, 0.550″, 3.000″, 4.000″
21. 0.1006″, 0.134″, 0.200″, 2.000″, 3.000″, 4.000″
23. 0.103″, 0.900″, 1.000″, 4.000″
25. 0.1008″, 0.149″, 0.550″

27. 1.003 mm, 1.07 mm, 2 mm, 10 mm
29. 1.09 mm, 5 mm, 50 mm, 90 mm
31. 1.007 mm, 1.7 mm, 1 mm, 40 mm
33. 1.08 mm, 1.3 mm, 1 mm, 80 mm
35. 1.06 mm, 1.8 mm, 1 mm, 10 mm
37. 1.001 mm, 1.07 mm, 4 mm
39. 1.009 mm, 1.09 mm, 7 mm, 30 mm
41. 1.005 mm, 6 mm, 60 mm
43. 1.007 mm, 1 mm
45. 1.03 mm, 2 mm, 20 mm, 80 mm, 90 mm
47. 1.004 mm, 1.7 mm, 10 mm
49. 1.005 mm, 1.05 mm, 1.5 mm, 2 mm, 50 mm

UNIT 25 ACHIEVEMENT REVIEW – SECTION 2

1. a. 7.5 ft
   b. 75 in

1. c. 28.8 ft
   d. 38 mm

1. e. 800 mm
   f. 21.8 cm

3. 5 complete lengths

5.

	Greatest Possible Error	ACTUAL LENGTH	
		Smallest Possible	Largest Possible
a.	0.01″	5.17″	5.19″
b.	0.00005″	0.83665″	0.83675″
c.	0.01 mm	46.15 mm	46.17 mm
d.	0.005 mm	15.345 mm	15.355 mm

7. a. 0.879″ ± 0.003″
   b. 5.2613″ ± 0.0006″
   c. 43.615 mm ± 0.015 mm
   d. 78.9135 mm ± 0.0045 mm
9. 16.56 mm
11. a. 0.06″
    b. 0.3″
    c. 0.52″
    d. 0.76″
    e. 0.94″
    f. 1.1″
    g. 1.32″

11. h. 1.44″
    i. 0.05″
    j. 0.19″
    k. 0.32″
    l. 0.49″
    m. 0.71″
    n. 1.02″
    o. 1.23″
    p. 1.47″
13. a. 3.781″
    b. 7.721″
    c. 2.888″

13. d. 26.16 mm
    e. 83.22 mm
    f. 29.60 mm
15. a. 0.1004″, 0.128″, 0.150″
    b. 0.1006″, 0.148″, 0.100″, 2.000″
    c. 0.1002″, 0.106″, 0.500″, 1.000″
    d. 0.1003″, 0.107″, 0.550″, 1.000″, 4.000″
    e. 0.1001″, 0.140″, 0.850″, 2.000″
    f. 0.1009″, 0.100″
    g. 0.1005″, 0.139″, 0.650″, 3.000″, 4.000″
    h. 0.1006″, 0.101″, 0.800″, 1.000″, 3.000″, 4.000″

SECTION 3 FUNDAMENTALS OF ALGEBRA

UNIT 26 SYMBOLISM

1. $6x + y$
3. $25 - b$
5. r/s
7. xy/m^2
9. a. $2 1/2R$
   b. $2 3/4R$

9. c. $2 1/4R$
   d. $6R$
11. $n - p - t$
13. a. 32
    b. 14
    c. 5

13. d. 3
    e. 1 1/6
15. a. 43
    b. 30
    c. 2 4/5
    d. 4

15. e. 45
17. a. 72 sq in
    b. 8.4852 in
19. a. 9 in
    b. 11.619 sq in
21. 31.5 sq in

23. 2400 mm^2
25. 62.8 sq in
27. a. 5 in
    b. 211.05 cu in

UNIT 27 SIGNED NUMBERS

1. a. (+)9
   b. (+)7
   c. (+)6
   d. (–)6
   e. (–)10
   f. (+)6
   g. (–)20
   h. (–)10
   i. (+)5
   j. (–)8
   k. (–)11
   l. (–)4
   m. (+)17.5
   n. (–)17.5
   o. (+)8.5
   p. (+)1.5
   q. (–)5 1/4
   r. (–)6 7/8
3. a. –22, –18, –1, 0, +2, +4, +17
   b. –21, –19, –5, –2, 0, +5, +13, +27
   c. –25, –10, –7, 0, +7, +10, +14, +25
   d. –14.9, –3.6, –2.5, 0, +0.3, +15, +17
   e. –16, –13 7/8, –4 3/8, +6, +14 1/8

5. a. 23
   b. 27
   c. 25
   d. –23
   e. –29
   f. 7
   g. –8
   h. –3
   i. –6
   j. –22
   k. –35
   l. –18.8
   m. –11.8
   n. –19.1
   o. –13
   p. –3 1/8
   q. –13 3/16
   r. –14.47
   s. 0.43
   t. 30.1
7. a. –24
   b. 24
   c. –20

7. d. 20
   e. –35
   f. 28
   g. 0
   h. –13
   i. 0.32
   j. 0.036
   k. –3/4
   l. 0
   m. –8
   n. –8
   o. 8
   p. 0
   q. 3
   r. 1
   s. –0.3
   t. –2 3/8
9. a. 4
   b. 8
   c. –8
   d. –27
   e. 16
   f. –32

9. g. 25
   h. –125
   i. 64
   j. 2.25
   k. –1.728
   l. 0.09
   m. –0.027
   n. –1/8
   o. 1/8
   p. –27/64
   q. 1/4
   r. 1/16
   s. –1/125
   t. –1/9.261
11. a. 2
    b. 9
    c. 2
    d. 3

11. e. –2
    f. 2
    g. –5
    h. 5
    i. 4
    j. 1/2
    k. 1/2
    l. 1/4
13. 14
15. 2
17. 21
19. 142
21. 9
23. 5 1/100
25. –19/100
27. 2
29. –0.5
31. 14

UNIT 28 ALGEBRAIC OPERATIONS OF ADDITION, SUBTRACTION, AND MULTIPLICATION

1. $21y$
3. $-22xy$
5. 0
7. $-8pt$
9. $15.2a^2 b$
11. $1\,1/4xy$
13. $2.91gh^3$
15. $6P$
17. $-1\,7/8xy$
19. $6.666\,M$
21. $5.5a^2 b^2 c$
23. $2a - 11m$
25. $3xy^2 + 3x^2 y$
27. $-2x^3 - 7x^2 + 4x + 12$
29. 0
31. $-0.4c + 3.6cd + 2.9d$

33. $2xy$
35. $-2xy$
37. $-10a^2$
39. $8mn^3$
41. $1\,1/4x^2$
43. $-13a + 7a^2$
45. $-2ax^2$
47. $d^2 t^2$
49. $3x - 18$
51. 0
53. $x^2 + 3xy$
55. 0
57. $3a^3 - 1.5a^2 + a$
59. $-d^2 - 2dt + dt^2 + 4$
61. $10.09e + 15.76f + 10.03$

63. x^3
65. $56a^4 b^3 c^3$
67. 0
69. $3d^8 r^4$
71. $0.12x^7 y^4$
73. 0
75. $-3.36bc$
77. $-4x^8 y^6$
79. $-49a^4 b^4$
81. $-x^4 y^2$
83. $-14x^3 y^5 + 21x^6 y^3$
85. $-8a^4 b^5 + 2a^3 b^4 + 8a^3 b^2$
87. $-4dt - 4t^2 + 4$
89. $3x^3 + 24x + 7x^2 + 56$
91. $10a^3 x^6 + 5ab^2 x^4 + 2a^2 bx^4 + b^3 x^2$

UNIT 29 ALGEBRAIC OPERATIONS OF DIVISION, POWERS, AND ROOTS

1. $2x$
3. -1
5. 0
7. $-6H$
9. 2.6
11. $5cd$
13. $8g^2 h$
15. xz^3
17. $4P^2 V$
19. $1/4FS^2$
21. $4x^2 + 6x$
23. $-3x^5 y + 2xy^3$
25. $-15a - 25a^4$
27. $-cd^2 + 5c^2 d + 1$
29. $3a^2 x + ax^2 - 2$
31. $4a - 6a^2 c - 8c^2$
33. $9a^2 b^2$
35. $8x^6 y^3$

37. $-27c^9 d^6 e^{12}$
39. $49x^8 y^{10}$
41. $a^6 b^3 c^9$
43. $-x^{12} y^{15} z^3$
45. $0.064x^9 y^3$
47. $10.24M^6 N^2 P^4$
49. $64a^8 b^{12} c^2$
51. $0.36d^6 e^6 f^{12}$
53. $9x^4 - 30x^2 y^3 + 25y^6$
55. $25t^4 - 60t^2 x + 36x^2$
57. $0.36d^6 t^4 - 0.24d^3 t^3 + 0.04t^2$
59. $4/9c^4 d^2 + c^3 d^3 + 9/16c^2 d^4$
61. $a^{16} b^4 + 2a^8 b^2 x^6 y^3 + x^{12} y^6$
63. $m^2 ns^3$
65. $9x^4 y^3$
67. $-3x^2 y^4$
69. $0.8a^3 c^4 f$
71. $1/4xy$

73. $-4d^2 t^3$
75. $2h^2$
77. $4a\sqrt[3]{c}$
79. $2/3a^2 b\sqrt{c}$
81. $-2a\sqrt[5]{b^3}$
83. $9b - 15b^2 + c - d$
85. $-ab - a^2 b + 6a$
87. $-16 - xy$
89. $1 - r$
91. $-2x + 18$
93. $6 + c^2 d$
95. $6a^2 - 6b$
97. $3b$
99. $7y^6 + 15$
101. $2\,2/3d$
103. $80a - 4a^4 b^6$
105. $5f^4 + 6f^2 h$

UNIT 30 INTRODUCTION TO EQUATIONS

1. 14
3. 11
5. 4
7. 9
9. 5

11. $0.5'', 1'', 3''$
13. 0.635 mm
15. 50 mm
17. $12°$
19. $36°$

21. $1/2''$
23. a. $3/4''$
    b. $1/2''$
    c. $1\,3/4''$

25. 6
27. 7
29. 6
31. 8

33. 84
35. 3
37. 48
39. 20

UNIT 31 SOLUTION OF EQUATIONS BY THE SUBTRACTION, ADDITION, AND DIVISION PRINCIPLES OF EQUALITY

1. 10
3. 19
5. 4
7. 36
9. -22
11. -53
13. 34
15. -50
17. 18.8

19. 24.09
21. 0
23. $-1\,5/8$
25. $-3\,1/4$
27. $-23\,1/8$
29. -17.101
31. $17''$
33. $13/16''$
35. 37.61 mm

37. $0.1004''$
39. $7\,11/32''$
41. $4.4286''$
43. $0.1653''$
45. -10
47. 137
49. 28
51. 83

53. 13
55. 78
57. 9.3
59. -3.69
61. -0.005
63. 0.09
65. -4.89
67. $1/8$

69. $-16\,5/32$
71. 19.0622
73. 48.1995
75. $4\,1/2''$
77. 53.3 mm
79. 806 mm
81. -3
83. 6

85. 5
87. 2.3
89. -19
91. 0
93. 20
95. -0.9
97. 19.75
99. 11

101. 32
103. -72
105. $-4\,1/2$
107. 0.2
109. $3/17$
111. $21.75°$
113. 120 mm
115. 120 r/min

UNIT 32 SOLUTION OF EQUATIONS BY THE MULTIPLICATION, ROOT, AND POWER PRINCIPLES OF EQUALITY

1. 20
3. 63
5. 27
7. 0
9. 36
11. 18
13. 23.4
15. -6
17. 0
19. 0.001

21. 0.0832
23. $3\,3/4$
25. 2
27. $3/8$
29. 0.9
31. 426.72 mm
33. $7.0711''$
35. $0.0325''$
37. 314.16 mm
39. 4

41. 9
43. 3
45. 12
47. 5
49. 100
51. $2/3$
53. $3/5$
55. $-1/2$
57. $4/5$
59. 0.3

61. 1.5
63. 0.4
65. 2.8
67. 0.3
69. a. 6 in
    b. $5/9$ ft
    c. 1.2 m
    d. 8 m
    e. 0.07 ft
71. 64

73. 1.44
75. 0.6724
77. 12.167
79. -0.001
81. 0
83. -32
85. -0.216
87. 0.001
89. $9/16$
91. $1/16$

93. $25/64$
95. $15\,5/8$
97. $5\,23/64$
99. $-27/1000$
101. a. 7.84 sq in
     b. 0.5625 sq ft
     c. 0.36 m^2
     d. $48\,400$ mm^2
     e. 166.41 sq in

UNIT 33 SOLUTION OF EQUATIONS CONSISTING OF COMBINED OPERATIONS AND REARRANGEMENT OF FORMULAS

1. 8
3. 7
5. 2
7. 0.2
9. 9
11. −0.67
13. 4.8
15. 7
17. 31.3
19. 3
21. 4

23. −1
25. 36
27. 3
29. 6
31. 0.630
33. 27,066.929
35. 5.000
37. 2327.586
39. 0.086
41. 143.239
43. 12.341

45. a. $a = \dfrac{A}{b}$
   b. $b = \dfrac{A}{a}$
   c. $a = \sqrt{d^2 - b^2}$
   d. $b = \sqrt{d^2 - a^2}$

47. a. $D_O = \sqrt{FW^2 + D^2}$
   b. $D = \sqrt{D_O{}^2 - FW^2}$
   c. $d = 2a + 2C - D_O$
   d. $C = (D_O + d - 2a)2$

49. a. $D = M + 1.5155P - 3W$
   b. $P = \dfrac{D + 3W - M}{1.5155}$
   c. $W = \dfrac{M - D + 1.5155P}{3}$

51. a. $D = \dfrac{L - 1.57d - 2x}{1.57}$
   b. $d = \dfrac{L - 1.57D - 2x}{1.57}$
   c. $x = \dfrac{L - 1.57D - 1.57d}{2}$

53. a. $S = \dfrac{Ca}{C - F}$
   b. $F = \dfrac{SC - Ca}{S}$

UNIT 34 RATIO AND PROPORTION

1. 2/5
3. 2/11
5. 6/23
7. 17/9
9. $a/2$
11. 4/3
13. 1/6
15. a. 2/1
   b. 2/3

15. c. 2/1
   d. 3/5
   e. 3/13
   f. 13/2
   g. 13/6
   h. 5/2
   i. 3/2
   j. 6/13

17. 1
19. 35
21. 12
23. 17.5
25. 2.2
27. 8.2
29. 4
31. 5/6

33. −31 1/2
35. 5.25
37. 0.5
39. 6
41. a. 16 in
   b. 1 1/8 in
   c. 72.9 mm
   d. 32.4 mm

43. a. 4.016 in
   b. 1.124 in
   c. 8.031 in
   d. 2.720 in
   e. 2.808 in
   f. 7.950 in
   g. 1.125 in
   h. 1.575 in

43. i. 5.300 in
   j. 1.686 in
   k. 9.450 in
   l. 2.040 in
   m. 3.150 in
   n. 0.843 in
   o. 8.031 in
   p. 3.744 in

UNIT 35 DIRECT AND INVERSE PROPORTIONS

1. a. 1.66 mm
   b. 2.59 mm
   c. 1.16 mm
   d. 2.33 mm
   e. 1.94 mm

3. a. 0.980 in
   b. 0.763 in
   c. 79.403 mm
   d. 12.966 mm
   e. 0.468 in

5. 1650 parts
7. 0.49 kg
9. a. 288 rpm
   b. 157.5 rpm

9. c. 28 teeth
   d. 25 teeth
   e. 154.3 rpm

UNIT 36 APPLICATIONS OF FORMULAS TO CUTTING SPEED, REVOLUTIONS PER MINUTE, AND CUTTING TIME

1. 60 fpm
3. 90 fpm
5. 100 fpm
7. 130 m/min
9. 106 m/min
11. 113 rpm

13. 43 rpm
15. 2292 rpm
17. 477 r/min
19. 3 626 r/min
21. 4.8 min
23. 12.0 min

25. 45 m/min
27. 146 fpm
29. 183 rpm
31. 3820 r/min
33. 87 rpm
35. 1.6 min

37. 271.3 min
39. 0.030 inch per revolution
41. 45 h
43. 14.05 h
45. 154 rpm

47. 1493 rpm
49. 230 rpm
51. 800 rpm
53. 414 rpm
55. 436 rpm
57. 343 rpm

UNIT 37 APPLICATIONS OF FORMULAS TO SPUR GEARS

1. 2
3. 0.6283 inch
5. 3.5714 inches
7. 0.1571 inch
9. 14 teeth
11. 8.2857 inches
13. 1.2047 inches
15. 0.1818 inch
17. 0.0785 inch
19. 0.0340 inch

21. 0.3082 inch
23. 0.1438 inch
25. 0.2222 inch
27. 15
29. 7
31. 26
33. 0.1429 inch
35. 0.0964 inch
37. 0.2857 inch
39. 2.7239 inches

41. 0.0143 inch
43. 23 teeth
45. 0.0023 inch
47. 0.0038 inch
49. 0.0060 inch
51. 6.7778 inches
53. 1.9375 inches
55. a. 108 mm
   b. 18.850 mm
   c. 120 mm

55. d. 6 mm
   e. 12 mm
   f. 9.425 mm
57. a. 25 mm
   b. 7.854 mm
   c. 30 mm
   d. 2.5 mm
   e. 5 mm
   f. 3.927 mm

59. a. 260 mm
   b. 31.417 mm
   c. 280 mm
   d. 10 mm
   e. 20 mm
   f. 15.708 mm
61. 30 teeth
63. 8.169 mm

UNIT 38 ACHIEVEMENT REVIEW — SECTION 3

1. a. $x + y - c$
   b. $ab \div c$
   c. $2M - P^2$
3. a. −39
   b. 16
   c. −14.4
   d. −54
   e. 0.78
   f. −6
   g. 64

3. h. 36
   i. −64
   j. −3
   k. 1/4
   l. −1.5
   m. 8
   n. −0.67
5. a. 16
   b. 38
   c. 14

5. d. 31.3
   e. −39.3
   f. −1.4
   g. −9
   h. 5.8
   i. 74.052
   j. −6.784
   k. −0.5
   l. 3
   m. 9

5. n. $\dfrac{-2}{3}$
   o. 49
   p. 125
   q. −0.3
   r. 27/64
7. a. 4.243
   b. 1.56
   c. 290.948
   d. 3.000
   e. 1.778

9. a. 18.6
   b. 10.8
   c. 1/3
   d. 3
   e. 32.5
   f. 0.72
   g. 5
   h. 3.5

11. a. 147 fpm
   b. 1194 r/min
   c. 3.82 min
   d. 3.5
   e. 0.5393 in

SECTION 4 FUNDAMENTALS OF PLANE GEOMETRY
UNIT 39 INTRODUCTION TO GEOMETRIC FIGURES

1. a. parallel
   b. perpendicular
   c. oblique
3. a. ‖
   b. ⊥
   c. °
   d. ′
   e. ″
5. 76°52′
7. 244°08′
9. 46°46′05″

11. 540°
13. 19°19′
15. 109°21′09″
17. 44°
19. 21°59′35″
21. 96°33′59″
23. 81°54′
25. 110°51′05″
27. 84°
29. 268°25′
31. 46°30′

33. 68°30′
35. 51°25′43″ (rounded)
37. 161°10′25″ (rounded)
39. 76°57′
41. 117°42′
43. 93°09′
45. 6°28′
47. 77°40′
49. 214°04′55″
51. 44°26′38″
53. 103°0′32″

55. 89°54′20″
57. 19°53′50″
59. 107.75°
61. 93.30°
63. 56.80°
65. 2.32°
67. 79.98°
69. 53.1792°
71. 98.3403°
73. 5.1047°
75. 61.2017°

UNIT 40 PROTRACTORS — SIMPLE AND CALIPER

1. ∠A = 25°
   ∠B = 42°
   ∠C = 54°
   ∠D = 77°
   ∠E = 93°
   ∠F = 11°
   ∠G = 23°
   ∠H = 46°

1. ∠I = 81°
   ∠J = 87°
3. The third angle measures 28°.
5. ∠1 = 29°
   ∠2 = 133°
   ∠3 = 29°
   ∠4 = 58°

5. ∠5 = 39°
   ∠6 = 27°
   ∠7 = 122°
   ∠8 = 31°
   ∠9 = 72°
   ∠10 = 103°
   ∠11 = 64°
   ∠12 = 48°

5. ∠13 = 150°
   ∠14 = 19°
7. 19°45′
9. 50°15′
11. 20°15′
13. 20°30′
15. a. 49°
    b. 14°

15. c. 73°
    d. 88°
    e. 22°11′
    f. 44°41′
    g. 71°38′
    h. 11°40′33″
    i. 30°59′1″

UNIT 41 ANGLES

1. a. ∠A, ∠BAF, ∠FAB
   b. ∠B, ∠ABC, ∠CBA
   c. ∠3, ∠BCD, ∠DCB
   d. ∠4, ∠CDE, ∠EDC
   e. ∠5, ∠DEF, ∠FED
   f. ∠6, ∠AFE, ∠EFA
3. a. acute
   b. right
   c. right
   d. acute
   e. obtuse
   f. acute
   g. straight
   h. acute

3. i. right
   j. straight
   k. reflex
5. a. ∠3 and ∠6, ∠4 and ∠5
   b. ∠1 and ∠6, ∠2 and ∠5, ∠3 and ∠8, ∠4 and ∠7
7. a. 152°
   b. 146°43′
9. a. ∠1, ∠2, ∠5, ∠7, ∠9, ∠11, ∠13, ∠15 = 107°
      ∠3, ∠4, ∠6, ∠8, ∠10, ∠12, ∠14 = 73°
   b. ∠1, ∠2, ∠5, ∠7, ∠9, ∠11, ∠13, ∠15 = 92°52′
      ∠3, ∠4, ∠6, ∠8, ∠10, ∠12, ∠14 = 87°08′
11. a. ∠2 = 68°, ∠3 = 112°
    b. ∠2 = 77°26′, ∠3 = 102°34′

UNIT 42 INTRODUCTION TO TRIANGLES

1. isosceles
3. scalene
5. right
7. equilateral
9. 180°
11. a. 28°
    b. 29°42′47″

13. a. 14.2″
    b. 14.2″
15. a. 81°30′
    b. 77°20′30″

17. a. 11°
    b. 46°
19. a. 48°
    b. 79°

21. a. ∠C
    b. ∠A
    c. ∠B

23. a. ∠D
    b. ∠E
    c. ∠2

UNIT 43 GEOMETRIC PRINCIPLES FOR TRIANGLES AND OTHER COMMON POLYGONS

1. Pairs A, B, D, and F
3. a. 56 mm
   b. 92.57 mm
5. a. 52°42′
   b. 37°18′

7. a. 77°
   b. 13°
   c. 13°
9. a. 4.909 in
   b. 2.640 in

11. a. 72.5 mm
    b. 113.6 mm
13. a. 115.70 mm
    b. 115.70 mm

15. a. 15 in
    b. 5 in
17. a. 960 mm
    b. 576 mm

19. a. 2.681 in
    b. 2.730 in
21. a. 63°
    b. 94°

UNIT 44 INTRODUCTION TO CIRCLES

1. a. Chord
   b. Diameter or Chord
   c. Radius
   d. Center
3. a. Sector
   b. Segment
   c. Radius
   d. Radius
   e. Chord
   f. Arc

5. a. 25.133 in
   b. 94.248 mm
   c. 116.868 mm
   d. 17.907 in
   e. 11.141 in
   f. 69.391 mm
   g. 49.656 mm
   h. 1.222 in
7. 243.47 mm

9. a. 1.75″
   b. 1.75″
11. a. 280 mm
    b. 162.286 mm
13. a. (1) 2.689″
       (2) 7.564″
    b. (1) 6.034″
       (2) 3.654″

15. a. 2.723″
    b. 249°49′
17. a. 12.00 in
    b. 12.800 in
19. a. 15.438 in
    b. 1.950 in
21. a. (1) 33°30′
       (2) 137.20 mm

21. b. (1) 67°38′
       (2) 207.70 mm
23. a. 59°
    b. 64°15′
25. a. 2.154 in
    b. 1.615 in

UNIT 45 ARCS AND ANGLES OF CIRCLES

1. 6.283 in
3. 243°30'
5. 14.921 in
7. a. (1) 73°
     (2) 34°
     (3) 134°
   b. (1) 68°
     (2) 39°
     (3) 104°

9. a. (1) 41°
     (2) 139°
   b. (1) 37°30'
     (2) 142°30'
11. a. (1) 94°
     (2) 16°
   b. (1) 51°30'
     (2) 35°46'

13. a. (1) 57°
     (2) 90°
   b. (1) 55°21'
     (2) 90°
15. a. (1) 47°30'
     (2) 27°30'
   b. (1) 45°
     (2) 28°

17. a. (1) 3°
     (2) 20°
   b. (1) 9°
     (2) 10°
19. a. 40.58 mm
   b. 82.40 mm
21. a. 1.911 in
   b. 1.844 in

23. a. 24°
   b. 31°14'
25. a. 75°
   b. 82°30'

UNIT 46 FUNDAMENTAL GEOMETRIC CONSTRUCTIONS

All problems are constructed.

UNIT 47 ACHIEVEMENT REVIEW – SECTION 4

1. a. 123°41'
   b. 60°18'
   c. 42°19'13"
   d. 13°16'
   e. 73°24'
   f. 22°11'9"
   g. 43°30'
   h. 25°50'
3. 145°38'51"

5. 64°8'31"
7. 104.3131°
9. a. 10°30'
   b. 19°45'
   c. 29°30'
11. a. 133°
   b. 86°42'
   c. 77°16'27"

13. a. (1) 39°43'
      (2) 1 ft
   b. (1) 60°
      (2) 9.6 in
      (3) 4.8 in
   c. (1) 17°30'
      (2) 72°30'

15. 198°
17. 114.59 mm
19. a. 4.962 in
   b. 10.300 in
21. a. 9.983 in
   b. 262.032°

23. Problem to be
   layed out using
   construction
   methods.

SECTION 5 TRIGONOMETRY

UNIT 48 INTRODUCTION TO TRIGONOMETRIC FUNCTIONS

1. r is hyp
   x is adj
   y is opp
3. a is opp
   b is adj
   c is hyp
5. a is hyp
   b is opp
   c is adj
7. d is hyp
   m is opp
   p is adj
9. h is opp
   k is hyp
   l is adj
11. m is opp
   p is hyp
   s is adj
13. m is hyp
   r is adj
   t is opp
15. f is adj
   g is hyp
   h is opp

17. $\sin \angle 1 = \dfrac{y}{r}$
    $\cos \angle 1 = \dfrac{x}{r}$
    $\tan \angle 1 = \dfrac{y}{x}$
    $\cot \angle 1 = \dfrac{x}{y}$
    $\sec \angle 1 = \dfrac{r}{x}$
    $\csc \angle 1 = \dfrac{r}{y}$

19. $\sin \angle 1 = \dfrac{h}{g}$
    $\cos \angle 1 = \dfrac{k}{g}$
    $\tan \angle 1 = \dfrac{h}{k}$
    $\cot \angle 1 = \dfrac{k}{h}$
    $\sec \angle 1 = \dfrac{g}{k}$
    $\csc \angle 1 = \dfrac{g}{h}$

21. $\sin \angle 1 = \dfrac{r}{s}$
    $\cos \angle 1 = \dfrac{p}{s}$
    $\tan \angle 1 = \dfrac{r}{p}$
    $\cot \angle 1 = \dfrac{p}{r}$
    $\sec \angle 1 = \dfrac{s}{p}$
    $\csc \angle 1 = \dfrac{s}{r}$

23. Group 1: a, b, d
   Group 2: a, c
   Group 3: a, c, d
25. 0.96126
27. 0.83910
29. 5.6713
31. 3.2361
33. 1.9912
35. 0.18795
37. 2.9887
39. 0.67129

41. 1.0249
43. 0.01455
45. 0.00291
47. 1.4101
49. 3°0'
51. 25°0'
53. 66°0'
55. 37°0'
57. 35°40'
59. 40°50'
61. 20°40'
63. 23°20'
65. 80°10'
67. 71°20'
69. 22°10'
71. 0°50'
73. 0.73817
75. 0.57271
77. 1.2557
79. 0.99705
81. 1.1854
83. 2.2372
85. 1.4404
87. 0.85097

89. 0.83564
91. 1.1960
93. 50°44'
95. 60°49'
97. 53°8'
99. 27°3'
101. 71°14'
103. 80°32'
105. 51°25'
107. 56°26'
109. 18°44'
111. 30°24'
113. 0.264 21
115. 0.122 91
117. 0.439 39
119. 0.399 87
121. 18.45°
123. 43.20°
125. 48.30°
127. 54.32°
129. 70.53°

UNIT 49 ANALYSIS OF TRIGONOMETRIC FUNCTIONS

1. a. side y and side r are almost the same length
   b. side x is very small compared to side r
   c. side x is very small compared to side y
3. a. side y is very small compared to side r
   b. side x and side r are almost the same length
   c. side x is very large compared to side y
5. a. 45°
   b. 1.000 . . .
   c. 1.000 . . .

7. a. 1.000 . . .
   b. 1.000 . . .
   c. 0
   d. 0
9. $\tan 18°$
11. $\cot 36°$
13. $\csc 22°$
15. $\cos 81°19'$
17. $\csc 39.25°$

19. $\sec 55°$
21. $\cos 41°$
23. $\csc 3°$
25. $\sec 39°$
27. $\cos 90°$
29. $\sin 77.8°$
31. $\csc 58°34'$
33. $\cos 84.11°$
35. $\tan 90°$

37. $\csc 45°31'$
39. $\cos 0°1'$
41. $\sin 40°$
43. $\cot 45°$
45. $\sec 43°$
47. $\cos 75°$
49. $\cot 89°10'$
51. $\csc 89.0°$

UNIT 50 BASIC CALCULATIONS OF ANGLES AND SIDES OF RIGHT TRIANGLES

1. $36°28'$
3. $42°2'$
5. $59.24°$
7. $54.10°$
9. a. $22°21'$
    b. $67°39'$

11. a. $29°30'$
    b. $60°30'$
13. 5.313 in
15. 50.465 in
17. 136.16 mm

19. 68.78 mm
21. a. 2.229 in
    b. 2.049 in
23. a. 7.285 in
    b. 2.480 in

25. a. $17°30'$
    b. 55.237 in
    c. 52.680 in
27. a. 93.49 mm
    b. 31.64 mm

27. c. $71.30°$
29. a. $16°30'$
    b. 1.579 in
    c. 5.559 in

31. a. $8°20'$
    b. 1.019 in
    c. 6.957 in

UNIT 51 SIMPLE PRACTICAL MACHINE APPLICATIONS

1. a. 4.2262 in
    b. 2.2778 in
    c. 5.9949 in
    d. 1.2764 in
    e. 4.7716 in

1. f. 6.9883 in
    g. 0.1164 in
    h. 0.4071 in
    i. 3.3929 in
3. $8°48'$

5. 13.12 mm
7. $51°25'$
9. 281.46 mm
11. 27.70 mm
13. 0.825 in

15. 10.32 mm
17. 2.381 in
19. 2.537 in
21. 54.28 mm
23. 259.05 mm

UNIT 52 COMPLEX PRACTICAL MACHINE APPLICATIONS

1. 3.748 in
3. $14°13'$

5. 0.564 in
7. $44°21'$

9. 298.85 mm
11. 4.499 in

13. 5.408 in
15. $37°26'$

17. 4.227 in
19. 0.667 in

21. $29°40'$
23. 1.399 in

UNIT 53 THE CARTESIAN COORDINATE SYSTEM

1. $\sin 125° = 0.81915$
    $\cos 125° = -0.57358$
    $\tan 125° = -1.4282$
    $\cot 125° = -0.70021$
    $\sec 125° = -1.7434$
    $\csc 125° = 1.2208$
3. $\sin 260° = -0.98481$
    $\cos 260° = -0.17365$
    $\tan 260° = 5.6713$
    $\cot 260° = 0.17633$
    $\sec 260° = -5.7588$
    $\csc 260° = -1.0154$

5. $\sin 300° = -0.86603$
    $\cos 300° = 0.50000$
    $\tan 300° = -1.7321$
    $\cot 300° = -0.57735$
    $\sec 300° = 2.0000$
    $\csc 300° = -1.1547$
7. $\sin 216°20' = -0.59248$
    $\cos 216°20' = -0.80558$
    $\tan 216°20' = 0.73547$
    $\cot 216°20' = 1.3597$
    $\sec 216°20' = -1.2413$
    $\csc 216°20' = -1.6878$

9. $\sin 146°10' = 0.55678$
    $\cos 146°10' = -0.83066$
    $\tan 146°10' = -0.67028$
    $\cot 146°10' = -1.4919$
    $\sec 146°10' = -1.2039$
    $\csc 146°10' = 1.7960$
11. $\sin 313.20° = -0.72897$
    $\cos 313.20° = 0.68455$
    $\tan 313.20° = -1.0649$
    $\cot 313.20° = -0.93906$
    $\sec 313.20° = 1.4608$
    $\csc 313.20° = -1.3718$

UNIT 54 OBLIQUE TRIANGLES: LAW OF SINES AND LAW OF COSINES

1. 3.627 in
3. 3.533 in
5. $33°54'$
7. $18°16'$
9. 128.73 mm

11. two solutions
13. one solution
15. one solution
17. two solutions
19. 122.13 mm

21. $71.48°$
23. $27°2'$
25. 36.96 mm
27. a. 1.163 in
    b. $42°19'$

29. a. 455.42 mm
    b. $14.52°$
31. a. $94°44'$
    b. $49°20'$
33. $71.98°$

35. 0.792 in
37. 97.36 mm
39. 1.212 in
41. $18.03°$

UNIT 55 ACHIEVEMENT REVIEW – SECTION 5

1. a. a is opp
       b is adj
       c is hyp
    b. a is adj
       b is opp
       c is hyp
    c. m is hyp
       s is opp
       p is adj

1. d. r is hyp
       x is adj
       y is opp
3. a. $45°50'$
    b. $26°30'$
    c. $80°10'$
    d. $41°24'$
    e. $50°36'$
    f. $80°32'$

5. a. $\cos 54°$
    b. $\cot 38°42'$
    c. $\sin 73°7'$
    d. $\tan 13.57°$
7. a. (1) 5.537 in
       (2) 3.075 in
    b. $6.60°$
    c. 118.936 mm
    d. 0.594 in

7. e. (1) $64.16°$
       (2) 177.776 mm
    f. 0.278 in
    g. $39°35'$
    h. $58.50°$
9. a. 309.999 mm
    b. $49°57'$
    c. (1) $110°30'$
       (2) 11.372 in

9. c. (3) 6.583 in
    d. 181.648 mm
    e. $125.10°$
    f. (1) 8.484 in
       (2) $69°22'$
       (3) $75°48'$
    g. 6.046 in
    h. $59.74°$
    i. $107°49'$

SECTION 6 COMPOUND ANGLES

UNIT 56 INTRODUCTION TO COMPOUND ANGLES

1. a. 3.680 in
    b. $65°57'$

3. a. 5.555 in
    b. $56°5'$

5. a. 1.897 in
    b. $65°4'$

UNIT 57 DRILLING AND BORING COMPOUND-ANGULAR HOLES: COMPUTING ANGLES OF ROTATION AND TILT USING GIVEN LENGTHS

1. a. $38°22'$
    b. $49°39'$

3. a. $43°9'$
    b. $49°6'$

5. a. $41°5'$
    b. $49°38'$

7. a.

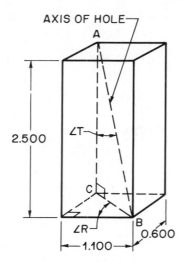

7. b. 28°37'
   c. 26°37'
9. a.

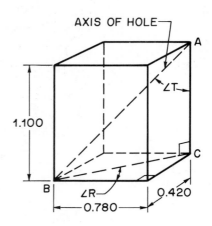

9. b. 28°18'
   c. 38°51'

UNIT 58 DRILLING AND BORING COMPOUND-ANGULAR HOLES: COMPUTING ANGLES OF ROTATION AND TILT USING GIVEN ANGLES

1. a. 37°52'
   b. 48°46'
3. a. 28.76°
   b. 34.43°
5. a. 32°31'
   b. 47°3'
7. a.

7. b. 30°49'
   c. 33°54'
9. a.

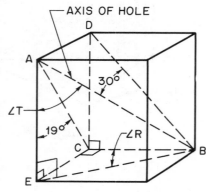

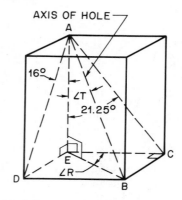

9. b. 36.40°
   c. 25.79°
11. a. 20°30'
    b. 42°52'

13. a. 21.89°
    b. 27.21°
15. a. 28.39°
    b. 31.84°

UNIT 59 MACHINING COMPOUND-ANGULAR SURFACES: COMPUTING ANGLES OF ROTATION AND TILT

1. a. 57°6'
   b. 49°0'
3. a. 58°8'
   b. 31°52'

5. a.

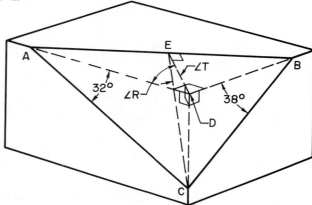

5. b. 51°21'
   c. 45°1'

7. a.

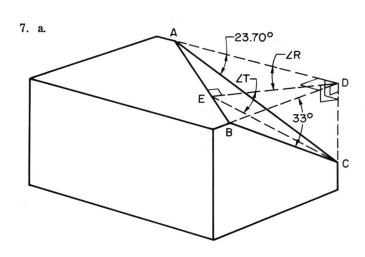

7. b. 55.94°
   c. 38.09°
9. a. 49°59′
   b. 53°31′
11. a. 46.23°
    b. 40.72°
13. a. 38°32′
    b. 40°28′

UNIT 60 COMPUTING ANGLES MADE BY THE INTERSECTION OF TWO ANGULAR SURFACES

1. a. 32°14′
   b. 37°18′
3. a. 32°53′
   b. 45°12′
5. a.

5. b. 31°24′
   c. 32°45′
7. a.

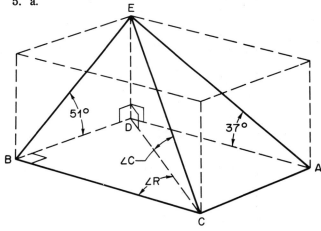

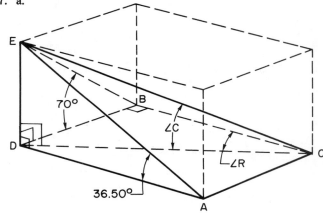

7. b. 15.07°
   c. 35.55°
9. a. 37°7′
   b. 30°5′

11. a. 44.24°
    b. 30.05°
13. a. 17°36′
    b. 15°40′

UNIT 61 COMPUTING COMPOUND ANGLES ON CUTTING AND FORMING TOOLS

1. 29°9′
3. 27°38′
5. a.

5. b. 25°14′
7. a.

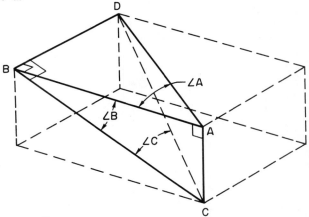

7. b. 29°27′
9. 34°4′
11. 25°20′

13. 30°33'
15. 32°24'
17. a.

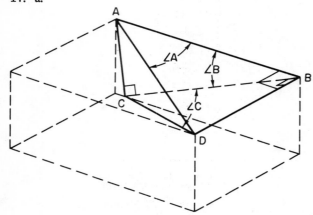

17. b. 30°52'
19. a.

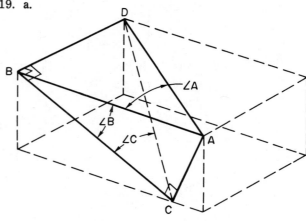

19. b. 32°54'
21. 30°45'
23. 36°25'

UNIT 62 ACHIEVEMENT REVIEW – SECTION 6

1. a. 30°4'   1. b. 44°57'   3. a. 55.54°   3. b. 50.00°   5. 31°36'

SECTION 7 NUMERICAL CONTROL
UNIT 63 INTRODUCTION TO NUMERICAL CONTROL

1. See Instructor's Guide   3. See Instructor's Guide

UNIT 64 POINT-TO-POINT PROGRAMMING ON TWO-AXIS MACHINES

1. a.

Hole	x	y
1	7.500"	−3.550"
2	9.600"	−0.250"
3	7.500"	1.625"
4	13.625"	2.450"
5	14.250"	−1.425"
6	14.250"	−3.550"

1. b.

Hole	x	y
1	7.500"	−3.550"
2	2.100"	3.300"
3	−2.100"	1.875"
4	6.125"	0.825"
5	0.625"	−3.875"
6	0	−2.125"

3. a.

Hole	x	y
1	160.00 mm	−75.38 mm
2	216.24 mm	3.94 mm
3	160 mm	49.00 mm
4	292.32 mm	69.16 mm
5	308.06 mm	−23.08 mm
6	308.06 mm	−75.38 mm

3. b.

Hole	x	y
1	160.00 mm	−75.38 mm
2	56.24 mm	79.32 mm
3	−56.24 mm	45.06 mm
4	132.32 mm	20.16 mm
5	15.74 mm	−92.24 mm
6	0	−52.30 mm

5. a.

Hole	x	y
1	−4.750"	4.750"
2	−0.654"	8.569"
3	3.514"	12.457"
4	8.074"	9.600"
5	13.776"	7.748"
6	17.500"	−0.100"
7	13.776"	1.752"
8	8.074"	−0.100"
9	4.550"	4.750"
10	−1.188"	−0.520"
11	−6.162"	−4.625"

5. b.

Hole	x	y
1	−4.750"	4.750"
2	4.096"	3.819"
3	4.168"	3.888"
4	4.560"	−2.857"
5	5.702"	−1.852"
6	3.724"	−7.848"
7	−3.724"	1.852"
8	−5.702"	−1.852"
9	−3.524"	4.850"
10	−5.738"	−5.270"
11	−4.974"	−4.105"

7. a.

Hole	x	y
1	-9.317″	18.412″
2	-3.793″	17.086″
3	5.885″	18.250″
4	11.400″	16.750″
5	18.935″	12.500″
6	5.885″	9.330″
7	-3.616″	8.075″
8	-13.541″	10.397″

7. b.

Hole	x	y
1	-9.317″	18.412″
2	5.524″	-1.326″
3	9.678″	1.164″
4	5.515″	-1.500″
5	7.535″	-4.250″
6	-13.050″	-3.170″
7	-9.501″	-1.255″
8	-9.925″	2.322″

9. a.

Hole	x	y
1	-237.06 mm	329.42 mm
2	-125.66 mm	302.84 mm
3	51.30 mm	329.30 mm
4	174.64 mm	302.18 mm
5	317.70 mm	222.50 mm
6	51.30 mm	163.00 mm
7	-136.30 mm	133.05 mm
8	-310.02 mm	187.28 mm

9. b.

Hole	x	y
1	-237.06 mm	329.42 mm
2	111.40 mm	-26.58 mm
3	176.96 mm	26.46 mm
4	123.34 mm	-27.12 mm
5	143.06 mm	-79.68 mm
6	-266.40 mm	-59.50 mm
7	-187.60 mm	-29.95 mm
8	-173.72 mm	54.23 mm

UNIT 65 BINARY NUMERATION SYSTEM AND BINARY CODED TAPE

1. $2(10^2) + 8(10^1) + 5(10^0)$
   $200 + 80 + 5 = 285$

3. $9(10^4) + 0(10^3) + 5(10^2) + 0(10^1) + 0(10^0)$
   $90,000 + 0 + 500 + 0 + 0 = 90,500$

5. $2(10^1) + 3(10^0) + 0(10^{-1}) + 2(10^{-2}) + 3(10^{-3})$
   $20 + 3 + 0 + 0.02 + 0.003 = 23.023$

7. $4(10^3) + 7(10^2) + 5(10^1) + 1(10^0) + 1(10^{-1}) + 0(10^{-2}) + 7(10^{-3})$
   $4000 + 700 + 50 + 1 + 0.1 + 0 + 0.007 = 4751.107$

9. $1(10^2) + 7(10^1) + 5(10^0) + 0(10^{-1}) + 7(10^{-2}) + 5(10^{-3}) + 3(10^{-4})$
   $100 + 70 + 5 + 0 + 0.07 + 0.005 + 0.0003 = 175.0753$

11. 1_{10}
13. 11_{10}
15. 15_{10}
17. 9_{10}
19. 21_{10}

21. 51_{10}
23. 0.5_{10}
25. 3.75_{10}
27. 2.125_{10}

29. 9.3125_{10}
31. 1100_2
33. 1010111_2
35. 101011_2

37. 1100110_2
39. 1_2
41. 110011_2
43. 0.1_2

45. 0.011_2
47. 1010001.11_2
49. 1101111.01_2
51. 10100011.111_2

#		DECIMAL NUMBER	VERTICAL BINARY-DECIMAL NUMBER	BINARY-CODE-DECIMAL TAPE $2^3 2^2 2^1 2^0$
53. 3173	10^3	3	11	
	10^2	1	1	
	10^1	7	111	
	10^0	3	11	
55. 803	10^2	8	1000	
	10^1	0	0	
	10^0	3	11	
57. 74,932	10^4	7	111	
	10^3	4	100	
	10^2	9	1001	
	10^1	3	11	
	10^0	2	10	
59. 1029	10^3	1	1	
	10^2	0	0	
	10^1	2	10	
	10^0	9	1001	

#		DECIMAL NUMBER	VERTICAL BINARY-DECIMAL NUMBER	BINARY-CODE-DECIMAL TAPE $2^3 2^2 2^1 2^0$
61. 8321	10^3	8	1000	
	10^2	3	11	
	10^1	2	10	
	10^0	1	1	
63. 429	10^2	4	100	
	10^1	2	10	
	10^0	9	1001	
65. 60,070	10^4	6	110	
	10^3	0	0	
	10^2	0	0	
	10^1	7	111	
	10^0	0	0	
67. 2097	10^3	2	10	
	10^2	0	0	
	10^1	9	1001	
	10^0	7	111	

UNIT 66 ACHIEVEMENT REVIEW — SECTION 7

1. Coordinates to be plotted

3. a.

Hole	x	y
1	7.275″	−1.100″
2	9.181″	2.510″
3	14.204″	2.510″
4	20.950″	2.510″
5	22.044″	−2.362″
6	22.044″	−4.482″
7	14.911″	−3.223″

b.

Hole	x	y
1	7.275″	−1.100″
2	1.906″	3.610″
3	5.023″	0
4	6.746″	0
5	1.094″	−4.872″
6	0	−2.120″
7	−7.133″	1.259″

5. a. 1
   b. 5
   c. 21
   d. 2.25
   e. 9.5625

7.

	Decimal Number		Decimal Number	Vertical Binary-Decimal Number	Binary-Code-Decimal Tape $2^3 2^2 2^1 2^0$
a.	5624	10^3	5	101	
		10^2	6	110	
		10^1	2	10	
		10^0	4	100	
b.	8078	10^3	8	1000	
		10^2	0	0	
		10^1	7	111	
		10^0	8	1000	
c.	1005	10^3	1	1	
		10^2	0	0	
		10^1	0	0	
		10^0	5	101	
d.	723	10^2	7	111	
		10^1	2	10	
		10^0	3	11	
e.	906	10^2	9	1001	
		10^1	0	0	
		10^0	6	110	
f.	36,840	10^4	3	11	
		10^3	6	110	
		10^2	8	1000	
		10^1	4	100	
		10^0	0	0	
g.	62	10^1	6	110	
		10^0	2	10	
h.	2057	10^3	2	10	
		10^2	0	0	
		10^1	5	101	
		10^0	7	111